Lecture Notes in Networks and Systems

Volume 255

The series "Lecture Notes in Networks and Systems" publishes the latest developments in Networks and Systems—quickly, informally and with high quality. Original research reported in proceedings and post-proceedings represents the core of LNNS.

Volumes published in LNNS embrace all aspects and subfields of, as well as new challenges in, Networks and Systems.

The series contains proceedings and edited volumes in systems and networks, spanning the areas of Cyber-Physical Systems, Autonomous Systems, Sensor Networks, Control Systems, Energy Systems, Automotive Systems, Biological Systems, Vehicular Networking and Connected Vehicles, Aerospace Systems, Automation, Manufacturing, Smart Grids, Nonlinear Systems, Power Systems, Robotics, Social Systems, Economic Systems and other. Of particular value to both the contributors and the readership are the short publication timeframe and the world-wide distribution and exposure which enable both a wide and rapid dissemination of research output.

The series covers the theory, applications, and perspectives on the state of the art and future developments relevant to systems and networks, decision making, control, complex processes and related areas, as embedded in the fields of interdisciplinary and applied sciences, engineering, computer science, physics, economics, social, and life sciences, as well as the paradigms and methodologies behind them.

Indexed by SCOPUS, INSPEC, WTI Frankfurt eG, zbMATH, SCImago.

All books published in the series are submitted for consideration in Web of Science.

More information about this series at http://www.springer.com/series/15179

Michal Choraś · Ryszard S. Choraś ·
Marek Kurzyński · Paweł Trajdos ·
Jerzy Pejaś · Tomasz Hyla
Editors

Progress in Image Processing, Pattern Recognition and Communication Systems

Proceedings of the Conference (CORES, IP&C, ACS) – June 28–30 2021

Springer

Editors
Michal Choraś
Institute of Computer Science
and Telecommunications
University of Science
and Technology
Bydgoszcz, Poland

Ryszard S. Choraś
Institute of Computer Science
and Telecommunications
University of Science
and Technology
Bydgoszcz, Poland

Marek Kurzyński
Faculty of Electronics
Wroclaw University of Science
and Technology
Wrocław, Poland

Paweł Trajdos
Faculty of Electronics
Wroclaw University of Science
and Technology
Wrocław, Poland

Jerzy Pejaś
West Pomeranian University of Technology
in Szczecin
Szczecin, Poland

Tomasz Hyla
West Pomeranian University of Technology
in Szczecin
Szczecin, Poland

ISSN 2367-3370 ISSN 2367-3389 (electronic)
Lecture Notes in Networks and Systems
ISBN 978-3-030-81522-6 ISBN 978-3-030-81523-3 (eBook)
https://doi.org/10.1007/978-3-030-81523-3

This Springer imprint is published by the registered company Springer Nature Switzerland AG
The registered company address is: Gewerbestrasse 11, 6330 Cham, Switzerland

Preface

This volume of LNNS contains the proceedings of the multi-conference consisting of three known and prestigious conferences with long tradition, namely:

- International Conference on Image Processing and Communications (IP&C 2021),
- International Conference on Computer Recognition Systems (CORES 2021),
- International Conference on Advanced Computer Systems (ACS 2021).

In 2021, we decided to hold a joint event in the city of Bydgoszcz at University of Science and Technology (UTP), but due to COVID-19 pandemic, the multi-conference is held virtually.

The multi-conference brought together the researchers, developers, practitioners and educators in the fields of image processing, computer recognition systems, networks, security and advanced computer systems.

The conference proceedings contains high-level papers which were selected through a strict review process. The presented papers are structured in several sections, namely:

- Advances in Pattern Recognition and Classifiers
- Progress in Image Processing and Computer Vision
- Practical Applications of Computer Recognition Systems
- Advances in ML for Security and Networks
- Progress in Advanced Computer Systems

Without the high-quality submissions from the authors, the success of the conference would not be possible. Therefore, we would like to thank all authors and also the reviewers for the effort they put into their submissions and evaluation.

We are also grateful to dr. Pawel Trajdos for editing this LNNS volume, as well as to Springer for publishing the IP&C, CORES and ACS proceedings in their Lecture Notes in Networks and Systems series.

Michal Choraś
Ryszard S. Choraś
Marek Kurzyński
Paweł Trajdos
Jerzy Pejaś
Tomasz Hyla

Organization

International Conference on Image Processing and Communications (IP&C 2021)

General Chairs

Ryszard Tadeusiewicz
Ryszard S. Choraś

Local Chair

Michał, Choraś

Program Committee

Łukasz Apiecionek
Kevin W. Bowyer
Dumitru Dan Burdescu
Mihai Carabas
Christophe Charrier
Leszek Chmielewski
Andrzej Dąbrowski
Andrzej Dobrogowski
Marek Domański
Ewa Grabska
Janusz Kacprzyk
Andrzej Kasiński
Andrzej Kasprzak
Joerg Keller
Rafał Kozik
Marek Kurzyński
Witold Malina
Andrzej Materka
Wojciech Mokrzycki
Sławomir Nikiel
Zdzisław Papir
Jens M. Pedersen
Jerzy Pejaś
Leszek Rutkowski
Khalid Saeed
Abdel-Badeeh M. Salem

International Conference on Computer Recognition Systems (CORES 2021)

Honorary Chair

Juliusz Lech Kulikowski

Chairs

Program Committee

Organization Committee Chair

Paweł Trajdos

Organization Committee

Paweł Trajdos
Robert Burduk
Szymon Wojciechowski

International Conference on Advanced Computer Systems (ACS 2021)

Conference Chairs

Tomasz Hyla
Imed El Fray
Jerzy Pejaś

Organization Committee

Witold Maćków
Sylwia Hardej
Luiza Fabisiak

Program Committee

Akira Imada
Albert Dipanda
Albert Sangrà
Aleksandr Cariow
Alexander Prokopenya
Alexander Schill
Andriy Luntovskyy
Andrzej Cader
Andrzej Kasiński
Andrzej Niesler
Andrzej Piegat
Anna Bartkowiak
Anna Grocholewska-Czuryło
Antoni Wiliński
Arkadiusz Orłowski
Bogdan Księżopolski
Costin Badica
Dariusz Frejlichowski
Elisabeth Rakus-Andersson
Eugeniusz Kuriata
Gisella Facchinetti
Grzegorz Bocewicz
Izabela Rejer
Jacek Pomykała
Jan Węglarz
Janusz Górski
Janusz Kacprzyk
Janusz Stokłosa
Jason T. J. Jung
Javier Lopez
Jerzy August Gawinecki
Jonathan Lawry
Jos Dumortier
Josef Pieprzyk
Khalid Saeed
Krzysztof Chmiel
Krzysztof Ciesielski
Kurosh Madani
Kurt Sandkuhl
Larisa Globa
Leon Bobrowski
Leszek Rutkowski

Marcin Paprzycki
Marcin Szpyrka
Marian Srebrny Marian Srebrny
Michelle Joab
Mieczysław Kula
Mirosław Kurkowski
Nabendu Chaki
Natalia Wawrzyniak
Nicolas Tadeusz Courtois
Oleg Finko
Özgür Ertuğ
Paweł Pawlewski
Piotr Andrzej Kowalski
Przemysław Mazurek
Robert Burduk
Ryszard S. Choraś
Ryszard Tadeusiewicz
Shinya Kobayashi
Sławomir Wierzchoń
Toru Yamaguchi
Vaclav Snašel
Valery Rogoza
Vincent Rijmen
William Steingartner
Witold Pedrycz
Władysław Homenda
Władysław Skarbek
Włodzimierz Bielecki
Zbigniew Adam Kotulski
Zbigniew Banaszak
Zenon Sosnowski

Contents

Advances in Pattern Recognition and Classifiers

Building an Ensemble of Classifiers via Randomized Models of Ensemble Members

Pawel Trajdos(✉) and Marek Kurzynski

Department of Systems and Computer Networks, Faculty of Electronics, Wrocław University of Science and Technology, Wybrzeże Wyspiańskiego 27, 50-370 Wrocław, Poland
{pawel.trajdos,marek.kurzynski}@pwr.edu.pl

Abstract. Many dynamic ensemble selection (DES) methods are known in the literature. A previously-developed by the authors, method consists in building a randomized classifier which is treated as a model of the base classifier. The model is equivalent to the base classifier in a certain probabilistic sense. Next, the probability of correct classification of randomized classifier is taken as the competence of the evaluated classifier. In this paper, a novel randomized model of base classifier is developed. In the proposed method, the random operation of the model results from a random selection of the learning set from the family of learning sets of a fixed size. The paper presents the mathematical foundations of this approach and shows how, for a practical application when learning and validation sets are given, one can determine the measure of competence and build a MC system with the DES scheme. The DES scheme with the proposed model of competence was experimentally evaluated on the collection of 67 benchmark datasets and compared in terms of eight quality criteria with two ensemble classifiers which use the previously-proposed concepts of randomized model. The proposed approach achieved the lowest ranks for almost all investigated quality criteria.

Keywords: Ensemble classification · Probabilistic model · Randomized classifier · Competence

1 Introduction

In ensemble classification methods the recognition result is obtained by combining the responses of ensemble members. The literature on MC systems has grown rapidly in the last 30 years and includes a variety of proposed and developed concepts, algorithms and techniques of ensemble classification methods as well as a wide range of their practical applications.

Among the different methods of operation of the MC systems, the dynamic ensemble selection (DES) scheme seems to be the most mature and promising approach. In the DES methods, base classifiers are dynamically (individually for

M. Choraś et al. (Eds.): CORES 2021, IP&C 2021, ACS 2021, LNNS 255, pp. 3–13, 2022.
https://doi.org/10.1007/978-3-030-81523-3_1

each recognized object) selected from the entire set (pool) and then classification results of selected classifiers are combined by weighted majority voting [8]. In most methods, the base classifiers are selected from the pool according to their accuracy measure (competence) in a local region of the feature space. It means that only an ensemble containing the most competent classifiers is selected to predict the label of a specific test sample. This approach is fully justified because usually, there are no classifiers in the pool that would be experts for all recognized objects, rather base classifiers are expert in different local regions.

In [3,4], we can find a comprehensive review of different DES methods containing the taxonomy and detailed characteristics of DES schemes and their comparative experimental analysis.

Although the presented methods differ in algorithms for determining classifier competence and in the interpretation of competence models justifying the division of methods into different categories (ranking-based, accuracy-based, probabilistic-based and behavior-based measures), they are dominated by solutions in which competence at a point is determined based on the properties of the tested classifier in terms of correct/incorrect classification of validation objects located in the neighborhood of the object being recognized.

Such techniques seem to be a justified approach if we notice that there are serious problems with the natural measure of classifier competence for the feature vector x, that is the probability of correct classification at this point. Firstly, this requires the adoption of a probabilistic model of the recognition task which in the case of many classifiers (e.g. artificial neural network etc.) is simply unjustified. Secondly, for deterministic base classifiers, the probability of correct classification at the point x is equal to either 0 or 1. Thus, the probability of correct classification in no way differentiates (in a continuous scale) base classifiers, but divides them into classifiers correctly/incorrectly classifying the object x.

In this study, we explicitly adopt the probability of correct classification at point x as a measure of classifier competence at this point, but we will do it indirectly through the randomized classifier being a probabilistic model of base classifier. The randomized classifier makes a decision randomly, depending on the value of a certain random variable, so its probability of correct classification at a point x is a number belonging to the $[0,1]$ interval. The randomized model is constructed in such a way that, in a certain probabilistic sense, it is equivalent to the base classifier. Therefore its probability of correct classification is taken as a point measure of the competence of the evaluated classifier.

This approach has already been considered by the authors in earlier papers [13,14]. The concept of the probabilistic model presented there leads to a randomized classifier whose values of classifying functions are the observed values of random variables with a probability distribution related to the values of classification functions produced by the evaluated base classifier. In the original probabilistic model, called Randomized Reference Classifier (RRC), the beta distribution was used. In the modified model, we applied the truncated Gaussian distribution. The developed concept for determining the competence of base classifiers through their randomized models proved to be a very effective approach. This is evidenced by the results of comparative studies presented in the source publications and above all the results presented in the review paper containing the state-of-the-art in the field of MC systems with DES scheme [4]. In experimental comparative studies of 30 MC systems using 30 benchmark databases, the RRC method took 3rd place in terms of average rank and 5th place in terms of average accuracy.

In this paper, a novel randomized model of a base classifier is proposed in which the random operation of the model results from a random selection of the learning set from the family of learning sets of a fixed size. This approach means that the concept of the base classifier is its randomized model determining the algorithm for processing features in the context of a training set before this set is randomly selected.

Although randomization has been applied for a long time in decision-making systems (see e.g. [1,5]), there are not many solutions in the literature that use a random mechanism to build multiple classifier systems. First of all, Kleinberg's method of stochastic discrimination (SD) must be mentioned here [7] and the random forest algorithm derived from SD. Noteworthy, there are bagging and boosting methods related to the construction of an ensemble of classifiers based on different training sets created in the random resampling process. In [10] interesting concept of building an ensemble of classifiers through randomization is presented, in which randomization procedure is associated with the random selection of the parameters of the base classifiers.

The method presented in the paper follows the trend of using randomization procedures to build a multiclassifier system. However, this is done originally and differently from the techniques listed above. Randomization in the presented approach allows determining the measure of competence, based directly on the probability of correct classification (at a given point) which evaluates the quality of the base classifier on a continuous scale.

The paper is organized as follows. Section 2 introduces the formal model of classification problem and gives an insight into the original method of determining competence via randomized model of classifier. Section 3 contains a description of the proposed method. Section 4 presents the multiclassifier system with DES scheme using the proposed method of determining competence measure. Section 5 provides results of experimental comparative analysis. Section 6 concludes the paper.

2 Competence of Base Classifier

2.1 Preliminaries

Let us consider a pattern recognition problem in which $x \in \mathcal{X} = \Re^d$ denotes feature vector of an object and $j \in \mathcal{M} = \{1, 2, 3, ..., M\}$. Is its class number. Let ψ be a trained classifier which maps the feature space $\mathcal{X}$ into the set of class numbers $\mathcal{M}$, viz.:

$$\psi(x) = i, \quad x \in \mathcal{X}, \; i \in \mathcal{M}. \tag{1}$$

We assume that ψ is described by the canonical model [9], i.e. for a given x, it first produces values of normalized classification functions (supports) $g_i(x), i \in \mathcal{M}$ ($g_i(x) \in [0,1], \sum g_i(x) = 1$) and then classifies an object according to the maximum support rule:

$$\psi(x) = i \Leftrightarrow g_i(x) = \max_{k \in \mathcal{M}} g_k(x). \tag{2}$$

Let $c(\psi|x)$ be a competence measure of classifier ψ at the point $x \in \mathcal{X}$. There exist several methods of building local competence measure $c(\psi|x)$ based on different paradigms. Their exhaustive review can be found in [4].

2.2 Competence via Randomization

It seems that the most natural competence measure of the classifier ψ for the feature vector x is the probability of correct classification $P_c(\psi(x))$. If ψ is a deterministic classifier, its operation at point x is strictly dichotomous: classifier correctly classifies x belonging to the j-th class if $\psi(x) = j$, or misclassifies x if $\psi(x) \neq j$. It means that probability of correct classification $Pc(\psi(x))$ of classifier ψ at the point x is either 0 or 1, viz.:

$$Pc(\psi(x)) = [\![\psi(x) = j]\!], \tag{3}$$

where $[\![\cdot]\!]$ denotes the Iverson bracket.

Therefore, a direct application of $Pc(\psi(x))$ as a competence measure $c(\psi|x)$ is not used in this paper.

In the proposed method, we first built a probabilistic model $\bar{\psi}$ of classifier ψ. The model $\bar{\psi}$ depends on random variable Y with probability distribution $P(Y)$. It means that the model $\bar{\psi}(Y)$ is the randomized classifier which makes classification randomly according to a probability distribution $P(\bar{\psi}(Y))$ dependent on $P(Y)$ and $\bar{\psi}$ as a function of Y [1]. Formally, $\bar{\psi}$ is also a function from $\mathcal{X}$ into $\mathcal{M}$ specifying how decisions are made given observation y of the random variable Y. It is obvious that distribution $P(Y)$ and construction of $\bar{\psi}$ should be proposed in such a way that the model is equivalent to ψ in a certain probabilistic sense.

The probability of correct classification of $P_c(\bar{\psi}(x))$ at point x belongs to the interval $[0,1]$. Therefore, we can use the probability of the correct classification of the randomized classifier as a measure of the competence of the classifier ψ at point x, i.e.:

$$c(\psi|x) = P_c(\bar{\psi}(x)). \tag{4}$$

The key problem of the proposed approach is related to the definition of probability distribution $P(Y)$ which results from the adopted relationship between the classifier ψ and its model $\bar{\psi}$.

Randomized Reference Classifier. RRC is a probabilistic classifier which classifying functions $\{\delta_j(x)\}_{j \in \mathcal{M}}$ are observed values of multidimensional random variable $Y = \{\Delta_j(x)\}_{j \in \mathcal{M}}$ fulfilling the following conditions: (i) $\Delta_i(x) \in$

$[0, 1]$, (ii) $\sum_{i \in \mathcal{M}} \Delta_i(x) = 1$, (iii) $\mathbf{E}[\Delta_i(x)] = g_i(x)$, where $\mathbf{E}$ is the expected value operator. Conditions (i) and (ii) arise from the normalization properties of class supports, whereas condition (iii) provides the equivalence of the randomized model $\bar{\psi}_{RRC}$ and base classifier ψ. Based on the latter condition, the RRC can be used to provide randomized model of any classifier that returns vector of class-specific supports $g_i(x), i \in \mathcal{M}$.

It is obvious that the probability of correct classification of an object x belonging to the class j using the RRC is as follows:

$$P_c(\bar{\psi}_{RRC}(x)) = P[\Delta_j(x) > \Delta_k(x), k \in \mathcal{M} \setminus j]. \tag{5}$$

The probability on the right side of (5) can be easily determined if we assume – as in the original work [14] – that random variables $\Delta_i(x)$ have the beta distribution.

Since $\bar{\psi}_{RRC}$ acts – on average – as the modeled base classifier ψ, the formula (4) is fully justified.

RRC with Gaussian Distribution. In the paper [13], we proposed the use of Gaussian distribution truncated to the interval $[0, 1]$ instead of the beta distribution suggested in the original RRC model. It means that in the proposed randomized model $\bar{\psi}_{GD}$ random variable Y is defined as in the RRC model. The expected value for each random variable $\Delta_i(x)$ is simply determined using formula (iii), but the standard deviation is calculated using a rescaled variance of the beta distribution with parameter that should be experimentally tuned in order to achieve the best classification accuracy.

3 The Proposed Model of Competence

3.1 The Mathematical Fundamentals

We suppose that classifier ψ is trained in the supervised learning procedure using a learning set

$$s_n = \{(x_1, j_1), (x_2, j_2), \ldots (x_n, j_n)\}, \tag{6}$$

containing n learning objects and their true classes. Let $\mathscr{S}_n$ be a family of all possible sets of a fixed size n.

Since the dependence of the classifier on the learning set is important for the proposed competence model, we will now treat classifier ψ as a function of two variables:

$$\psi(x, s_n) = i, \quad x \in \mathcal{X}, \; s_n \in \mathscr{S}_n, \; i \in \mathcal{M}. \tag{7}$$

We assume that over the family $\mathscr{S}_n$, random variable S_n with probability distribution $P(S_n)$ is defined. Learning set s_n is the observed value of S_n. It means that set s_n is randomly selected from the family $\mathscr{S}_n$ according to $P(S_n)$.

In the proposed method, we adopt $Y = S_n$. As a consequence, we get the model $\bar{\psi} = \psi(S_n)$.

The difference between $\psi(x, s_n)$ and its model $\psi(x, S_n)$ is obvious. $\psi(x, S_n)$ is a random variable presenting the concept of classifier construction or the algorithm of its activity describing how to process object features x in the context of variables S_n to obtain the classification result. In other words, $\psi(x, S_n)$ refers to the general classifier concept before the training set s_n has been drawn from $\mathscr{S}_n$. Thus, $\psi(x, S_n)$ is randomized classifier whose classification depends on the result of the draw s_n from the family $\mathscr{S}_n$. In turn, $\psi(x, s_n)$ is a classifier trained using a randomly selected learning set s_n which classifies object x into the specific class i from the set $\mathcal{M}$. As it results from the above description, in the proposed method the probability of correct classification of the concept $\psi(x, S_n)$ averaged over the family $\mathscr{S}_n$ is used as the competence of the classifier $\psi(x, s_n)$ being the realization of this concept.

Since random variable $[\![\psi(x, S_n) = j]\!]$ has binary (Bernoulli) distribution (see formula (3)) with success and failure probability equal to P_c and $1 - P_c = P_e$, respectively. We have the following formula for the probability of correct classification of the model $\bar{\psi}$ at the point x:

$$P_c(\bar{\psi}(x)) = P_c(\psi(x, S_n)) = E_{S_n}[\![\psi(x, S_n) = j]\!] \tag{8}$$

$$= \int_{\mathscr{S}_n} [\![\psi(x, s_n) = j]\!] \, dP(s_n). \tag{9}$$

If probability distribution of the draw of set s_n from $\mathscr{S}_n$ is uniform, then we have

$$P_c(\bar{\psi}(x)) = P_c(\psi(x, S_n)) = \frac{\|s_n : P_c(\psi(x, s_n)) = 1\|}{\|\mathscr{S}_n\|}, \tag{10}$$

where $\|A\|$ denotes Lebesgue measure of A.

3.2 Empirical Case

Suppose now that the finite set of learning sets of size n is available:

$$(s_n^{(1)}, s_n^{(2)}, ..., s_n^{(K)}). \tag{11}$$

Then estimation of the average correct classification probability (10) can be calculated as follows:

$$P_c(\psi(x, S_n)) \approx \frac{K_c}{K}, \tag{12}$$

where K_c denotes the number of learning sequences for which $P_c(\psi(x, s_n)) = 1$.

In this case, we have empirical competence measure of classifier ψ at the point x which is the estimated value of probability of correct classification of ψ at this point averaged over learning sets of the fixed size n.

3.3 Competence Set

To calculate (12), the true class of the object x must be known. Thus, we assume that a validation set

$$\mathcal{V} = \{(x_1, j_1), (x_2, j_2), \dots (x_N, j_N)\}. \tag{13}$$

containing pairs of feature vectors and their true class labels is available. In this case, from (4), (12) and (13) we have (in the feature space $\boldsymbol{\mathcal{X}}$) the set of N validation objects with competences of classifier ψ:

$$\mathcal{C}_\psi = \{(x_1, c(\psi|x_1)), (x_2, c(\psi|x_2)), \dots (x_N, c(\psi|x_N))\}. \tag{14}$$

Set $\mathcal{C}_\psi$ is called the competence set of classifier ψ.

3.4 Generalization Method – Potential Function Model

In the second step, based on the competence set (14), the competence function (measure) $c(\psi|x)$ is determined. In other words, information contained in the set $\mathcal{C}_\psi$, i.e. values of competence for validation points $x_k \in \mathcal{V}$, is generalized to the whole feature space $\boldsymbol{\mathcal{X}}$. In the potential function method of generalization, the feature vectors x_k are considered to be the locations of the competence sources $c(\psi|x_k)$ that influence the entire feature space $\mathcal{X}$ creating a competence field. The competence at x is a result of the cumulative influence of the sources, where the influence of each source is proportional to $c(\psi|x_k)$ and it decreases as the distance between x_k and x increases. This interpretation allows using the normalized Gaussian potential function model [15] to construct the competence function as follows:

$$c(\psi|x) = \frac{1}{D} \sum_{x_k \in \mathcal{V}} c(\psi|x_k) \exp(-\text{dist}(x_k, x)^2), \tag{15}$$

where D denotes the normalizing factor and $\text{dist}(x, y)$ is the Euclidean distance between x and y.

4 Multiclassifier System with des Scheme

In the multiclassifier system (MCS), a set of trained classifiers $\Psi = \{\psi_1, \psi_2, \dots, \psi_L\}$ called base classifiers is given. We assume that all classifiers from the set Ψ meet the assumptions presented in Subsect. 2.1 and 3.1. For each classifier $\psi_l \in \Psi$, the competence function $c(\psi_l|x)$ is calculated according to (4), (12) and (15).

The proposed multiclassifier system uses a dynamic ensemble selection (DES) strategy which consists of two steps. In the first step, an ensemble of competent classifiers is selected from the entire set Ψ for a given x

$$\Psi(x) = \{\psi_{l1}, \psi_{l2}, ..., \psi_{lx} : c(\psi_{lm}|x) > \alpha\}, \tag{16}$$

where the threshold value α is arbitrary, but it is usually assumed that $\alpha = 1/M$. This step eliminates inaccurate classifiers (worse than random guessing) and keeps the ensemble relatively diverse.

The selected classifiers are combined using a weighted vector of class supports, where the weights are equal to the competences, viz.:

$$g_j^{(MCS)}(x) = \sum_{\psi_l \in \Psi(x)} c(\psi_l|x)\, g_j^{(\psi_l)}(x). \tag{17}$$

Finally, the MCS system classifies object x using the maximum rule:

$$\psi^{(MCS)}(x) = i \Leftrightarrow g_i^{(MCS)}(x) = \max_{j \in \mathcal{M}} g_j^{(MCS)}(x). \tag{18}$$

It should be emphasized that the computationally complex procedures for determining the source competence measure for validation objects – thanks to the concept of the competence set – is fully implemented in the learning phase. In the classification phase, only a simple (and quick) procedure of determining the classifying functions of the ensemble classifier and making decisions according to the maximum principle remains to be implemented.

5 Experiments

The experimental study is aimed at comparing the proposed method of classifier selection with the original approach proposed in [14], and its modification using the Gaussian Distribution [13]. This paper does not contain a comparison between RRC based DES algorithms and the state-of-the-art methods of building classifier ensembles. This is because that kind of comparison has already been done [4].

5.1 Experimental Setup

During the experimental evaluation the following heterogeneous ensemble classifiers were compared: ψ_{C} – DES scheme with the proposed model of competence, ψ_{B} – DES scheme proposed in [14], ψ_{N} – DES scheme proposed in [13].

The following base classifiers were used to constitute each of the ensembles: ψ_{NB} – Naive Bayes classifier with kernel density estimation, ψ_{J48} – Weka version of the C4.5 algorithm with Laplace smoothing, ψ_{NC} – Nearest centroid (Nearest Prototype), ψ_{SVM} – SVM classifier with linear kernel (no kernel), ψ_{LDA} – LDA classifier, ψ_{QDA} – QDA classifier, ψ_{KNN} – Nearest neighbour classifier.

Each of the above-mentioned base classifiers was trained using 31 bootstrap samples from the training dataset. The outcomes of the classifiers built using bootstrap samples were used to calculate the correct classification probability using formula (12).

The experimental code was implemented using WEKA framework. The source code of the algorithms is available online[1].

To evaluate the proposed methods, the following classification-loss criteria were used [11]: macro-averaged and micro-averaged FDR (1-precision), FNR (1-recall), F_1, Matthews correlation coefficient (MCC). Macro and micro averaged criteria were used because those families of criteria differ significantly in scoring majority and minority classes. That is, macro averaged criteria are more sensitive to performance for the minority class. Whereas micro averaged criteria are more sensitive to majority class performance.

Following the recommendations of [6], the statistical significance of the obtained results was assessed using the two-step procedure. The first step is to perform the Friedman test [6] for each quality criterion separately. Since the multiple criteria were employed, the familywise errors (FWER) should be controlled [2]. To do so, the Bergman-Hommel [2] procedure of controlling FWER of the conducted Friedman tests was employed. When the Friedman test shows that there is a significant difference within the group of classifiers, the pairwise tests using the Wilcoxon signed-rank test [6] were employed. To control FWER of the Wilcoxon-testing procedure, the Bergman-Hommel approach was employed [2]. For all tests the significance level was set to $\alpha = 0.05$.

The experimental evaluation was conducted on the collection of the 67 benchmark datasets taken from the Keel repository. Benchmark datasets are available online[2].

During the preprocessing stage, the datasets underwent a few transformations. First, all nominal attributes were converted into a set of binary variables. The transformation is necessary whenever the distance-based algorithms are employed [12]. To reduce the computational burden and remove irrelevant information, the PCA procedure with the variance threshold set to 95% was applied. The features were also normalized to have zero mean value and unit variance.

5.2 Results and Discussion

To compare multiple algorithms on multiple benchmark sets, the average ranks approach [6] is used. Due to the page limit, full results are published online[3].

The numerical results are given in Table 1. The table is divided into eight sections – one section is related to a single evaluation criterion. The first row of each section is the name of the quality criterion investigated in the section. The second row shows the p-value of the Friedman test. The third one shows the average ranks achieved by algorithms. The following rows show p-values resulting from pairwise Wilcoxon test. Above each section the names of the investigated algorithms are placed.

[1] https://github.com/ptrajdos/rrcBasedClassifiers/tree/develop.

[2] https://github.com/ptrajdos/MLResults/blob/master/data/KeelData.tar.xz.

[3] https://github.com/ptrajdos/MLResults/raw/a3b4168a0b0aabee7ef8cd1056baf4a8578a9f6d/RandomizedClassifiers/CORES2021.zip.

Table 1. Statistical evaluation. Wilcoxon test results.

	ψ_C	ψ_B	ψ_N	ψ_C	ψ_B	ψ_N	ψ_C	ψ_B	ψ_N	ψ_C	ψ_B	ψ_N
Nam.	MaFDR			MaFNR			MaF1			Macro Matthews		
Frd.	1.000e+00			2.523e−01			3.470e−01			1.000e+00		
Rank	1.955	2.061	1.985	1.795	1.992	2.212	1.803	2.000	2.197	1.939	1.932	2.129
ψ_C		.457	.457		.174	**.036**		.150	**.007**		.236	.224
ψ_B			.652			.148			.154			.224
Nam.	MiFDR			MiFNR			MiF1			Micro.Matthews		
Frd.	1.000e+00			1.000e+00			1.000e+00			1.000e+00		
Rank	1.871	1.985	2.144	1.871	1.985	2.144	1.871	1.985	2.144	1.864	1.992	2.144
ψ_C		.411	.411		.411	.411		.411	.411		.386	.370
ψ_B			.411			.411			.411			.370

According to the table, the proposed approach achieved the lowest ranks for all the investigated quality criteria except macro-averaged MCC. For the macro-averaged MCC, the proposed approach is the second one; however, the difference between the average ranks is very small.

On the other hand, the approach based on the Gaussian distribution achieves the lowest rank in terms of all quality criteria except macro-averaged FDR.

However, the conducted statistical tests show that there are almost no significant differences between the investigated algorithms. The significant differences are only between the proposed method and the Gaussian-based RRC-DES. These differences show that the proposed approach is better in terms of macro averaged FNR and macro-averaged F_1 measure. It means that the proposed approach is better at discovering minority class objects. This is done without lowering precision. This property may be useful for the imbalanced classification task. The results show that using the data-driven distribution for the RRC model may improve the classification quality.

6 Conclusions

In this paper, we proposed a data-driven approach to building the RRC-based dynamic-ensemble-selection classifier. The classifier was compared with the original RRC-based approach and the RRC-method built using Gaussian distribution. The conducted experimental study shows that there are almost no significant differences between the proposed method and the previously-proposed algorithms. The only observed differences may suggest that using the proposed approach in imbalanced classification problems may allow a significant improvement over the previously mentioned methods. However, this result is achieved at the cost of training an ensemble of classifiers using bootstraped datasets. Consequently, future research is aimed at building a distribution-driven RRC approach with the reduced computational burden.

Acknowledgment. This work was supported by the statutory funds of the Department of Systems and Computer Networks, Wroclaw University of Science and Technology.

References

1. Berger, J.O.: Statistical Decision Theory and Bayesian Analysis. Springer, New York (1985). https://doi.org/10.1007/978-1-4757-4286-2
2. Bergmann, B., Hommel, G.: Improvements of general multiple test procedures for redundant systems of hypotheses. In: Bauer, P., Hommel, G., Sonnemann, E. (eds.) Multiple Hypothesenprüfung/Multiple Hypotheses Testing, vol. 70, pp. 100–115. Springer, Heidelberg (1988). https://doi.org/10.1007/978-3-642-52307-6_8
3. Britto, A.S., Sabourin, R., Oliveira, L.E.: Dynamic selection of classifiers–a comprehensive review. Pattern Recognit. **47**(11), 3665–3680 (2014). https://doi.org/10.1016/j.patcog.2014.05.003
4. Cruz, R.M., Sabourin, R., Cavalcanti, G.D.: Dynamic classifier selection: recent advances and perspectives. Inf. Fusion **41**, 195–216 (2018). https://doi.org/10.1016/j.inffus.2017.09.010
5. DeGroot, M.H.: Optimal Statistical Decisions. Wiley, New York (2004). https://doi.org/10.1002/0471729000
6. Demšar, J.: Statistical comparisons of classifiers over multiple data sets. J. Mach. Learn. Res. **7**, 1–30 (2006)
7. Kleinberg, E.: Stochastic discrimination. Ann. Math. Artif. Intell. **1**(1–4), 207–239 (1990). https://doi.org/10.1007/bf01531079
8. Ko, A.H., Sabourin, R., Britto, A.S., Jr.: From dynamic classifier selection to dynamic ensemble selection. Pattern Recognit. **41**(5), 1718–1731 (2008). https://doi.org/10.1016/j.patcog.2007.10.015
9. Kuncheva, L.I.: Combining Pattern Classifiers. Wiley, New York (2014). https://doi.org/10.1002/9781118914564
10. Santucci, E., Didaci, L., Fumera, G., Roli, F.: A parameter randomization approach for constructing classifier ensembles. Pattern Recognit. **69**, 1–13 (2017). https://doi.org/10.1016/j.patcog.2017.03.031
11. Sokolova, M., Lapalme, G.: A systematic analysis of performance measures for classification tasks. Inf. Process. Manag. **45**(4), 427–437 (2009). https://doi.org/10.1016/j.ipm.2009.03.002
12. Tian, Y., Deng, N.: Support vector classification with nominal attributes. In: Hao, Y., et al. (eds.) Computational Intelligence and Security, pp. 586–591. Springer, Heidelberg (2005). https://doi.org/10.1007/11596448_86
13. Trajdos, P., Kurzynski, M.: Randomized reference classifier with Gaussian distribution and soft confusion matrix applied to the improving weak classifiers. In: Burduk, R., Kurzynski, M., Wozniak, M. (eds.) Progress in Computer Recognition Systems. Advances in Intelligent Systems and Computing, vol. 977, pp. 326–336. Springer, Cham (2019). https://doi.org/10.1007/978-3-030-19738-4_33
14. Woloszynski, T., Kurzynski, M.: A probabilistic model of classifier competence for dynamic ensemble selection. Pattern Recogn. **44**(10–11), 2656–2668 (2011). https://doi.org/10.1016/j.patcog.2011.03.020
15. Woloszynski, T., Kurzynski, M., Podsiadlo, P., Stachowiak, G.W.: A measure of competence based on random classification for dynamic ensemble selection. Inf. Fusion **13**(3), 207–213 (2012). https://doi.org/10.1016/j.inffus.2011.03.007

Hybrid Learning Model for Syntactic Pattern Recognition

Mariusz Flasiński[1(✉)], Janusz Jurek[1], and Tomasz Peszek[2]

[1] Information Technology Systems Department, Jagiellonian University, ul. prof. St. Łojasiewicza 4, 30-348 Cracow, Poland
mariusz.flasinski@uj.edu.pl

[2] School of Banking and Management, al. Kijowska 14, 30-079 Cracow, Poland

Abstract. The novel hybrid learning model based on neural networks and grammatical inference is proposed in the paper. The model is used within the multi-derivational parsing of vague language methodology. The foundations of the methodology and the learning algorithms are presented. The model has been used for the implementation of the short term electrical load forecasting system.

1 Introduction

There are two main subareas in pattern recognition: the decision-theoretic subarea [3,5] and the syntactic/structural one, which - in turn - can be divided into the structural approach [4], the algebraic approach [22] and the syntactic approach, SPR, [12,14,15]. The signal structure analysis (e.g. ECG, EEG and industrial equipment-generated) has been one of the main applications of SPR since the early 1970s [14]. The efficient parsing of DPLL(k) grammars [10,18] was used for signal analysis in high energy physics [2], otolaryngology (audiometry-based diagnosis) [11], fetal palates diagnosis [21] and electrical load forecasting [13]. In the case of the last application area, new signal structural patterns occur relatively often. Therefore, the self-learning of the underlying grammar by the SPR system, called *SPRELP* (*Syntactic Pattern Recognition-based Electrical Load Prediction system*) [13], has appeared to be the crucial requirement. Since the methodology used in *SPRELP* is based on both Kohonen neural networks and the parsing system [13], the hybrid model of learning based on the NNs learning and the grammatical inference for DPLL(k) grammars has been developed. This hybrid learning model is presented in the paper. The scheme of the *SPRELP* system is introduced in Sect. 2. The idea of the learning of the so-called vague patterns by the Kohonen NNs is presented in Sect. 3. The fourth section contains the grammatical inference rules. The application results are discussed in Sect. 5, whereas the final section includes the conclusions.

M. Choraś et al. (Eds.): CORES 2021, IP&C 2021, ACS 2021, LNNS 255, pp. 14–22, 2022.
https://doi.org/10.1007/978-3-030-81523-3_2

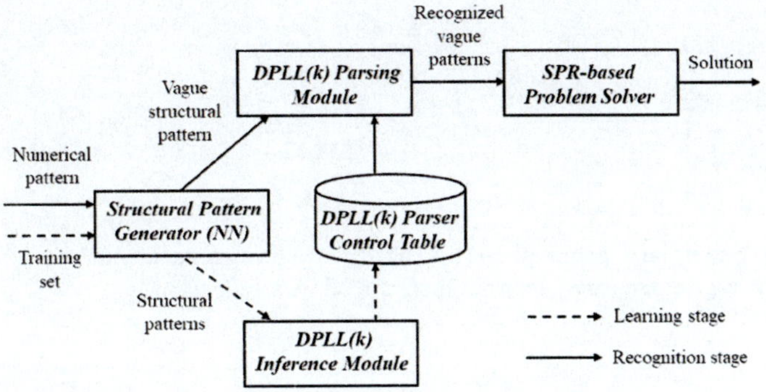

Fig. 1. The scheme of the vague DPLL(k) language SPR system.

2 Model of Vague Language Parsing

The losing of information on the fuziness/uncertainty of objects/phenomena during the transforming of their numerical representations into structural (idealized) patterns in one of the main open problems in syntactic pattern recognition [14,15]. Therefore, we have proposed the methodology which is based on the two-stage recognition scheme in [13].

A general scheme of a syntactic pattern recognition system which is constructed according to this methodology is shown in Fig. 1. The so-called vague structural patterns are generated by a neural network at the first stage. A *vague structural pattern* consists of vague primitives which represent the fuzzy/uncertain nature of patterns. (This stage is discussed in Sect. 3.) Vague structural patterns are analyzed/recognized by the DPLL(k) parsing module at the second stage. Instead of determining one "correct" (idealized) pattern, the parsing module can identify b best vague patterns which are acceptable as approximations of the "idealized" template. These best structural patterns are attributed with some numerical measures. These measures can be of the form of distance, probability, fuzziness measures, etc. and they are, finally, used for computing the "averaged" pattern by the syntactic pattern recognition-based problem solver. This recognition path of our approach has been presented in [13] in a more detailed way.

As we have mentioned in Introduction, if new structural patterns occur relatively often, a syntactic pattern recognition system should allow us to learn a grammar automatically [14,15]. Therefore, the learning path has been constructed for our scheme. At the first stage of this path, the neural network is trained and it generates structural patterns which, in turn, are used as the input to the DPLL(k) inference module. At the second stage of the learning path, this modules generates the underlying DPLL(k) grammar which constitutes the DPLL(k) parser control table.

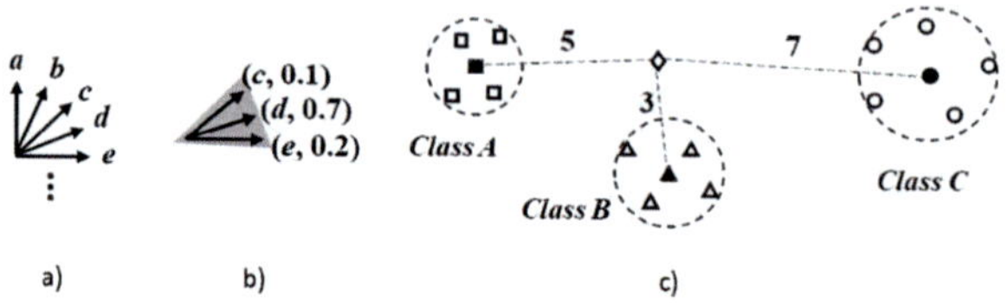

Fig. 2. (a) Exemplary primitives, (b) the exemplary vague primitive, (c) ascribing weights to a vague primitive according to the NN classification.

3 Neural Network as Vague Structural Pattern Generator

As we have mentioned in the previous section, a vague structural pattern consists of vague primitives which are defined as follows. Let Σ_T be a t-element set of terminal symbols (of a grammar). A *vague primitive* is a vector $\mathbf{v} = (a_{k_1}\mu_{k_1}, a_{k_2}\mu_{k_2}, \ldots, a_{k_j}\mu_{k_j})$, $1 \leq j \leq t$, where: $a_{k_1}, a_{k_2}, \ldots, a_{k_j} \in \Sigma_T$ are different symbols, $\mu_{k_1}, \mu_{k_2}, \ldots, \mu_{k_j}$ are the measures, called *attributes*, ascribed to $a_{k_1}, a_{k_2}, \ldots, a_{k_j}$, correspondingly. We have also assumed that attributes can be of the form of distance, probability, fuzziness measures, etc. For example, if a subset of a set of primitives is defined as in Fig. 2a, then we can define a vague primitive as it is shown in Fig. 2b. The Euclidean metric is used for attributing vague primitives in the present version of our model. In order to generate fuzzy primitives on the basis of numerical patterns, the Kohonen neural network is applied. Its neurons represent the classes. (The classes correspond to the terminal symbols (from Σ_T) of the underlying grammar.)

The activation function of the neuron is based on the metric criterion. Strictly speaking, it is calculated on the basis of the reciprocal of the distance between the unknown pattern presented and the centroid of the cluster represented by this neuron. The activations of the neurons are normalized to the unit vector. For example, if the unknown pattern is placed in the feature space as it is shown in Fig. 2c (the numbers are the distances), then the attribute for the class/structural primitive A is calculated as $\mu_A = \frac{1/5}{1/5+1/3+1/7}$. During the recognition stage, after the presenting of the unknown numerical pattern, only the neurons with the highest activations are identified in order to generate a fuzzy primitive.

During the unsupervised learning stage, the maximum diameter of the class (neuron) is set. Let $N_A(j)$ be the cluster corresponding to the class/neuron A after the j-th iteration, $N_A(j) = \{\mathbf{V}_1, \ldots, \mathbf{V}_u\}$. Let $\mathbf{C}_A(j) = \frac{\mathbf{V}_1 + \ldots + \mathbf{V}_u}{u}$ be the centroid of $N_A(j)$. Let $\mathbf{X}$ be the unknown pattern presented at the $(j+1)$-th iteration. The learning rules are defined in the following way.

$$N_A(j+1) = \begin{cases} N_A(j), & \mathbf{X} \text{ has not been assigned to class } A, \\ N_A(j) \cup \{\mathbf{X}\}, & \mathbf{X} \text{ has been assigned to class } A, \end{cases}$$

and

$$\mathbf{C}_A(j+1) = \begin{cases} \mathbf{C}_A(j), & \mathbf{X} \text{ has not been assigned to class } A, \\ \frac{u \cdot \mathbf{C}_A(j) + \mathbf{X}}{u+1}, & \mathbf{X} \text{ has been assigned to class } A. \end{cases}$$

4 DPLL(k) Grammar Inference Scheme

As we have mentioned, the class of DPLL(k) grammars [10] is used in our approach. Let us introduce this class in the following way.

Definition 4.1. A *dynamically programmed (context-free) grammar* is a quintuple $G = (\Sigma_N, \Sigma_T, P, S)$, where: Σ_N is a set of nonterminal symbols; Σ_T is a set of terminal symbols; P is a set of n productions of the form: $p_i = (\pi_i, L_i, R_i, A_i, DCL_i)$, in which $\pi_i : \bigcup_{k=1,\ldots,n} DCL_k \to \{TRUE, FALSE\}$ is the predicate of applicability of p_i; $L_i \in \Sigma_N$ and $R_i \in \Sigma^*$ are the left- and right-hand sides of p_i, respectively; A_i is the sequence of actions *add, read, move* performed over $\bigcup_{k=1,\ldots,n} DCL_k$; DCL_i is the derivation control tape for p_i; S is the start symbol (axiom), $S \in \Sigma_N$. □

Definition 4.2. Let $G = (\Sigma_N, \Sigma_T, P, S)$ be a dynamically programmed context-free grammar. The grammar G is called a *DPLL(k) grammar*, if the following two conditions are fulfilled.

1. The Stearns's condition of the LL(k) grammars [23]. (The top-down left-hand side derivation is deterministic if it is allowed to look at k input symbols to the right of the current position of the input head in the string).
2. There exists a certain number σ such that after the application of at most σ productions we get at least one new terminal symbol. □

The grammatical inference rules for DPLL(k) grammars operate on the so-called *polynomial specification of the language* [20]. Let us introduce the following definition.

Definition 4.3. Let T be the set of terminal symbols which occur in the sample, V be a subset of positive integers. *Polynomial specification of a language* is of the form $L(T, V) = S_i^{w_j(n_k)}$, where w_j is a polynomial of a variable $n_k \in V$. S_i is called a *polynomial structure* and it is defined in a recursive way as follows: (1) $S_i = (a_{i_1} ... a_{i_r})$, where $a_{i_j} \in T$ (Then S_i is called a *basic* polynomial structure.) or (2) $S_i = (S_{i_1}^{w_{i_1}(n_{i_1})} ... S_{i_r}^{w_{i_r}(n_{i_r})})$, where S_{i_k} is defined as in (1) or (2). (Then S_i is called a *complex* polynomial structure.) □

For example, let $T = \{a, b, c\}$, $V = \{n\}$ be a set of integer variables. Then, $L(T, V) = ((ab)^n c^{n^2+2})^n$ is the example of a polynomial specification of a language. The following polynomial structures are defined for $L(T, V)$.

$S_1 = (ab)^n c^{n^2+2}$	$w_1(n) = n$
$S_{1_1} = ab$	$w_{1_1}(n) = n$
$S_{1_2} = c$	$w_{1_2}(n) = n^2 + 2$

Now, we can define the rules of grammatical inference for DPLL(k) grammars, according to the model presented in [19].

The DPLL(k) Grammar Inference Rules. Let $L(T, V)$ be a polynomial specification of a language L. The DPLL(k) grammar $G = (\Sigma_N, \Sigma_T, P, S)$ generating the language L is defined in the following way.

1) $\Sigma_T := T$.
2) For each $n \in V$, positive integer variables v_n and d_n are defined.
3) For each polynomial structure of $L(T, V)$ which is denoted by S, the nonterminal $X_S \in \Sigma_N$ and positive integer variables c_S and e_S are defined.
4) For each polynomial structure of $L(T, V)$ which is denoted by S, the following productions of P are defined.

- if S is of the form of $(a_1 ... a_r)$ then the productions are defined as follows:

l	π_l	$L_l \longrightarrow R_l$	A_l
1_S	$c_S = 0$	$X_S \longrightarrow a_1 ... a_r X_S$	$c_S := 1; e_S := w(v_n)$
2_S	$c_S < e_S$	$X_S \longrightarrow a_1 ... a_r X_S$	$c_S := c_S + 1$
3_S	$(c_S = e_S)$ and $(d_n = true)$	$X_S \longrightarrow \lambda$	$c_S := 0;$
4_S	$(c_S = e_S)$ and $(d_n = false)$	$X_S \longrightarrow \lambda$	$c_S := 0; d_n := true$
5_S	$(c_S = e_S)$ and $(d_n = false)$	$X_S \longrightarrow a_1 ... a_r X_S$	$c_S := c_S + 1; v_n := v_n + 1;$ $e_S := w(v_n)$

- if S is of the form of $(S_1^{w_1(n_1)} ... S_r^{w_r(n_r)})$ then the productions are defined as follows:

l	π_l	$L_l \longrightarrow R_l$	A_l
1_S	$c_S = 0$	$X_S \longrightarrow X_{S_1} ... X_{S_r} X_S$	$c_S := 1; e_S := w(v_n)$
2_S	$c_S < e_S$	$X_S \longrightarrow X_{S_1} ... X_{S_r} X_S$	$c_S := c_S + 1$
3_S	$(c_S = e_S)$ and $(d_n = true)$	$X_S \longrightarrow \lambda$	$c_S := 0;$
4_S	$(c_S = e_S)$ and $(d_n = false)$	$X_S \longrightarrow \lambda$	$c_S := 0; d_n := true$
5_S	$(c_S = e_S)$ and $(d_n = false)$	$X_S \longrightarrow X_{S_1} ... X_{S_r} X_S$	$c_S := c_S + 1; v_n := v_n + 1;$ $e_S := w(v_n)$

5) Let X_{S_1} be the nonterminal defined for the first polynomial structure. The initial production of P is defined as follows:

l	π_l	$L_l \longrightarrow R_l$	A_l
1	$true$	$S \longrightarrow X_{S_1}$	$c_{S_1} := 0; ...; c_{S_q} := 0;\ e_{S_1} := -1; ...; e_{S_q} := -1$ $v_{n_1} := 1; ...; v_{n_t} := 1;\ d_{n_1} := false; ...; d_{n_t} := false$

Let us consider the following example of the grammatical inference according to the rules defined. We infer the DPLL(k) grammar for the exemplary polynomial specification of the language $L(T, V) = ((ab)^n c^{n^2+2})^n$ introduced above. Since there is only one variable $n \in V$, we define two corresponding variables v_n and d_n.

There are three polynomial structures: S_1, S_{1_1}, S_{1_2}. Therefore, we define three corresponding nonterminals: X_1, X_{11}, X_{12} in Σ_N, and six variables: c_1, c_{11}, c_{12}, e_1, e_{11}, e_{12}. Let us notice that S_{1_1} and S_{1_2} are basic polynomial structures, and S_1 is a complex polynomial structure. One can easily notice that the following grammar productions are defined according to the inference rules.

l	π_l	$L_l \longrightarrow R_l$	A_l
1_{11}	$c_{11} = 0$	$X_{11} \longrightarrow abX_{11}$	$c_{11} := 1; e_{11} := v_n$
2_{11}	$c_{11} < e_{11}$	$X_{11} \longrightarrow abX_{11}$	$c_{11} := c_{11} + 1$
3_{11}	$(c_{11} = e_{11})$ and $(d_n = true)$	$X_{11} \longrightarrow \lambda$	$c_{11} := 0;$
4_{11}	$(c_{11} = e_{11})$ and $(d_n = false)$	$X_{11} \longrightarrow \lambda$	$c_{11} := 0; d_n := true$
5_{11}	$(c_{11} = e_{11})$ and $(d_n = false)$	$X_{11} \longrightarrow abX_{11}$	$c_{11} := c_{11} + 1; v_n := v_n + 1;$ $e_{11} := v_n$
1_{12}	$c_{12} = 0$	$X_{12} \longrightarrow cX_{12}$	$c_{12} := 1; e_{12} := v_n * v_n + 2$
2_{12}	$c_{12} < e_{12}$	$X_{12} \longrightarrow cX_{12}$	$c_{12} := c_{12} + 1$
3_{12}	$(c_{12} = e_{12})$ and $(d_n = true)$	$X_{12} \longrightarrow \lambda$	$c_{12} := 0;$
4_{12}	$(c_{12} = e_{12})$ and $(d_n = false)$	$X_{12} \longrightarrow \lambda$	$c_{12} := 0; d_n := true$
5_{12}	$(c_{12} = e_{12})$ and $(d_n = false)$	$X_{12} \longrightarrow cX_{12}$	$c_{12} := c_{12} + 1; v_n := v_n + 1;$ $e_{12} := v_n * v_n + 2$
1_1	$c_1 = 0$	$X_1 \longrightarrow X_{11}X_{12}X_1$	$c_1 := 1; e_1 := v_n$
2_1	$c_1 < e_1$	$X_1 \longrightarrow X_{11}X_{12}X_1$	$c_1 := c_1 + 1$
3_1	$(c_1 = e_1)$ and $(d_n = true)$	$X_1 \longrightarrow \lambda$	$c_1 := 0;$
4_1	$(c_1 = e_1)$ and $(d_n = false)$	$X_1 \longrightarrow \lambda$	$c_1 := 0; d_n := true$
5_1	$(c_1 = e_1)$ and $(d_n = false)$	$X_1 \longrightarrow X_{11}X_{12}X_1$	$c_1 := c_1 + 1; v_n := v_n + 1;$ $e_1 := v_n$

Finally, the initial production of P is defined as follows.

l	π_l	$L_l \longrightarrow R_l$	A_l
1	*true*	$S \longrightarrow X_1$	$c_1 := 0; c_{11} := 0; c_{12} := 0$ $e_1 := -1; e_{11} := -1; e_{12} := -1$ $v_n := 1; d_n := false$

5 Application for STLF

According to the scheme of the vague DPLL(k) language SPR system with the learning path described in the previous sections, the short-term (electrical) load forecasting (STLF) system has been constructed. The *SPRELP* (*Syntactic Pattern Recognition-based Electrical Load Prediction system*) has been developed with one of the biggest Polish electricity distribution companies (approx. 50 TWh of electricity supplied to ca 5.5 million customers per year). The forecast is generated on the basis of the input numerical pattern which consists of: temperature (hourly forecast), insolation (hourly forecast) and day characteristics. During the learning stage, such data from the previous periods together with the previous STLFs constitute the historical data used for both learning of the Kohonen neural network and then the DPLL(k) grammar inference.

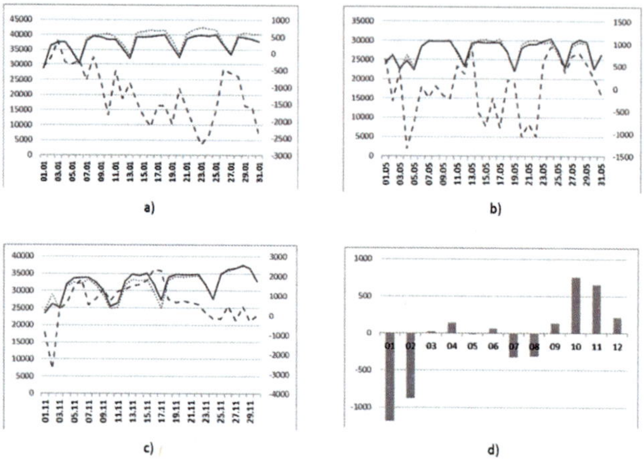

Fig. 3. The daily actual load in MWh (solid line) vs *SPRELP* forecast (dotted line) on the left scale and the forecast error (dashed line) on the right scale for typical months of: winter (a), spring (b) and autumn (c); the monthly forecast error (d).

The exemplary both the daily actual load and the *SPRELP* forecast are shown in Figs. 3 (a)–(c). One can easily notice that the forecast errors are different each from other in various seasons. Whereas during springs the forecast errors are small (the same holds for summers), they increase during winters and autumns (which results from less precise weather predictions) as it is shown in Fig. 3(d).

Table 1. The comparison of the STLF methods

Method	MAPE	MAE	RMSE
Amjady & Keynia 2009 [1]	1.68–2.02	87–124	*NA*
Fan & Hyndman 2012 [6]	1.88	110	*NA*
Yang et al. 2013 [26]	5.13	97	*NA*
Hong & Wang 2014 [17]	2.81	*NA*	*NA*
Wang et al. 2016 [25]	3.86	*NA*	*NA*
Hermias et al. 2017 [16]	5.92	454	579
Tian et al. 2018 [24]	3.96	692	1134
Our method	2.84	370	399

The performance of our hybrid NN/syntactic pattern recognition model has been compared with the performance of the STLF methods published recently. For the forecast accuracy the following metrics has been compared: MAPE (Mean Absolute Percentage Error - the metrics which is used commonly to evaluate the performance of STLF methods), MAE (Mean Absolute Error) and RMSE (Root Mean Square Error). The comparison results are shown in Table 1. It can be noticed that our model generates reasonable good forecasts in terms of these criteria.

6 Concluding Remarks

The solution of the one of the main open problems in syntactic pattern recognition, that is the losing of information on the fuziness/uncertainty of objects/phenomena during the transforming of their numerical representations into structural patterns [14,15] has been proposed. We have focused on the learning stage in the syntactic pattern recognition scheme in the paper. The novel concept of the vague languages has been introduced as the basis paradigm of the solution proposed. During the learning stage, the vague languages are generated by the Kohonen neural networks. Then, the vague word is treated as the element of the sample by the grammatical inference algorithm. This hybrid learning model has been used for the implementing of the learning path in the syntactic pattern recognition framework. The framework has been applied in the area of the short-term electrical load forecasting and it has given satisfactory results. It seems that the performance of the model could be improved, if the forecast was not generated for a big area as a whole but for its several subareas separately. The graph model could be then used for the representing of divided areas and the graph parsing [7–9] could be used for the structural analysis. The research into this problem is to be launched and its results can be the subject of further publications.

References

1. Amjady, N., Keynia, F.: Short-term load forecasting of power systems by combination of wavelet transform and neuro-evolutionary algorithm. Energy **34**, 46–57 (2009)
2. Behrens, U., Flasiński, M., Hagge, L., Jurek, J., Ohrenberg, K.: Recent developments of the ZEUS expert system ZEX. IEEE Trans. Nucl. Sci. **NS-43**, 65–68 (1996)
3. Bishop, C.M.: Pattern Recognition and Machine Learning. Springer, New York (2006)
4. Bunke, H., Sanfeliu, A. (eds.): Syntactic and Structural Pattern Recognition - Theory and Applications. World Scientific, Singapore (1990)
5. Duda, R.O., Hart, P.E., Stork, D.G.: Pattern Classification. Wiley, New York (2001)
6. Fan, S., Hyndman, R.J.: Short-term load forecasting based on a semi-parametric additive model. IEEE Trans. Power Syst. **27**, 134–141 (2012)
7. Flasiński, M.: Parsing of edNLC-graph grammars for scene analysis. Pattern Recogn. **21**, 623–629 (1988)
8. Flasiński, M.: Distorted pattern analysis with the help of node label controlled graph languages. Pattern Recogn. **23**, 765–774 (1990)
9. Flasiński, M.: On the parsing of deterministic graph languages for syntactic pattern recognition. Pattern Recogn. **26**, 1–16 (1993)
10. Flasiński, M., Jurek, J.: Dynamically programmed automata for quasi contexts sensitive languages as a tool for inference support in pattern recognition-based real-time control expert systems. Pattern Recogn. **32**, 671–690 (1999)

11. Flasiński, M., Reroń, E., Jurek, J., Wójtowicz, P., Atłasiewicz, K.: On the construction of the syntactic pattern recognition-based expert system for Auditory Brainstem Response analysis. In: Kurzyński, M., Puchała, E., Woźniak, M., Żołnierek, A. (eds.) Computer Recognition Systems. Advances in Soft Computing, vol. 30, pp. 503–510. Springer, Heidelberg (2005). https://doi.org/10.1007/3-540-32390-2_59
12. Flasiński, M.: Syntactic pattern recognition: paradigm issues and open problems. In: Chen, C.H. (ed.) Handbook of Pattern Recognition and Computer Vision, chap. 1, 5th edn, pp. 3–25. World Scientific, New Jersey (2016)
13. Flasiński, M., Jurek, J., Peszek, T.: Application of syntactic pattern recognition methods for electrical load forecasting. In: Burduk, R., Jackowski, K., Kurzyński, Woźniak, M., Żołnierek, A. (eds.) CORES 2015. Advances in Intelligent Systems and Computing, vol. 403, pp. 599–608. Springer, Cham (2016). https://doi.org/10.1007/978-3-319-26227-7_56
14. Flasiński, M.: Syntactic Pattern Recognition. World Scientific, New Jersey (2019)
15. Fu, K.S.: Syntactic Pattern Recognition and Applications. Prentice Hall, Englewood Cliffs (1982)
16. Hermias, J.P., Teknomo, K., Monje, J.C.N.: Short-term stochastic load forecasting using autoregressive integrated moving average models and hidden Markov model. In: 2017 International Conference on Information and Communication Technologies (ICICT), Karachi, pp. 131–137 (2017)
17. Hong, T., Wang, P.: Fuzzy interaction regression for short term load forecasting. Fuzzy Optim. Decis. Making **13**, 91–103 (2014)
18. Jurek, J.: On the linear computational complexity of the parser for quasi context sensitive languages. Pattern Recogn. Lett. **21**, 179–187 (2000)
19. Jurek, J.: Syntactic Pattern Recognition with the GDPLL(k) Grammars. Habilitation Dissertations Series, vol. 365. Jagiellonian University Publishers, Cracow (2005). (in Polish)
20. Jurek, J.: Generalisation of a language sample for grammatical inference of GDPLL(k) grammars. In: Kurzyński, M., Puchała, E., Woźniak, M., Żołnierek, A. (eds.) CORES 2007. Advances in Soft Computing, vol. 45, pp. 282–288. Springer, Heidelberg (2007). https://doi.org/10.1007/978-3-540-75175-5_35
21. Jurek, J., Wójtowicz, W., Wójtowicz, A.: Syntactic pattern recognition-based diagnostics of fetal palates. Pattern Recogn. Lett. **133**, 144–150 (2020)
22. Kulikowski, J.L.: Algebraic Methods in Pattern Recognition. Springer, Wien (1971). https://doi.org/10.1007/978-3-7091-2884-8
23. Lewis, P.M., II., Stearns, R.E.: Syntax-directed transduction. J. ACM **15**, 465–488 (1968)
24. Tian, C., Ma, J., Zhang, C., Zhan, P.: A deep neural network model for short-term load forecast based on long short-term memory network and convolutional neural network. Energies **11**, 3493 (2018)
25. Wang, P., Liu, B., Hong, T.: Electric load forecasting with recency effect: a big data approach. Int. J. Forecast. **32**, 585–597 (2016)
26. Yang, Y., Wu, J., Chen, Y., Li, C.: A new strategy for short-term load forecasting. Abstract Appl. Anal. 208964 (2013)

Distance Metrics in Clustering and Weighted Scoring Algorithm

Jakub Klikowski and Robert Burduk(✉)

Department of Systems and Computer Networks,
Wroclaw University of Science and Technology, Wroclaw, Poland
{jakub.klikowski,robert.burduk}@pwr.edu.pl

Abstract. One of the current challenges for the supervised classification methods is to obtain acceptable values of the performance measures for an imbalanced dataset. There is a significant disproportion in the number of objects from different class labels in datasets with a high imbalanced ratio. This article analyzes the clustering and weighted scoring algorithm based on estimating the number of clusters that consider the minimum number of objects from the minority class label in each cluster. This algorithm uses the distance metric when determining the value of the score function. Therefore, this article aims to analyze the impact of selecting the distance metric on the six classification performance measures' value. The performed experiments show that the Euclidean distance allows obtaining the best classification results for imbalanced datasets.

Keywords: Imbalanced data · Ensemble of classifiers · Class imbalance · Decision boundary · Scoring function

1 Introduction

One of the trends of machine learning is supervised classification. The dataset in which one attribute identifies the class label can be used for supervised classification. Otherwise, the dataset may be used by the unsupervised classification algorithm. In real-world recognition problems, it is very rare that the number of objects in each class label is equal. In an imbalanced dataset, the number of objects available for different class labels is unequal e.g. 90% objects are from one class label, and 10% are for the other class label. Such a dataset is also called a skewed dataset. Many practical problems concern skewed dataset because they arise directly from the problem's characteristics and the available training data [10]. Examples of practical problems with imbalanced dataset include, among others: network intrusion detection [2,15], source code fault detection [6], or in general fraud detection [1].

There are various approaches to deal with the imbalanced dataset problem. In general, they are classified into two categories as follows: data-level [7,14,24] and algorithm-level [25,28] methods. The data-level methods use a resampling

M. Choraś et al. (Eds.): CORES 2021, IP&C 2021, ACS 2021, LNNS 255, pp. 23–33, 2022.
https://doi.org/10.1007/978-3-030-81523-3_3

process that can be performed by oversampling, undersampling, and hybrid in nature. The algorithm-level methods are based on either modifying the state-of-the-art classification algorithms in order to adapt them to the skewed datasets or proposing specific ones [9,13,16].

This article deals with the algorithm-level approach and uses a previously developed by authors clustering and weighted scoring based on estimating the number of clusters algorithm that consider the minimum number of objects from the minority class label in each cluster [12]. In particular, in this article, we focused on the problem of selecting the distance metric. The experimental studies used a wide spectrum of imbalanced datasets and eight distance metrics. The experimental research aims to select the distance metric, which in the tested algorithm allows obtaining the best results for the classification of an imbalanced datasets.

The paper is structured as follows: the Sect. 2 describes clustering and weighted scoring based on estimating the number of clusters algorithm. In the Sect. 3 the experiments that were carried out are presented. The obtained results and the discussion are included in the Sect. 4. Finally, we conclude the paper in the Sect. 5.

2 Clustering and Weighted Scoring Based on Estimating the Number of Clusters Algorithm

One frequently used approach to data classification is to build ensembles of classifiers. Such an ensemble consists of multiple classifiers, usually those with quite poor predictive ability. However, when these models make a joint decision this ability improves significantly, allowing solving problems with difficult characteristics. When designing such an ensemble, two important aspects should be taken into account. First, it is necessary to ensure adequate diversification of the ensemble. This means that models should be trained and selected so that they have different properties [19]. Then each of them will have its own area of competence to make decisions. Another important aspect is a properly designed decision rule. It heavily depends on whether the built ensemble correctly classifies data and whether the joint decision improves models' predictive quality. A successful rule construction significantly improves the classifiers' competence by coherently combining the decisions and thereby selecting the most important values from them. Classifier ensembles are a very good choice in regard to classifying difficult data. In particular, they are suitable for imbalanced problems, which can be proved by numerous articles devoted to this topic [20,22,29].

The Clustering and Weighted Scoring with Estimating the Number of Clusters *CWS-ENC* is an extension of the CWS method developed by Ksieniewicz and Burduk [17]. It is an ensemble of classifiers approach, where the models diversification is ensured by separating problem classes into clusters. Additionally, an important aspect is that the method implements a decision rule based strongly on this segmentation's geometric characteristics. The main strategy adopted in the proposed method based on the homogeneous [27] ensemble with linear base classifiers, where each model is trained on a combination of [8] data class clusters computed with the *K-Means* [5] segmentation algorithm.

Compared to the previous proposal, the changes made the aim to extend the original approach with a procedure to select the number of clusters automatically. However, to present the motivation of the introduced changes, it is necessary to explain the main idea of this method step by step. In essence, this ensemble's base concept is to build a predictive method by combining multiple base classification models $\Psi_1, \Psi_2, \ldots, \Psi_K$, where K is the number of classifiers. Models are learned on sub-spaces that are determined by a segmentation algorithm for data from a different class. For this, the K-means algorithm is used. In fact, for the binary problem, each classifier will be built on a data cluster from two classes. This means that the number of K heavily depends on how many clusters are formed.

In the original idea, this number was selected in a deterministic way at the beginning of the experiments. The main change that will be made to this method is automatically selecting the number of clusters. This modification's main motivation is that data describing various problems have very different characteristics of distributions and class sizes. This also applies to imbalanced datasets, where the differences in the number of objects of a particular class are significantly higher. At this point, finding one ideal number of clusters into which we want to divide the data appears an issue. If the majority data is divided into more clusters, it will be sort of data balancing by segmentation. However, when the minority data is divided into the same number of subsets, the problem returns to the original level of imbalance or even worsens in some cases because eventually there is no control over a single cluster's size.

This raises the idea that it is best to determine these numbers for each class separately. However, doing this manually would require much work and it would not be a good solution from the perspective of using this method for further development. Hence, an idea emerged to do some automation, which would allow, in a dynamic way, to make a fast and relatively successful optimization of the number of clusters for each set of objects from a given class. Such choices are made possible by using a metric that provides some measurable indicator of the clusters' quality. For this purpose, the Silhouette Coefficient [11] metric was used. This is the maximum value of the mean Silhouette Value $sv(x)$ [23] over the entire dataset and it is described by the Eq. 1:

$$sv(x) = \frac{b(x) - a(x)}{max\{a(x), b(x)\}}, \tag{1}$$

where x is an object from cluster C, $b(x)$ is a mean dissimilarity measure of x to cluster C and $a(x)$ is a mean dissimilarity measure of x to other clusters. In addition to the metric itself, some procedure is needed to search for the optimal solution. The idea for this is to explore different solutions in a predefined range. In such a way, we have some upper limit on the number of clusters, however, it is possible to adapt the method to the problem under consideration easily. The different number of clusters in the range from 2 to K_{max} are created and then evaluated with the Silhouette Coefficient metric. The K_{max} value equal to 5 was selected during preliminary experiments.

The imbalanced data, characterized as the minority class, has very small sample counts relative to the majority class. This low population may also be very small without reference to the majority class. For example, one may encounter datasets with only a few minority class samples for the entire set. Keeping this possibility in mind, we introduced some heuristic rules to the above idea of partitioning data into clusters. This rule is designed to prevent the division of samples from a class when their abundance is extremely low. This approach will prevent additional imbalance when it is already significantly high, which causes the classifier's performance to worsen. The 25 samples were set as the default value of this threshold, determined in the preliminary experiments.

The CWS-ENC method relies on using linear base classifiers. Such a condition is due to the method's nature, which builds the predictive ability of models based on relationships in geometric space. Usually, a classifier ensemble makes a joint decision based on the predictions that models provide. This is an approach known as majority voting, which could be done in a weighted or equally important fashion. Another way is called support accumulation, which is to draw a decision based on the aggregation of the derived probabilities of samples fit to each model. This method uses a very similar approach to the latter. Estimating the class membership of a given sample "x" is determined mainly by its location in the geometric space. The function $sf_l(x)$ used to determine the distance from the decision boundaries of models (Ψ_l) forming an ensemble of classifiers to clusters' centroids (Eq. 2).

$$sf_l(x) = \|\psi_l(x)\| + \sum_{c=1}^{C} d_c, \tag{2}$$

where d_c is the distance from the object x to clusters' centroids expressed by any distance metric which is parameter of the method. The scoring function that calculates model supports based on the distances is represented by the Eq. 3:

$$wsf_l(x) = 1 - \frac{sf_l(x)}{\sum_{l=1}^{L} sf_l(x)}, \tag{3}$$

To better explain how the proposed method works, a pseudocode has been prepared (Algorithm 1). Firstly, subsets containing samples belonging to the same classes are extracted from the dataset. The symbol D_i denotes these subsets. Next, for each of them, the number of objects is calculated. Here, a special rule is used to prevent very small datasets from being divided into smaller ones. When the calculated number is less than the assumed threshold S_{min}, the algorithm goes to the next step without performing segmentation. When this condition is met, the procedure of splitting the data into additional clusters begins. First, the method tries to determine the best number of clusters. It is done in a manner similar to a bruteforce. Further sets of clusters are created using the K-means algorithm for different values of K from 2 to K_{max}, where this value is the method's input parameter. For each setup with K^{M_i} clusters, the Silhouette Coefficient metric is calculated and the setup with the best score is selected.

Algorithm 1: CWS-ENC – for binary problem

Input: D – Learning set
x – object
K_{max} – maximum number of clusters
S_{min} – minimum number of samples

Output: The ensemble classifier decision

1 Divide D into D_i subsets of data, where $i \in 1, 2$ is the number of class labels.
2 If the size of D_i is greater than S_{min} determine the best number of clusters M_i from 2 to K_{max} using the Silhouette Coefficient metric for each class. Otherwise, M_i is equal 1.
3 Divide D_i into K^{M_i} clusters using the K-means clustering algorithm separately for each $i - th$ class.
4 Find the cluster centroids $C_1^i, \ldots, C_{M_i}^i$ as the means of the points in the respective clusters.
5 Train base classifier $\Psi_1, \ldots, \Psi_L$ using each combination of clusters from different class labels, i.e. one cluster from each class label, $L = M_1 * M_2$.
6 Calculate weighted scoring functions for the object x:

$$ws f_l(x) = 1 - \frac{s f_l(x)}{\sum_{l=1}^{L} s f_l(x)},$$

where

$$s f_l(x) = \|\Psi_l(x)\| + \sum_{c=1}^{2} d_c.$$

7 The ensemble classifier decision:

$$\widehat{\Psi}(x) = sign\left(\sum_{l=1}^{L} ws f_l(x) \Psi_l(x)\right),$$

where $\Psi(x)$ is the prediction returned by base classifier $\Psi(x) \in \{-1, 1\}$.

---

3 Experimental Evaluation

During prediction, the distance between instances and centroids is calculated according to the Eq. 3. A very important part of this formula is the way in which this distances are determined. There are actually many different metrics to compute the distance between two points. Choosing the right approach can have an impact on the performance quality. This causes a certain problem that should be taken up during the further development of the proposed method. It is also important to consider whether these differences are statistically significant. Therefore, the main goal of this work is to investigate different ways of measuring the distance between using metrics. Correctly performed experimental analysis should allow us to clearly determine potentially the best option. In the following, the research questions are formulated:

RQ1: Does the selection of a distance metric have a statistically significant impact on the predictive performance of the CWE-ENS method?
RQ2: Which distance metric achieves the best possible classification quality of the CWE-ENS method?

3.1 Setup

The evaluation procedure applied in this work is based on the usage of *Stratified K-fold Cross-validation*, where each fold maintains the same imbalance ratio. Because some of the datasets have small size, the number of folds was set as 5, to avoid forming very small subsets that will complicate obtaining good predictive ability of the models. As the main objective of this work is to analyze the impact and select the best distance metric for the CWS-ENC method, the experimental studies were performed on the following variants of the method using different distance metrics:

- *Euclidean* distance (EUC)
- *Manhattan* distance (MAN)
- *Chebyshev* distance (CHE)
- *Jensen-Shannon* distance (JSH)
- *Bray-Curtis* distance (BRY)
- *Canberra* distance (CAN)
- *Correlation* distance (COR)
- *Square Euclidean* distance (SQE)

This research was done on 58 selected datasets where all problems were dichotomies. The names of these sets, the imbalance ratio, the number of samples, and the number of features are described in the Table 1. The achieved performance of tested method variants was measured by six selected metrics of classification quality:

- *balanced accuracy score* (BAC),
- *F1-score* (F-1)
- *G-mean* (GMN)
- *precision* (PRE)
- *recall* (REC)
- *specificity* (SPE)

Next, based on these metric results, the statistical analysis was made using the *Wilcoxon rank test* with the significance level *alpha* = 0.05 [4]. The experimental environment was implemented using the Python language. A large contribution to the majority of the project implementation are libraries: *scikit-learn* [21] with which the cross validation was performed and the base classifiers used; *scipy* [26] which provided different ways to calculate the distance metric between two points; *stream-learn* [18] which mainly has been used to compute various classification metrics. Furthermore, the obtained results and the full implementation can be found on the Github repository[1].

[1] Repository link: https://github.com/w4k2/cws-enc-cores21.

Table 1. Overview of real datasets used in experimental evaluation (KEEL [3])

Dataset name	IMB. RATIO	SAMPLES	FEATURES	Dataset name	IMB. RATIO	SAMPLES	FEATURES
abalone-21_vs_8	40	581	8	*glass4*	15	214	9
abalone-3_vs_11	32	502	8	*glass5*	23	214	9
abalone9-18	16	731	8	*glass6*	6.4	214	9
cleveland-0_vs_4	13	177	13	*led7digit-0-2-4-5-6-7-8-9_vs_1*	11	443	7
dermatology-6	17	358	34	*lymphography-normal-fibrosis*	24	148	18
ecoli-0-1-3-7_vs_2-6	39	281	7	*new-thyroid1*	5.1	215	5
ecoli-0-1-4-6_vs_5	13	280	6	*newthyroid2*	5.1	215	5
ecoli-0-1-4-7_vs_2-3-5-6	11	336	7	*poker-9_vs_7*	30	244	10
ecoli-0-1-4-7_vs_5-6	12	332	6	*shuttle-6_vs_2-3*	22	230	9
ecoli-0-1_vs_2-3-5	9.2	244	7	*shuttle-c2-vs-c4*	20	129	9
ecoli-0-1_vs_5	11	240	6	*vowel0*	10	988	13
ecoli-0-2-3-4_vs_5	9.1	202	7	*winequality-red-3_vs_5*	68	691	11
ecoli-0-2-6-7_vs_3-5	9.2	224	7	*winequality-red-8_vs_6*	35	656	11
ecoli-0-3-4-6_vs_5	9.2	205	7	*winequality-red-8_vs_6-7*	46	855	11
ecoli-0-3-4-7_vs_5-6	9.3	257	7	*winequality-white-3_vs_7*	44	900	11
ecoli-0-3-4_vs_5	9	200	7	*winequality-white-9_vs_4*	33	168	11
ecoli-0-4-6_vs_5	9.2	203	6	*yeast-0-2-5-6_vs_3-7-8-9*	9.1	1004	8
ecoli-0-6-7_vs_3-5	9.1	222	7	*yeast-0-2-5-7-9_vs_3-6-8*	9.1	1004	8
ecoli-0-6-7_vs_5	10	220	6	*yeast-0-3-5-9_vs_7-8*	9.1	506	8
ecoli2	5.5	336	7	*yeast-0-5-6-7-9_vs_4*	9.4	528	8
ecoli3	8.6	336	7	*yeast-1-2-8-9_vs_7*	31	947	8
ecoli4	16	336	7	*yeast-1-4-5-8_vs_7*	22	693	8
glass-0-1-4-6_vs_2	11	205	9	*yeast-1_vs_7*	14	459	7
glass-0-1-5_vs_2	9.1	172	9	*yeast-2_vs_4*	9.1	514	8
glass-0-1-6_vs_2	10	192	9	*yeast-2_vs_8*	23	482	8
glass-0-1-6_vs_5	19	184	9	*yeast3*	8.1	1484	8
glass-0-4_vs_5	9.2	92	9	*yeast4*	28	1484	8
glass-0-6_vs_5	11	108	9	*yeast5*	33	1484	8
glass2	12	214	9	*yeast6*	41	1484	8

4 Results

Figure 1 shows the results that were obtained from the experimental study. It contains the summary results after the statistical analysis. In this graph, one can notice a slight advantage of EUC distance measure over MAN distance measure and a significant advantage over other metrics. It can be clearly seen that MAN for the scores obtained with the SPE metric obtains higher quality. However, the other metrics give an advantage for the EUC measure. It is also worth noting that for the REC metric, distance measures such as CHE, BRY, CAN, COR perform much better. Nevertheless, the results for the other metrics are significantly low, which may imply getting a method with a strong tendency to classify too biased towards the minority class.

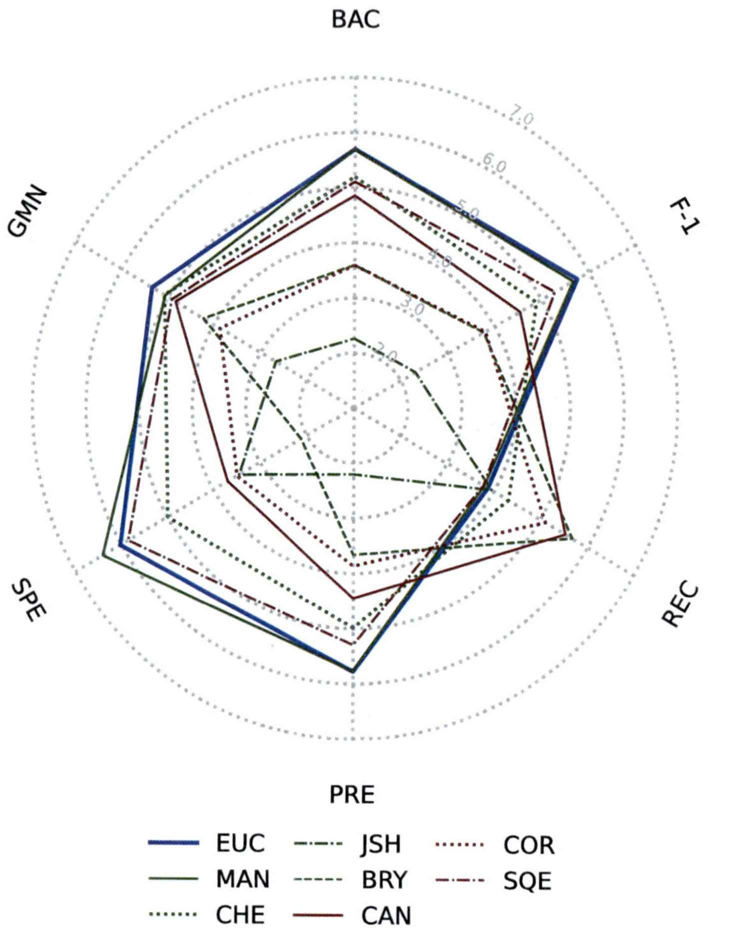

Fig. 1. Radar plot of mean ranks obtained by the Friedman test.

Complementing this radar plot, the Table 2 presents the same results but in tabular form. This reading allows to look more closely at the results and highlights those with statistical differences. The columns contain names of the tested variants of the method with specified numbers, and the rows contain names of the selected metrics for which the predictive performance was calculated. In each row, there are two lists of values. The former (above) are the average rank scores calculated based on the given metric, and the latter (below) are numbers indicating the advantage with statistical significance over other methods.

For example, the BAC metric, it can be seen that the method variant using the EUC distance measure has a statistical advantage over the JSH, BRY, CAN, COR, SQE measures. It is now noticeable that the EUC outperforms most of the other variants with statistical significance. The exceptions are MAN and

CHE. However, despite the absence of statistically significant differences, the EUC measure scores better anyway.

Table 2. Results for mean ranks and statistical significance

	EUC (1)	MAN (2)	CHE (3)	JSH (4)	BRY (5)	CAN (6)	COR (7)	SQE (8)
BAC	5.698	5.690	5.198	2.259	3.595	4.853	3.595	5.112
	4,5,6,7,8	4,5,6,7	4,5,7	–	4	4,5,7	4	4,5,7
F-1	5.750	5.664	4.888	2.310	3.767	4.543	3.802	5.276
	3,4,5,6,7,8	3,4,5,6,7	4,5,7	–	4	4,5,7	4	4,5,7
GMN	5.750	5.664	4.888	2.310	3.767	4.543	3.802	5.276
	3,4,5,6,7,8	3,4,5,6,7	4,5,7	–	4	4,5,7	4	4,5,7
PRE	5.776	5.767	4.983	2.207	3.664	4.448	3.862	5.293
	3,4,5,6,7,8	3,4,5,6,7	4,5,7	–	4	4,5	4	4,5,7
REC	3.871	3.793	4.310	3.948	5.672	5.526	5.103	3.776
	–	–	–	–	1,2,3,4,8	1,2,3,4,8	1,2,4,8	–
SPE	6.017	6.397	4.991	3.448	2.121	3.690	3.491	5.845
	3,4,5,6,7	3,4,5,6,7	4,5,6,7	5	–	5	5	3,4,5,6,7

4.1 Lessons Learned

In summary, the research questions stated above will be answered:

RQ1: Does the selection of a distance metric have a statistically significant impact on the predictive performance of the CWE-ENS method?
The experimental analysis conducted allows us to unequivocally state that the differences between the results are statistically significant for the selected distance metrics on imbalanced datasets. This is evident at first glance in the radar plot, and is fully confirmed by the tabular results. Not every variant achieved a result with statistical significance. However, the advantage of some metrics over the others is definitely noticeable.

RQ2: Which distance metric achieves the best possible classification quality of the CWE-ENS method?
It is noticeable that the best performance is achieved by a variant of the CWS-ENS method using the Euclidean distance measure. However, the answer to the question cannot be obvious. Although the favorite variant has been chosen, it should be remembered that these results show a specific problem under investigation. For the datasets that were used, these results are the best. However, one should remember that different data may require a reanalysis of the parameters. Additionally, it is worth noting that the distance measured using the Manhattan measure obtains very similar results, with no statistically significant differences. It is one of the variants that cannot be rejected and is worth considering in the next stages of the method development.

5 Conclusions

This paper has analyzed the distance metric's selection in the supervised learning algorithm with clustering feature space. The authors previously proposed the analyzed algorithm. It is a fusion algorithm of linear base classifiers that uses feature space partitioning and the object's distance from the decision boundary to determine the score function's value. In experimental studies, we tested eight distance metrics using unbalanced datasets. The obtained results were analyzed using appropriate non-parametric statistical tests and six classification performance measures. Overall, the best results were obtained for the Euclidean distance metric, while the Manhattan metric obtains very similar results.

Acknowledgements. This work was supported by the Polish National Science Centre under the grant No. 2017/25/B/ST6/01750 as well as by the statutory funds of the Department of Systems and Computer Networks, Faculty of Electronics, Wroclaw University of Science and Technology.

References

1. Abdallah, A., Maarof, M.A., Zainal, A.: Fraud detection system: a survey. J. Netw. Comput. Appl. **68**, 90–113 (2016)
2. Abdulhammed, R., Faezipour, M., Abuzneid, A., AbuMallouh, A.: Deep and machine learning approaches for anomaly-based intrusion detection of imbalanced network traffic. IEEE Sens. Lett. **3**(1), 1–4 (2018)
3. Alcalá-Fdez, J., et al.: Keel data-mining software tool: data set repository, integration of algorithms and experimental analysis framework. J. Multiple-Valued Log. Soft Comput. **17** (2011)
4. Alpaydin, E.: Introduction to Machine Learning. MIT Press, Cambridge (2014)
5. Basu, S., Banerjee, A., Mooney, R.: Semi-supervised clustering by seeding. In: Proceedings of 19th International Conference on Machine Learning (ICML-2002). Citeseer (2002)
6. Choraś, M., Pawlicki, M., Kozik, R.: Recognizing faults in software related difficult data. In: International Conference on Computational Science, pp. 263–272. Springer (2019)
7. Fotouhi, S., Asadi, S., Kattan, M.W.: A comprehensive data level analysis for cancer diagnosis on imbalanced data. J. Biomed. Inf. **90**, 103,089 (2019)
8. Fred, A., Lourenço, A.: Cluster ensemble methods: from single clusterings to combined solutions. In: Supervised and Unsupervised Ensemble Methods and Their Applications, pp. 3–30. Springer (2008)
9. Galar, M., Fernandez, A., Barrenechea, E., Bustince, H., Herrera, F.: A review on ensembles for the class imbalance problem: bagging-, boosting-, and hybrid-based approaches. IEEE Trans. Syst. Man Cybern. Part C (Appl. Rev.) **42**(4), 463–484 (2011)
10. Haixiang, G., Yijing, L., Shang, J., Mingyun, G., Yuanyue, H., Bing, G.: Learning from class-imbalanced data: review of methods and applications. Expert Syst. Appl. **73**, 220–239 (2017)
11. Kaufmann, L., Rousseeuw, P.J.: Finding Groups in Data: An Introduction to Cluster Analysis. Wiley, New York (1990)

12. Klikowski, J., Burduk, R.: Clustering and weighted scoring algorithm based on estimating the number of clusters. In: International Conference on Computational Science. Springer (2021, accepted)
13. Klikowski, J., Ksieniewicz, P., Woźniak, M.: A genetic-based ensemble learning applied to imbalanced data classification. In: International Conference on Intelligent Data Engineering and Automated Learning, pp. 340–352. Springer (2019)
14. Koziarski, M., Woźniak, M., Krawczyk, B.: Combined cleaning and resampling algorithm for multi-class imbalanced data with label noise. arXiv preprint arXiv:2004.03406 (2020)
15. Kozik, R., Choras, M., Keller, J.: Balanced efficient lifelong learning (B-ELLA) for cyber attack detection. J. UCS **25**(1), 2–15 (2019)
16. Krawczyk, B., Woźniak, M., Schaefer, G.: Cost-sensitive decision tree ensembles for effective imbalanced classification. Appl. Soft Comput. **14**, 554–562 (2014)
17. Ksieniewicz, P., Burduk, R.: Clustering and weighted scoring in geometric space support vector machine ensemble for highly imbalanced data classification. In: International Conference on Computational Science, pp. 128–140. Springer (2020)
18. Ksieniewicz, P., Zyblewski, P.: stream-learn-open-source python library for difficult data stream batch analysis. arXiv preprint arXiv:2001.11077 (2020)
19. Kuncheva, L.I.: Combining Pattern Classifiers: Methods and Algorithms. Wiley, Hoboken (2004)
20. Lopez-Garcia, P., Masegosa, A.D., Osaba, E., Onieva, E., Perallos, A.: Ensemble classification for imbalanced data based on feature space partitioning and hybrid metaheuristics. Appl. Intell. **49**(8), 2807–2822 (2019)
21. Pedregosa, F., et al.: Scikit-learn: machine learning in Python. J. Mach. Learn. Res. **12**, 2825–2830 (2011)
22. Rodríguez, J.J., Diez-Pastor, J.F., Arnaiz-Gonzalez, Kuncheva, L.I.: Random balance ensembles for multiclass imbalance learning. Knowl.-Based Syst. **193**, 105,434 (2020). https://doi.org/10.1016/j.knosys.2019.105434. https://www.sciencedirect.com/science/article/pii/S0950705119306598
23. Rousseeuw, P.J.: Silhouettes: a graphical aid to the interpretation and validation of cluster analysis. J. Comput. Appl. Math. **20**, 53–65 (1987)
24. Szeszko, P., Topczewska, M.: Empirical assessment of performance measures for preprocessing moments in imbalanced data classification problem. In: IFIP International Conference on Computer Information Systems and Industrial Management, pp. 183–194. Springer (2016)
25. Trajdos, P., Kurzynski, M.: A correction method of a base classifier applied to imbalanced data classification. In: International Conference on Computational Science, pp. 88–102. Springer (2020)
26. Virtanen, P., et al.: SciPy 1.0 contributors: SciPy 1.0: fundamental algorithms for scientific computing in Python. Nat. Methods **17**, 261–272 (2020). https://doi.org/10.1038/s41592-019-0686-2
27. Woźniak, M., Graña, M., Corchado, E.: A survey of multiple classifier systems as hybrid systems. Inf. Fusion **16**, 3–17 (2014)
28. Zhang, C., et al.: Multi-imbalance: an open-source software for multi-class imbalance learning. Knowl.-Based Syst. **174**, 137–143 (2019)
29. Zyblewski, P., Ksieniewicz, P., Woźniak, M.: Classifier selection for highly imbalanced data streams with minority driven ensemble. In: Rutkowski, L., Scherer, R., Korytkowski, M., Pedrycz, W., Tadeusiewicz, R., Zurada, J.M. (eds.) Artificial Intelligence and Soft Computing, pp. 626–635. Springer, Cham (2019). https://doi.org/10.1007/978-3-030-20912-4_57

Exploration of Hardware Acceleration Methods for an XNOR Traffic Signs Classifier

Dominika Przewlocka-Rus(✉), Marcin Kowalczyk, and Tomasz Kryjak

AGH University of Science and Technology in Krakow, Krakow, Poland
{dominika.przewlocka,marcin.kowalczyk,tomasz.kryjak}@agh.edu.pl

Abstract. Deep learning algorithms are a key component of many state-of-the-art vision systems, especially as Convolutional Neural Networks (CNN) outperform most solutions in the sense of accuracy. To apply such algorithms in real-time applications, one has to address the challenges of memory and computational complexity. To deal with the first issue, we use networks with reduced precision, specifically a binary neural network (also known as XNOR). To satisfy the computational requirements, we propose to use highly parallel and low-power FPGA devices. In this work, we explore the possibility of accelerating XNOR networks for traffic sign classification. The trained binary networks are implemented on the ZCU 104 development board, equipped with a Zynq UltraScale+ MPSoC device using two different approaches. Firstly, we propose a custom HDL accelerator for XNOR networks, which enables the inference with almost 450 fps. Even better results are obtained with the second method – the Xilinx FINN accelerator – enabling to process input images with around 550 frame rate. Both approaches provide over 95% accuracy on the test set.

Keywords: XNOR · FPGA · Traffic signs recognition · Hardware acceleration · FINN

1 Introduction

The problem of traffic sign detection and recognition is important for many applications, such as advanced driver assistance and safety systems, autonomous vehicles (self-driving cars), or traffic signs maintenance (i.e., detection of damage or incorrect positions). Due to the wide application and specificity of the problem, three aspects should be considered: (1) The algorithm has a direct impact on the safety of both its user (e.g. driver) and other road users (e.g. pedestrians). Therefore, we propose an approach based on machine learning. In particular, we use deep convolutional neural networks, which are very well suited for solving classification problems and characterised by high accuracy; (2) We expect the system to process images in real-time, since this has a direct impact on the so-called 'reaction time' and therefore also safety. (3) We would like to

M. Choraś et al. (Eds.): CORES 2021, IP&C 2021, ACS 2021, LNNS 255, pp. 34–45, 2022.
https://doi.org/10.1007/978-3-030-81523-3_4

keep the power consumption as low as possible, because the energy budget of modern cars (especially electric) is very limited. Therefore, it is justified to use programmable heterogeneous systems that enable designing almost any computational architecture, also one that will process data in parallel and pipelined, with minimal energy overhead.

Since the hardware implementation of such computationally complex solutions is not straightforward, for the presented work, we (1) decided to use XNOR networks; (2) explore the possibilities of hardware acceleration with custom implementation as well as using open-source library from Xilinx – FINN [1], [2].

The main contributions of this paper are: (1) the proposed energy efficient and real-time accurate traffic sign classifier run on the ZCU 104 development board with a heterogeneous Zynq UltraScale+ MPSoC device with clock frequency of 100 MHz; (2) design space exploration and comparison of different acceleration techniques (custom vs FINN).

The remainder of this paper is organised as follows: in Sect. 2 we briefly summarise the previous work concerning the research topic. Section 3 describes the proposed network architectures as well as classification results. In Sect. 4 we present the hardware accelerators' experiments – in Sect. 4.1 for the custom one, in Sect. 4.2 for the FINN-based. Section 5 discusses the obtained results and shows a comparative analysis of both solutions. Finally, in Sect. 6 we summarise our work and discuss possible future research directions.

2 Previous Work

Traffic sign recognition or classification is a particularly interesting topic due to its application in autonomous vehicles (AV) or advanced driver assistant systems (ADAS). One of the most popular databases is GTSRB (German Traffic Sign Recognition Benchmark), which was first presented for the challenge during the IJCNN 2011 conference [12]. The winners with the highest classification accuracy on the test set were: (1) 99.46% [14], (2) 98.31% [15] and (3) 96.14% [16]. The first two were based on convolutional neural networks, while the third used a random forests classifier. The human performance in that task was estimated as 98.84%. Due to the popularity of the topic, several other papers on classification of traffic signs were presented later. The state-of-the-art classification accuracy using convolutional neural networks is about 99.71% [13] for GTSRB.

The described in the literature hardware implementations of traffic sign recognition mainly focus on solutions without the use of deep convolutional neural networks. In [3] and [4] the authors propose a software-hardware implementation of traffic sign detection based on colour information in the HSV space and pattern matching based classification. Article [5] describes a software-hardware solution based on the HOG (Histogram of Oriented Gradients) algorithm. In the work [6] a hardware implementation of a deep convolutional neural network that classifies signs was proposed. The implementation of binary neural networks in FPGAs is the subject of several articles. In [9] the authors summarise the first

HLS binary network accelerator targeting architecture for CIFAR-10 dataset classification. Comparing to other hardware implementations (for standard convolutional networks), the binary one was far more efficient in terms of resources utilisation and power consumption, achieving even 6 times more GOPS/Watt. In [1,2] the authors present the FINN tool for fast implementation of quantised neural networks and test their solution on common benchmarks, such as MNIST, CIFAR-10 or SVHN. [7] describes a solution for CIFAR-10 benchmark that achieves 332 classifications per second with accuracy of 86.06%. Authors of [8] propose a decomposition of the input maps (multichannel image) to 1-bit precision, which boosts the calculations of the first convolutional layer. [10] presents an implementation based on Resource-Aware Model Analysis and evaluates the accelerator on MNIST, CIFAR-10 and AlexNet benchmarks.

3 Traffic Sign Classification with XNOR Networks

For the classification of traffic signs, the GTSRB (German Traffic Sign Recognition Benchmark) database was used [12]. It consists of images of German traffic signs representing 43 classes. 12630 images are used as the test set and 39209 as the training set, however, the number of training samples belonging to each class is not balanced. Each image is in the RGB colour space of size varying between 15×15 up to 250×250 pixels. At the same time, not all are square, differ in illumination and quality. Due to the above described issues, the images had to be standardised before training. The preprocessing of the input data was split into two parts: (1) standardisation of images with CLAHE (Contrast Limited Adaptive Histogram Equalisation - [17]) for the Y channel from the YCbCr colour space and (2) data augmentation to balance class representations. Data augmentation was performed using rotation, translation, shear, and scaling operations, so finally each class is represented by 10000 examples.

Candidates for traffic signs are classified by an XNOR-type neural network, described in detail in the work [11]. The network is characterised by binary weights and activations: $W \in \{-1, 1\} \wedge f(Wx) \in \{-1, 1\}$ (with W as weights, x as layer's input and $f()$ as activation function). The input image is represented by integer or floating point values, which means that the first convolutional layer is not fully binary.

The binary operations can vastly reduce the computational complexity of the network, which can be crucial in the case of embedded systems and particularly for the low-power ones. For a binary convolutional layer, the output weighted sum can be calculated in three stages:

1. XNOR operation between the vectorised input window I_{in} and the vectorised filter W – $I_{XNOR} = I_{in} \bar{\oplus} W$.
2. Bit count the obtained vector – $P = count(I_{XNOR})$.
3. Calculation of the convolution result based on $C = 2 * P - N$, where N is the size of the window.

Table 1. Neural network's architecture for custom accelerator

Layer	Input size	Output size	Kernel size
Conv-1	$32 \times 32 \times 3$	$28 \times 28 \times 64$	$5 \times 5 \times 64$
Max-1	$28 \times 28 \times 64$	$14 \times 14 \times 64$	2×2
Conv-2	$14 \times 14 \times 64$	$10 \times 10 \times 128$	$5 \times 5 \times 128$
Max-2	$10 \times 10 \times 128$	$5 \times 5 \times 128$	2×2
FC-1	$5 \times 5 \times 128$	512	$1 \times 1 \times 512$
FC-2	512	43	$1 \times 1 \times 43$

Next, the real-valued bias is added. The activation is a binarisation function: $Pix_{out} = max(0, min(1, \frac{Pix_{in}+1}{2}))$.

Subsections 3.1 and 3.2 describe the proposed networks accelerated using the ZCU 104 development board – respectively, for the custom accelerator and the Xilinx FINN based one.

3.1 Custom Architecture

The proposed network architecture is presented in Table 1. The convolutional block consists of the following layers (1) convolutional, (2) maxpooling, (3) batch normalisation, (4) activation. The dense (fully connected) block consists of the following layers (1) fully connected, (2) batch normalisation, (3) activation. The last dense block does not have an activation layer – the vector obtained from the batch normalisation layer is the output vector (network classification result).

For training the network, we used the code provided by [11]. The input images' pixels are normalised to range $[-1, 1]$ and the Square Hinge loss is used.

3.2 FINN Architecture

Since the FINN tool is still under strong development, several features are not yet available. The architecture proposed in Sect. 3.1 cannot be transformed into hardware modules using FINN (for today). It is required that the consecutive layers are designed in such way, that the output of the last convolutional one has a size of 1×1. The input images are normalised to the range $[-1, 1]$. In Table 2 the adjusted architecture is shown.

Table 3 summarises the results of both networks – for the custom and FINN accelerator. For the evaluation metrics, the accuracy, precision, recall, and F1 score were used, calculated as a weighted average with relation to the number of test examples for each class. The high precision and recall scores for both networks show that the predictions are most likely to be true positives – however, a more detailed analysis of such statistical metrics for each class should be continued in the future.

Table 2. Neural network's architecture for FINN tool

Layer	Input size	Output size	Kernel size
Conv-1	$32 \times 32 \times 3$	$28 \times 28 \times 64$	$5 \times 5 \times 64$
Max-1	$28 \times 28 \times 64$	$14 \times 14 \times 64$	2×2
Conv-2	$14 \times 14 \times 64$	$10 \times 10 \times 128$	$5 \times 5 \times 128$
Max-2	$10 \times 10 \times 128$	$5 \times 5 \times 128$	2×2
Conv-3	$5 \times 5 \times 128$	$5 \times 5 \times 513$	$1 \times 1 \times 512$
FC-1	512	43	$1 \times 1 \times 43$

Table 3. Evaluation of the proposed TSR Networks (C-Net for custom accelerator, F-Net – FINN accelerator)

	Accuracy	Precision	Recall	F1 score
C-Net	96.38%	96.4%	96.28%	96.23%
F-Net	95%	95.21%	95%	95%

4 Hardware Acceleration

The hardware acceleration targets a heterogeneous Zynq UltraScale+ MPSoC device, available on the ZCU 104 development board, both for custom and FINN accelerator. The target frequency is set to 100MHz (this value can be increased if necessary). Subsection 4.1 describes the developed accelerator as well as the obtained results for the proposed traffic sign classification network. Subsection 4.2 summarises the experiments with the FINN tool.

4.1 Custom Accelerator

The simplified accelerator scheme is presented in Fig. 1. The proposed solution can be used for the acceleration of virtually any simple convolutional network (within the available resources limitation) using the developed Verilog code generator. The key components of the proposed accelerator are the Convolutional Block and Dense Block parts, as well as several BRAM (Block RAM) sets which store the input, intermediate results and fully connected layers' weights. The overall organisation of operations is based on several crucial ideas:

1. The input and intermediate feature maps are organised in BRAM sets in such way that each BRAM stores one channel, which ease the process of writing the output feature maps.
2. The data is read in a sequence that enables the calculation of convolutional and maxpooling layers without additional memory usage.

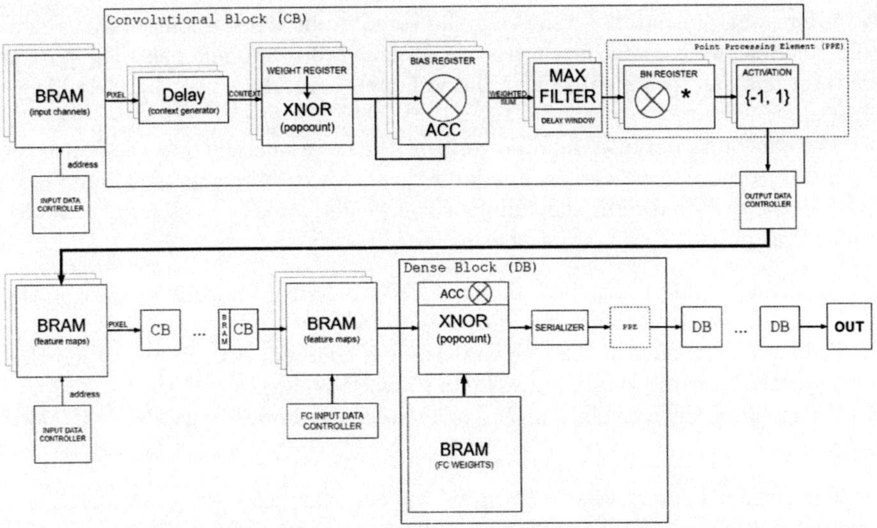

Fig. 1. A simplified scheme of the custom accelerator. The proposed architecture consists of three main components: Convolutional Blocks, Dense Blocks and BRAM sets for inputs and outputs from layers. The operations in layers are streamlined.

3. The weights of the convolutional layers are stored in the on-board registers for better access.
4. The weights of the fully convolutional layers are stored in BRAM. For enhanced parallelisation of the calculations, the memory is organised in a way to enable reading at once the weights connected with the one, selected input.

The Convolutional Block (CB) consists of modules responsible for layers: convolutional, maxpooling, batch normalisation, and binary activation. In the case of the first convolutional layer, the network's input is a greyscale (one channel – this is applicable for the described TSR network) or colour image (three channels) – each channel is stored in a separate BRAM memory (each 16 Kb). In the case of subsequent layers, the BRAM set stores the output (feature maps) from the previous layer. The filter weights are stored in the on-board registers, therefore the utilisation of BRAM memory is reduced, and the time needed to read the weights is shortened. Reading pixels, from both the input image and the feature maps between successive convolutional blocks, is organised in a specific order that allows the implementation of the maxpooling operation without the use of additional memory. The data is read as follows:

1. Read one context along all channels, in relation to the same central pixel.
2. Move the filter window and repeat the previous step – further contexts are selected in such way, that in the result of successive convolutions, a context ready for the maxpooling operation is obtained.

All filters work in parallel, receiving the same input data – contexts in a specific position (with a specific central pixel) along all input channels. The partial convolutions are accumulated and the weighted sum along the channels is calculated.

The convolutional filter module varies with the input type. For the first layer – real-valued input – it is an accumulator that adds or subtracts incoming pixels to/from the weighted sum, depending on the weight value (-1 or 1). For a binary input, the convolution is performed in three stages:

1. Vector of weights ('flattened' filter) is XNORed with the context (also in the form of a vector) – the operation takes 1 clock cycle.
2. The number of ones in the received vector is counted. Bit counting operation is performed using Hamming weights (5 clock cycles).
3. The result of the convolution C is calculated based on $C = 2 * P - N$ (1 clock cycle).

The obtained partial results (from subsequent channels) are accumulated and the bias is added. The maxpooling module accepts consecutive pixels and when ready acts as maximum filter. The output value is finally subjected to batch normalisation and binary activation (standard arithmetic operations). The batch normalisation formula is simplified to $Pix_{out} = A \cdot Pix_{in} + B$, where $A = \gamma \cdot \frac{1}{\sigma}$ and $B = -\gamma \cdot \frac{1}{\sigma} \cdot \mu + \beta$, in order to reduce the resource utilisation (DSP instances). The results are written in parallel from all filters, to different, appropriate BRAM instances (corresponding output map).

Implementation of a fully connected layer (Dense Block (DB)) assumes the use of a larger BRAM memory due to the significantly higher number of weights (in relation to the convolutional layer). The input pixels are read in a serial way from the feature map of the previous convolutional layer. Before receiving a pixel, a set of weights corresponding to the connection for all neurons is read from the BRAM memory – one word in memory stores the weights of a given connection for all the layer's neurons. The input is XNORed with each weight (i.e., the equivalent of multiplication in networks with full precision is performed) and then the results are sent to the respective accumulators representing the neurons. The weighted sum of all the layer's neurons is calculated in parallel. Then the data (outputs from all neurons) is serialised, bias is added, and values are normalised, if needed.

The post-implementation resource usage is presented in Table 4. The frame rate of network's inference is around 449.25 (calculated basing on clock frequency and number of clock cycles needed for input processing, as well as tested in hardware), with power consumption of 4.396 W (estimated using Vivado).

Table 4. Post-implementation resources usage for binary convolutional neural network on ZCU104

Resource	Utilisation	Available	Utilisation %
LUT	79302	230400	34.42
LUTRAM	350	101760	0.34
FF	44170	460800	9.59
BRAM	152	312	48.72
DSP	902	1728	52.20

4.2 FINN Accelerator

Acceleration of neural networks using the FINN tool requires several well-defined steps. Given the model, preferably trained using the Brevitas [18] tool, in the ONNX (Open Neural Network Exchange) format, one has to perform a set of operations from streamlining transformations, converting to HLS (High-Level Synthesis) layers and dataflow partitioning, to adjusting folding to maximise the performance. All but the last is done in a rather automatic manner and for more information we suggest consulting the FINN documentation. The folding influences the frame rate of the network and the usage of resources. Finally, the HLS IP per layer can be generated, as well as the stitched design and Vivado/Vitis project ready to be synthesised and run in hardware. If the accelerated network matches the framework's capabilities, the process is a rather smooth one.

To explore the possibilities of FINN, we decided to perform several experiments with folding, following the official examples – 'there are several folding factors for each layer, controlled by the PE (parallelisation over outputs) and SIMD (parallelisation over inputs) parameters (...) the higher the PE and SIMD values are set, the faster the generated accelerator will run, and the more FPGA resources it will consume'. The experiments' parameters, as well as the final frame rate of the accelerated network and energy consumption are shown in Table 5 – for each layer, the folding parameters (PE, SIMD, FIFO depth) are presented. Table 6 shows the resources' usage for the defined experiments. It is easily observed that with higher folding parameters, the network process the input faster – the difference between experiments 1 and 4 is more than 500 frames per second. One can also notice the rather slight increase of the used resources – perhaps for more complex (in terms of architecture) networks the results would vary more.

Table 5. Folding experiments for the TSR network with the FINN accelerator – parameters, final throughput and energy consumption (estimated with Vivado).

No	Conv 1	Conv 2	Conv 3	FC 1	FPS	Power [W]
1	(4, 1, 32)	(4, 4, 32)	(4, 4, 32)	(1, 1, 32)	64.51	3.521
2	(4, 1, 128)	(4, 4, 128)	(4, 4, 128)	(1, 1, 128)	64.51	3.523
3	(8, 1, 128)	(8, 8, 128)	(8, 8, 128)	(1, 1, 128)	212.41	3.522
4	(16, 1, 128)	(16, 16, 128)	(4, 16, 128)	(1, 1, 128)	582.22	3.547

Table 6. Post-implementation resources usage for the FINN accelerated TSR network in ZCU104

Resource	Total available	E1 [%]	E2 [%]	E3 [%]	E4 [%]
LUT	230400	5.81	5.89	5.75	6.14
LUTRAM	101760	1.62	1.67	1.59	1.70
FF	460800	4.29	4.29	4.4	4.62
BRAM	312	22.92	22.92	22.28	21.96
DSP	1728	–	–	–	–

5 Comparison of the Considered Approaches

Firstly, it should be noted that when starting the work on the custom XNOR accelerator, we were not familiar with the FINN tool, as it was in an early stage of development back then. After completing our implementation, we have read about FINN and only then decided to compare both approaches. The analysis of the results shall be divided into two parts – the binary networks and their accuracy, and the hardware acceleration.

In Sect. 3 we have proposed two network architectures with reduced precision for traffic sign recognition on the GTSRB dataset – the 'original' and FINN-adapted. Both networks achieved similar and relatively good accuracy, despite the use of different tools. It should be noted that the F-Network's accuracy is lower, despite the similar training conditions. Better results could be achieved using more parameters (filters or layers) or other architecture, e.g. residual networks. Since the FINN accelerator proved itself to be area-efficient (in terms of resource usage) for the proposed architecture, a slight increase in computational complexity should not be an issue for real-time and energy efficient processing. We have compared our networks to other traffic sign classifiers and placed ourselves among the existing top CNN-based classifiers – Table 7. In [13] and [14] the authors proposed networks with significantly more parameters. In [15] the presented solution employs layer skipping to provide different scales of receptive fields to the classifier, which also increases the model complexity. All mentioned solutions work with floating point precision – when comparing the accuracy of our XNOR models with other proposals, the 'size issue' and hardware acceleration suitability should be also taken into account.

Table 7. Comparison of our traffic sign classifier to the state-of-the-art solutions. C stands for the network developed with the custom accelerator, F – for FINN

Network	Accuracy	
Garcia et al. [13]	99.71%	
Ciresan et al. [14]	99.46%	1st at IJCNN 2011
Sermanet et al. [15]	99.17%	2nd at IJCNN 2011
Human performance	98.84%	
Our XNOR C-Net	96.28%	
Zaklouta et al. [16]	96.14%	3rd at IJCNN 2011
Our XNOR F-Net	95%	

The results presented in Sect. 4 basically prove the lead of the FINN accelerator, both in terms of energy consumption and the used resources (Table 5 and 6) for the same test conditions – the ZCU104 device and clock frequency of 100 MHz. The proposed custom accelerator needs over 4.396W, which is over 25% of the total power consumption of the FINN-generated one. The disproportion in resource usage is also noticeable. The inference speed of the FINN-accelerated network (Table 5) depends on the folding parameters. However, achieving the frame rate close to the one from the custom accelerator neither is a problem nor does it cause a significant increase in power demand – which is still lower for the FINN-based accelerator.

Perhaps it is worth noting what is the cause of such results – experiment E4 (see Table 5 and 6) is chosen as a reference point. Since the resources' usage and energy consumption are closely related, a comparative analysis of the architectures of these two accelerators should help to understand the differences. Firstly, one can observe a great difference in the BRAM usage (48.72% vs 21.96%). The main cause shall be storing the outputs of the subsequent convolutional blocks in case of the custom accelerator. The FINN architecture is fully streamlined and consecutive outputs are immediately passed for processing to the next layer (without any buffering). Next, one can notice the lack of usage of DSP elements in FINN accelerator. For the custom one, DSP are used as neurons in the fully connected layers – in case of the FINN tool it seems that the accumulation is implemented with popcount operation. The last factor influencing the number of used LUTs (34.42% vs 6.14%) and FFs (9.59% vs 4.62%) could be in the batch normalisation module, which is implemented as threshold in the FINN accelerator – in contrary to the arithmetic operations in the custom one.

6 Summary

The starting point for this work was the development of an energy-efficient and real-time traffic sign classification system. The proposed neural networks are characterised by high accuracy and at the same time relatively low computational complexity – both in terms of the architectures' size and the ability to be inferenced with effective binary operations. As a consequence, the hardware acceleration on the ZCU 104 developement board with Zynq UltraScale+ MPSoC device was possible.

The exploration of acceleration methods was focused on two approaches: using the designed custom accelerator and the FINN tool from Xilinx. The conducted experiments showed the FINN tool superiority, due to the use of some advanced optimisations, which we did not apply in our custom solution. However, it should be once more emphasised that our research goal was not to propose a better solution than FINN, as we did not know of its existence or any other open-source XNOR accelerator (at the beginning).

Eventually, using the FINN accelerator, we achieved the set goal, which was an energy-efficient real-time classifier. The developed classifier is ready to use in the complete embedded traffic sign recognition system. Because it is characterised by high frame rate, it is also suitable for UHD input video stream. The development of the system for both recognition and classification for high resolution camera input (up to 4K) shall be continued in future work.

Acknowledgements. The work presented in this paper was supported by the National Science Centre project no. 2016/23/D/ST6/01389 entitled "The development of computing resources organisation in latest generation of heterogeneous reconfigurable devices enabling real-time processing of UHD/4K video stream".

References

1. Blott, M., et al.: FINN-R: an end-to-end deep-learning framework for fast exploration of quantized neural networks. ACM Trans. Reconfig. Technol. Syst. (TRETS) **11**(3), 1–23 (2018)
2. Umuroglu, Y., et al.: FINN: a framework for fast, scalable binarized neural network inference. In: Proceedings of the 2017 ACM/SIGDA International Symposium on Field-Programmable Gate Arrays, FPGA 2017, pp. 65–74 (2017)
3. Han, Y., Oruklu, E.: Real-time traffic sign recognition based on Zynq FPGA and ARM SoCs. In: IEEE International Conference on Electro/Information Technology, pp. 373–376, June 2014
4. Farhat, W., Faiedh, H., Souani, C., Besbes, K.: Real-time hardware/software codesign of a traffic sign recognition system using Zynq FPGA. In: 2016 11th International Design Test Symposium (IDT), pp. 302–307, December 2016
5. Zhou, Y., Chen, Z., Huang, X.: A pipeline architecture for traffic sign classification on an FPGA. In: 2015 IEEE International Symposium on Circuits and Systems (ISCAS), pp. 950–953, May 2015
6. Yao, Y., Zhang, Z., Yang, Z., Wang, J., Lai, J.: FPGA-based convolution neural network for traffic sign recognition. In: 2017 IEEE 12th International Conference on ASIC (ASICON), pp. 891–894, October 2017
7. Zhou, Y., Redkar, S., Huang, X.: Deep learning binary neural network on an FPGA. In: 2017 IEEE 60th International Midwest Symposium on Circuits and Systems (MWSCAS), pp. 281–284, August 2017
8. Shimoda, M., Sato, S., Nakahara, H.: All binarized convolutional neural network and its implementation on an FPGA. In: 2017 International Conference on Field Programmable Technology (ICFPT), pp. 291–294, December 2017

9. Zhao, R., et al.: Accelerating binarized convolutional neural networks with software-programmable FPGAs. In: Proceedings of the 2017 ACM/SIGDA International Symposium on Field-Programmable Gate Arrays, ser. FPGA 2017 (2017)
10. Liang, S., Yin, S., Liu, L., Luk, W., Wei, S.: FP-BNN. Neurocomputing **275**(C), 1072–1086 (2018). https://doi.org/10.1016/j.neucom.2017.09.046
11. Courbariaux, M., Bengio, Y.: BinaryNet: training deep neural networks with weights and activations constrained to +1 or −1 CoRR, vol. abs/1602.02830 (2016). http://arxiv.org/abs/1602.02830
12. Stallkamp, J., Schlipsing, M., Salmen, J., Igel, C.: The German traffic sign recognition benchmark: a multi-class classification competition. In: IEEE International Joint Conference on Neural Networks, pp. 1453–1460 (2011)
13. Arcos-García, Á., Álvarez-García, J.A., Soria-Morillo, L.M.: Deep neural network for traffic sign recognition systems: an analysis of spatial transformers and stochastic optimisation methods. Neural Netw. **99**, 158–165 (2018)
14. Cireşan, D., Meier, U., Masci, J., Schmidhuber, J.: Multi-column deep neural network for traffic sign classification. Neural Netw. **32**, 333–338 (2012)
15. Sermanet, P., LeCun, Y.: Traffic sign recognition with multi-scale convolutional networks. In: The 2011 International Joint Conference on Neural Networks, pp. 2809–2813 (2011)
16. Zaklouta, F., Stanciulescu, B., Hamdoun, O.: Traffic sign classification using K-D trees and random forests. In: The 2011 International Joint Conference on Neural Networks, pp. 2151–2155 (2011)
17. Zuiderveld, K.: Contrast limited adaptive histogram equalization. In: Graphics Gems IV, pp. 474–485. Academic Press Professional Inc. (1994)
18. Pappalardo, A.: Xilinx/brevitas Zenodo. https://doi.org/10.5281/zenodo.3333552

ALEA: An Anonymous Leader Election Algorithm for Synchronous Distributed Systems

Deepanjan Mitra[1(✉)], Agostino Cortesi[2], and Nabendu Chaki[1]

[1] University of Calcutta, Kolkata 700106, WB, India
nabendu@ieee.org
[2] Ca' Foscari University, Via Torino, 30172 Venice, Italy
cortesi@unive.it

Abstract. In a distributed system of nodes, leader election is the task of designating one of the nodes as the leader. In this work, we introduce ALEA, an anonymous leader election algorithm for distributed system with synchronous assumptions. The algorithm does not require a node to have knowledge of the identities of other nodes present in the network, ensuring privacy in the whole election process. Proof of other desirable properties such as uniqueness, agreement and termination are detailed as well. The overall number of exchanged messages is $O(n^2)$, n being the number of nodes, and its actual performance is experimentally evaluated and discussed.

Keywords: Distributed system · Anonymity · Leader election

1 Introduction

In distributed systems, multiple processes across different machines communicate among each other so that the complete system appears as a single coherent system to the end-user. One of the main problems of these processes is election of a leader. A leader election algorithm is a basis for more complex and high-level algorithms. As for example, this could be a step towards agreeing on the identity of a leader: whether to commit a transaction to a database, or state machine replication. In real world applications like smart power grids, estimation of state(s), control of UAVs, synchronization in the blockchain operations, etc. often rely on consensus algorithms.

In the rest of this section, we describe briefly about existing leader election algorithms. The study and analysis leads to a comprehensive understanding of the domain before the formulation of the problem that has been solved in this paper.

1.1 Related Work

The Leader election algorithms find application in many fields. Sometimes these are used in consensus algorithms like Raft consensus algorithm [8] which was

M. Choraś et al. (Eds.): CORES 2021, IP&C 2021, ACS 2021, LNNS 255, pp. 46–58, 2022.
https://doi.org/10.1007/978-3-030-81523-3_5

developed to make a simpler consensus algorithm as compared to Paxos algorithm [6]. Leader election in Raft, takes place when a follower node times out while waiting for a heartbeat message from the leader node. The timed-out node changes its state to candidate, votes for itself and broadcasts vote requests to obtain a majority and becomes the leader. An algorithm for leader election may vary in communication mechanism, network topology and size of the network. The variant could be process names too - i.e., whether they have a unique identity or are anonymous. Several leader election algorithms have been developed based on these properties. Leader election is a basic symmetry breaking operation in distributed computing. The paper [3] discusses the leader election process, in any arbitrary network with unlabelled nodes and agents, that exploit asymmetries in the initial configuration of agents due to its topology and port labelling. Several methods can be used to detect asymmetry. For instance, in [3] the concept of port awareness is used to introduce some asymmetry. The use of a particular model of computation [2] is also used for leader election. Itai and Rodeh [5] proposed a probabilistic leader election algorithm for anonymous unidirectional rings, where each process selects a random identity from a finite domain, and processes with the largest identity start a new election round if they detect a name clash, with the assumption that the size of the ring is known to all processes, so that each process can recognize its own message.

1.2 Contribution

In this paper, we adopt a different strategy to elect a leader which is different from the methodologies described above, that make use of local numbering of edges incident at a vertex or that are harnessing an underlying topology. Our algorithm uses the concept of solving challenges in the leader election process. The candidate nodes have to solve some challenges posed by follower nodes and the one which solves the maximum number of challenges should be elected leader. The motivation of solving challenges is that, in some cases, for instance in a peer to peer distributed transaction management system, a node may assign some task or part of a bigger task to several other nodes. Such tasks can be used for the concept of challenges that are used in the leader election process.

The rest of the paper is structured as follows. The ALEA algorithm is described in Sect. 2. In Sect. 3, the uniqueness, agreement, termination properties of the algorithm are described, followed by experimental evaluation and discussion in Sects. 4 and 5 respectively. Section 6 concludes.

2 The ALEA Algorithm

In ALEA, the nodes, in the network, except the leader, are in the follower state as long as they receive a heartbeat message from the leader node. When the followers do not receive any heartbeat message from leader for a period of time, they turn candidates and broadcast vote requests. The follower nodes broadcast challenges in response to the vote requests. Candidates solve those challenges and the one that solves the maximum is elected leader. However, ties are natural in any election process in which case ALEA resorts to a break-tie procedure which takes help of signatures to break the tie.

2.1 Assumptions

a) The network is anonymous. The nodes do not have any knowledge about the identities of other nodes present in the system or about the total number of nodes present in the system.
b) The communication channels between nodes are reliable with no delays. No messages are lost during transmission and messages are received by intended receiver within a finite period of time.
c) Each candidate node solves at least one challenge.

2.2 Terminology Used in the Algorithm

Each node operates in one of the following states:

- Follower: A passive node, which only responds to messages received and will not initiate any communications.
- Candidate: An active node which is attempting to become a leader, they initiate RequestVote messages.
- Leader: An active node which is currently leading the network of nodes.

The algorithm divides time into 'rounds' or 'terms'. Each round or term is defined by a number that monotonically increases with every round. Each node maintains the following information (Table 1):

Table 1. Annotation of terms used in algorithm

Term	Meaning
CTN	Current term number
ET	Election timer
ST	Solve-Timer value that takes a random value chosen from a certain range [a, b] where 'a' and 'b' are integers
WT	Wait-Timer value
RC	Record of received challenges
NC	Number of challenges solved [initialized to zero at the beginning of each term]
DS	Digital signature
MAX	Maximum of the number of challenges solved by any node, initialized to some negative value

2.3 The ALEA Algorithm: Step by Step

Initialization: Initially, all the nodes, except the leader, are in the follower state and remain in that state as long as they receive a heartbeat message from the leader node. Each node maintains an Election-Timer (ET). Since the Election-Timer is a random number, this makes it less probable that all nodes become candidates at the same time.

Step 1) **Broadcast vote_request:** When a node does not receive any heartbeat message for a period equal to ET, it switches its state to candidate, increments current term number and broadcasts a VOTE_REQUEST (term) message.

Step 2) **Receipt of vote request and broadcast of challenge:** A node N on receipt of a VOTE_REQUEST (term) message, updates its current term with the term number received in the message and takes the following actions:
- (i) If N is a candidate, then it silently discards the message.
- (ii) If N is a non-candidate, it broadcasts a message CHALLENGE (term, challenge).

Step 3) **Receipt of challenge message:** On receipt of a CHALLENGE (term, challenge) by a node N, two things can occur:
- (i) If N is not a candidate node, it silently discards the message.
- (ii) If N is a candidate, then it performs the following actions: If it receives challenge for first time, it stores the received challenge in the RC list, starts a timer called Solve-Timer (ST) and begins to solve the received challenge. Otherwise, if the Solve-Timer (ST) has not expired, and the node is busy solving some other challenge, then it appends the received challenge in the RC list. If the node has solved a particular challenge and ST has not expired, it takes up the next challenge from the RC list (if the list is not empty).

When ST times out, the node broadcasts a message SOLVED (n, SIGN(n), PK) to all other nodes. Here n is the number of challenges solved by N, SIGN is a digital signature that node N computes on value of n and PK denotes the public key of that node.

Step 4) **Receipt of solved message:** A node N on receipt of the first SOLVED(n, SIGN(n), PK) message, performs the following actions:
- (i) The Node N starts a timer Wait-Timer(WT) (initialized to a constant value).
- (ii) If N is candidate and ST timer has not yet expired, then the node stops the timer and broadcasts the SOLVED(n, SIGN, PK) to all other nodes. (Let us denote the received signature as SIGN_recvd and the signature that it has broadcast as SIGN_sent. The candidate node maintains other information in addition to MAX. It also records the signature value corresponding to the n value with which MAX is updated). Here, P is a pool of digital signatures corresponding the candidate nodes that have solved the same number of challenges n

and value of n is the maximum of the number of challenges solved by any candidate node.
If MAX < n:

MAX = n

DS = (SIGN_recvd, PK)

Make P empty

Add (SIGN_recvd, PK) to P

If MAX = n:

Add (SIGN_rcvd, PK) to P

Else Discard the message.

Step 5) **Change state of node:** When the Wait-Timer times out, node N if a candidate, first checks if it has solved number of challenges equal to the MAX value recorded. If so, it adds the signature that it had broadcast, to P. A node checks if P is a singleton set or not, if not, it uses the sub-procedure break-tie on P. (The break-tie procedure has been described in Subsect. 2.4)
If $|P| = 1$ then

return DS

Else

call Procedure break-tie (P, MAX)

Then, the node compares m (number of challenges solved by it) and the value of MAX. Three cases may occur:

Case 1: If m < MAX then node N changes its state to follower.

Case 2: If m > MAX then node N changes its state to leader.

Case 3: If m = MAX then node N checks the following:

If DS = SIGN_sent

Node N changes its state to leader.

Else

Node N changes its state to follower.

Step 6) **Broadcast of heartbeat message by leader:** When a node N that has changed its state to leader, it broadcasts a HEARTBEAT (n, SIGN(n)) message containing the value n digitally signed using PK i.e. the public key of the node. The follower node on receipt of the heartbeat message verifies if it comes from the leader using the public key PK received and the digital signature that it has recorded, and it also checks if MAX = n. If the above conditions hold true, the follower nodes agree upon the elected leader. As long as the elected node remains leader it periodically broadcasts heartbeat messages.

2.4 Resolving Ties Among the Candidate Nodes

Problem occurs when more than one candidate happens to solve the same number of challenges and equal to the maximum of the number of challenges solved, in which case there would be a tie. Since there can be only one elected leader in a term, so the following procedure break-tie (P, L) is used to resolve the problem. The meaning of the terms used in the sub-procedure break-tie is described in the Table 2. The sub-procedure break-tie is called when a node finds that it has solved more than one challenge and the number of challenges solved is equal to the maximum of the number of challenges solved by all nodes of the network. This sub-procedure takes two arguments as input namely P and L where P is the pool of digital signatures corresponding nodes that have solved number of challenges equal to MAX value and L denotes the maximum of the number of challenges solved as recorded by the node from receipt of solved messages.

Table 2. Annotations of terms used in sub procedure break-tie

Term	Meaning
P	Pool of digital signatures corresponding the candidate nodes that have solved the same number of challenges n and value of n is the maximum of the number of challenges solved by any candidate node
T	Sum of the digital signatures in P. The size of digital signature computed by each node is equal
res	This variable is used to store the maximum value of remainder computed and is initialized to some small value
SIGN	A digital signature in the pool P
DS	The digital signature value returned by the procedure after tie is resolved
Q	A list containing signatures corresponding which modulo values computed are same. In case tie occurs again, the procedure is called with Q as argument recursively until the tie is broken

Listing 1. Pseudocode

```
procedure break-tie(P, L)
begin
    M = | P | i.e. cardinality of P
    T = $\sum_{i=1}^{M} sign_i$
    G = $\lceil \frac{T}{M*L} \rceil$
    res = -1
    DS = null
    Q = empty
    for each (SIGN_i, PK_i) in P:
        val = SIGN_i mod G
        if res < val:
           res = val
           DS = (SIGN_i, PK_i)
           make Q empty and add DS to Q
        else if res = val:
           add DS to list Q
       end if
    end for
    if | Q | > 1:
       call procedure break-tie(Q, L)
    else if | Q | = 1
       return DS
    end if
end procedure
```

The main idea used in the above sub-procedure is that we take a sum of the digital signature values present in P. Let us denote this sum as T. We consider another variable G which is computed as the greatest integer of T when divided by the product of the cardinality of P and value of L. Now each signature in the pool P is taken and the remainder of the division of signature by the value of G is considered. The digital signature corresponding the maximum value of remainder is considered the result of the sub-procedure. If two or more signatures happen to produce the same remainder, then the signatures are added to the pool and the sub-procedure is called recursively on this new pool until a result is obtained. Now, question may arise whether the above procedure will converge or not. If no two signatures give the same remainder then the sub-procedure converges. On the other hand, if two or more signatures produce the same remainder values equal to the maximum remainder then recursive calls are made. However, the number of such signatures decreases with each recursive call. Hence, we can say that the sub-procedure break-tie will eventually converge to return a result to the algorithm.

3 ALEA's Main Properties

A leader election algorithm is expected to satisfy the following properties:

- Uniqueness: there is exactly one node that is elected as a leader.
- Agreement: all other processors should know who the leader is, and they should be able to distinguish messages coming from the leader.
- Termination: the election algorithm should terminate within finite time.

Theorem 3.1. *ALEA satisfies the uniqueness property.*

Proof. Let us assume that, there are two nodes N and M that have been elected leaders and let n and m be the number of challenges solved by them respectively. As both N and M have been elected leaders, so, the number of challenges solved by them is greater than MAX recorded by nodes N and M. Since we had initially made the assumption that the communication channels are reliable and that no messages are lost during transmission, (assumptions (b), in Sect. 2.1), each SOLVED message will be received by each node of the network and each node will duly record the MAX value (as described in step 4 of the algorithm). Therefore, the maximum of n and m will be the value of MAX recorded by the nodes. Now, two cases may occur.

a) If $n \neq m$: then one of nodes N or M will become the leader and the other a follower, based on the actions described in cases 1 and 2 of step 5 in the algorithm.
b) If $n = m$: In this case, the nodes N and M will take actions in accordance with the actions described in case 3 of step 5 in the algorithm as a result of which one of nodes N and M will become the leader. When two or more nodes happen to solve same number of challenges and equal to the maximum of number of challenges solved by any other node, then the tie will be broken using the break-tie procedure. The digital signature DS value recorded along with the MAX value will be used to decide on the leader. Thus, only one of the nodes N and M will emerge as the leader node.

Therefore, our initial assumption, that in a given term two nodes N and M have been elected leaders, is not correct. Thus, the uniqueness property holds true.

Theorem 3.2. *ALEA satisfies the agreement property.*

Proof. In an anonymous system, this property may be redefined as, once a node is elected leader, all the other nodes should agree on the elected leader. From step 6 of the algorithm we observe that, once a leader gets elected, it broadcasts a HEARTBEAT(n, SIGN(n)) message containing the number of challenges solved by it digitally signed. Let us assume that there exists a follower node N for which $n \neq MAX$ or $SIGN(n) \neq DS.SIGN$. Since we had initially made the assumption that the communication channels are reliable and that messages are not lost during transmission, each SOLVED message will be received by each node in the network and each node will duly record the MAX value (as described in step 4 of the algorithm). Therefore, for every follower node, the MAX value should be exactly equal to the value of n (number of challenges solved by the leader) propagated in the HEARTBEAT message, and the SIGN value received in the heartbeat message must also agree with the SIGN value recorded in DS. Thus, we arrive at a contradiction. Hence our assumption, that there exists a follower node N for which $n \neq MAX$ or $SIGN(n) \neq DS$. SIGN, was wrong. Therefore, all follower nodes agree upon the elected leader. Thus, the algorithm satisfies the agreement property.

Theorem 3.3. *ALEA satisfies the termination property.*

Proof. Let us assume that a node N has been elected leader and broadcasts the heartbeat message containing the number of challenges solved by it digitally signed. Therefore, all other nodes in the network should be in the follower state. Let us consider some node other than the leader node N, is not in the follower state. Let us denote this node as M. Since a node can be in any one of three states. So, this node M can either be a candidate or a leader. We have already shown that in a given term, there can be only one elected leader (uniqueness property). Thus, node M cannot be in the leader state. Now let us check if node M can be in candidate state. Let us go back to step 5 of the algorithm. Since node M cannot be a leader, it does not satisfy case 2 of step 5. So, we have to investigate cases 1 and 3 of step 5 in the algorithm. Either the node M must have m (number of challenges solved by M) $<$ MAX and hence node M must change its state to follower, or the node M must have m $=$ MAX (in which case the node M must have been a candidate node in the election process and assuming that at least one candidate node solves at least one challenge). In the latter case, the node M will compare the digital signature value recorded from the receipt of SOLVED messages from other nodes (as described in step 4) and that it had broadcast i.e. SIGN_sent. Since a leader has already been elected, then the recorded DS value must not be equal to SIGN_sent. Thus, the node must change its state to follower. Therefore, our initial assumption that node M is not a follower when there is an elected leader is wrong. Hence, once a node has been elected leader, all other nodes are in follower state and the leader election terminates in finite time.

Theorem 3.4. *ALEA preserves anonymity of the nodes.*

Sketch of the Proof. We can see that whenever nodes need to send messages to other nodes, it uses broadcasting because the identity of other nodes are unknown to it. All messages that are exchanged between nodes are broadcast messages and do not contain the identity of intended recipient nodes. The digital signatures being exchanged between the nodes do not reveal any information about the private key of nodes or some other information about its identity because, recovering the private key from a digital signature is computationally difficult. Thus, no node will be able to know the identity of other nodes present in the network.

4 Experimental Evaluation

4.1 Scenario

The algorithm was simulated using the Contiki-Cooja simulator and conducted on an Intel Core i3-6006U with 4 GB RAM. The underlying topology is a wireless communication network. Each node can communicate directly with all other nodes by broadcasting messages. Moreover, we assume that no messages get lost during transmission. The two important parameters in this algorithm are the time taken and the total number of message transmission involved in the election process. It is needed to analyse the effects on time taken, on increasing the number of nodes in the network. Moreover, since we consider a model of anonymous distributed computation where nodes communicate by broadcasting messages, the total number of messages involved in the election process as the number of nodes in the network increases, is analysed.

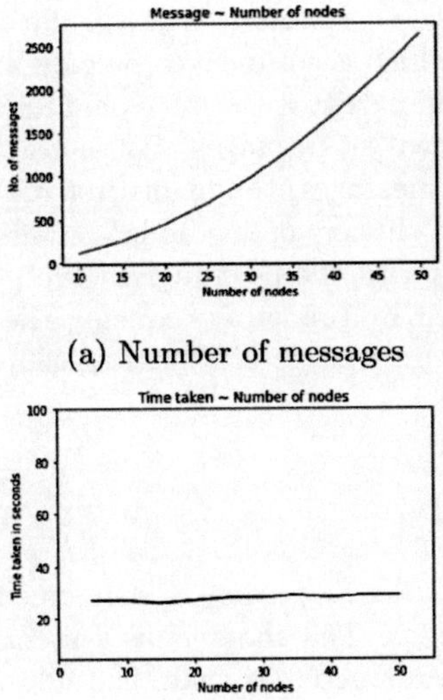

(a) Number of messages

(b) Time Taken

Fig. 1. Simulation results

4.2 Experimental Results

The algorithm was run on a network of ten to fifty nodes and the two parameter values were recorded as an average over the different cases based on number of candidates and number of challenges solved. It is seen in the Fig. 1b that the time taken remains almost the same when the number of nodes is increased from ten to fifty. This is subject to the assumption that the messages broadcast are directly received by all other nodes without any intermediate hops. This is

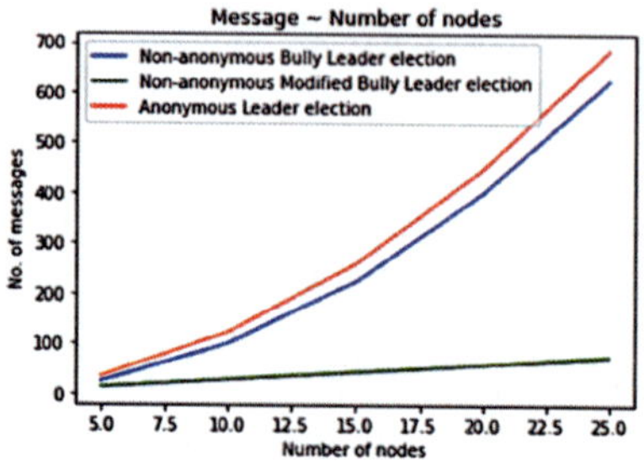

Fig. 2. Comparison of message complexity

because we have considered synchronous assumptions in the algorithm and hence hop by hop message transmission cannot be considered. On the other hand, in Fig. 1a it is seen that the total number of messages increases in a quadratic fashion as the number of nodes increases.

The nature of the message-node curve is due to the broadcast of messages taking place in the network which is assumed to be completely connected topology. This quadratic message complexity is the price paid for anonymity in the network which necessitates broadcast of messages. Given the synchronous assumption and the large number of messages, the algorithm may be used in a hierarchical network topology with clusters of few nodes performing localised operations and results being passed on to the next higher level. The number of messages exchanged may be reduced by randomly reducing the number of active candidates in the network. We propose to use these measures to reduce the message complexity in future work.

5 Discussion

Anonymity is desirable as far as privacy is concerned and is inevitable when nodes do not have unique ids. The anonymous systems however, are not developed with an objective of running faster than their non-anonymous counterparts. The objective of the paper is to present an algorithm that uses the concept of solving challenges. However the concept of challenge here, is at an abstract level. We do not mention any specific challenge. The challenges can be thought of as certain tasks that are needed to be performed. The nature of these tasks vary according to the application scenarios. For instance, in distributed transaction management system, a node may assign some task or part of a bigger task to several other nodes. These may be modelled as challenges by the nodes in the network and help the nodes choose their next coordinator when needed. We have compared the performance of ALEA with non-anonymous leader election algorithm to show that it is no worse than some popular non-anonymous leader election algorithms. We consider the Bully leader election algorithm. ALEA has a similar message complexity of quadratic order as that of the Bully leader election algorithm. Figure 2 shows a comparison graph where the proposed algorithm's message complexity is compared with that of the Bully algorithm. Next we consider, the modified Bully leader election [1]. The graph shows that both ALEA and the Bully algorithm are having curve shape that describes the message complexity as $O(n^2)$, while the modified Bully algorithm is having linear growth. The reader may wonder why can't the modification applied to Bully Leader election be also applied to this algorithm in an attempt to reduce message complexity. In Bully algorithm [4], the modification is done to remove duplicate or parallel election when a failed leader node becomes alive. This reduces the number of messages. However, ALEA is designed under the assumption that node failures do not occur. In the modified Bully leader election in [7], when the old leader node or process comes alive after it had crashed, a coordinator message is immediately sent to take over the leadership by bullying all the processes. However such a modification requires that nodes should know the identity of the coordinator node which is not possible in the anonymous setting. As mentioned before, the case of node failure is not considered. In order to ensure that a failed node on recovery can join the existing network, some actions may be undertaken. Let us denote a node that has recovered from failure as R. Now, node R can recover in:

a. Recovery in regular time: If node R is the leader, then if it does not recover within the election timer expiration, the other nodes having not received a heartbeat message, will start an election procedure as described in the algorithm. Any node irrespective of the fact whether it had been or had not been a leader before failure, on recovery during regular time will join the network as a follower node. Receiving heartbeat messages from the leader.
b. Recovery in election time: A node R on recovery during election time, if receives vote-request then it will participate in election process, otherwise if it receives a solved message when it has sent no vote-request, it stays in the network passively as a follower until participation in future leader election.

6 Conclusions

In this work, a leader election algorithm has been discussed, which maintains the anonymity of nodes in the network. Apparently, the proposed algorithm has a higher message complexity as compared to some other existing leader election algorithms. This is the price to pay to maintain anonymity of nodes in the network.

The algorithm is uniform, i.e. the nodes in the network do not need to know the total number of nodes present in the network. Being an uniform algorithm, ALEA allows scalability and transparency, and processes can join or leave the network without causing any hindrance to the other processes.

We are aware that the proposed algorithm has the limitation of lying on synchronous assumptions, which may restrict its adoption in real world scenarios. Another limitation is the fragility with respect to internal attacks: a malicious candidate may falsely propose to have solved a large number of challenges with the aim of winning the election. Such attacks need to be identified. In future, further improvements of the ALEA algorithm will be devised to properly address these open issues.

References

1. Beaulah, S.P.: Efficient algorithms for enhanced performance in distributed systems (2014). https://shodhganga.inflibnet.ac.in/bitstream/10603/73526/25/09_chapter%204.pdf
2. Boldi, P., Shammah, S., Vigna, S., Codenotti, B., Gemmell, P., Simon, J.: Symmetry breaking in anonymous networks: characterizations. In: ISTCS, pp. 16–26 (1996)
3. Dereniowski, D., Pelc, A.: Leader election for anonymous asynchronous agents in arbitrary networks. Distrib. Comput. **27**(1), 21–38 (2014)
4. Garcia-Molina, H.: Election in a distributed computing system. Trans. Comput. **31**, 48–59 (1982)
5. Itai, A., Rodeh, M.: Symmetry breaking in distributive networks. In: 22nd Annual Symposium on Foundations of Computer Science (SFCS 1981), pp. 150–158. IEEE (1981)
6. Lamport, L.: The Part-Time Parliament, pp. 277–317. Association for Computing Machinery, New York (2019). https://doi.org/10.1145/3335772.3335939
7. Murshed, M., Allen, A.R., et al.: Enhanced bully algorithm for leader node election in synchronous distributed systems. Computers **1**(1), 3–23 (2012)
8. Ongaro, D., Ousterhout, J.: In search of an understandable consensus algorithm. In: 2014 {USENIX} Annual Technical Conference ({USENIX}{ATC} 14), pp. 305–319 (2014)

Progress in Image Processing and Computer Vision

Comparing Concepts of Quantum and Classical Neural Network Models for Image Classification Task

Rafał Potempa(✉) and Sebastian Porebski

Department of Cybernetics, Nanotechnology and Data Processing, Faculty of Automatic Control, Electronics and Computer Science, Silesian University of Technology, Gliwice, Poland
rafal.potem@gmail.com, sebastian.porebski@polsl.pl

Abstract. While quantum architectures are still under development, when available, they will only be able to process quantum data when machine learning algorithms can only process numerical data. Therefore, in the issues of classification or regression, it is necessary to simulate and study quantum systems that will transfer the numerical input data to a quantum form and enable quantum computers to use the available methods of machine learning. This material includes the results of experiments on training and performance of a hybrid quantum-classical neural network developed for the problem of classification of handwritten digits from the MNIST data set. The comparative results of two models: classical and quantum neural networks of a similar number of training parameters, indicate that the quantum network, although its simulation is time-consuming, overcomes the classical network (it has better convergence and achieves higher training and testing accuracy).

Keywords: Quantum neural network · Quantum computing · Quantum circuit · Image recognition · Quantum data representation

1 Introduction

The paper shows that it is possible to unlock the potential of computational methods dedicated to the Turing-Church-Deutsch machines [4] although quantum computers are still in the initial phase of their development [18]. It is still unclear which of the physical implementations of quantum bits, i.e., by electrons [7,14,16], atomic nuclei [13], photons [19,20], or various physical effects, e.g., [9,12], would be the final one.

The paper shares the knowledge gathered in the research of the quantum neural networks, which is the extension of exploratory results published in [17]. The work was performed using the TensorFlow Quantum [2] library and similarly to [17], it was inspired by the ideas presented in [6] and [8], providing the extension of quantum neural network for solution of MNIST classification from [2] to all the ten digits. Besides, by using appropriate representation techniques,

M. Choraś et al. (Eds.): CORES 2021, IP&C 2021, ACS 2021, LNNS 255, pp. 61–71, 2022.
https://doi.org/10.1007/978-3-030-81523-3_6

it was possible to keep the input image larger. Nonetheless, a more efficient way of quantum representation of images, like FRQI [10] might have been used, but this particular example may require considering the use of a recurrent network or sampling layers to store the outcomes of the consecutive measurements on the interaction of quantum and classical network parts, hence increasing the complexity of the solution. Nevertheless, the technique of dividing the image into four parts helped to reduce the computational effort and made it possible to outperform the classical neural network in comparative experiments. Thus the work successfully simulates the learning of a quantum neural network and shows that quantum learning has a great potential for the development of machine learning.

2 Materials and Methods

The study deals with a popular MNIST Handwritten Digits Database [11]. The original data set contains images of digits from 0 to 9 represented by matrices of integer values from 0 to 255, of size 28 × 28, representing the grayscale color values of the image pixels. Total number of images is $N = 70\ 000$. Database authors [11] divided it into training and testing subsets of size $\frac{6}{7}N$ and $\frac{1}{7}N$, respectively, hence the data set is prepared for hold-out verification.

2.1 Quantum Computing

The quantum bit, or shortly *qubit*, is a conceptual quantum analogy of a classical bit. While a classical bit can be in either of states called 0 and 1, a qubit can resemble the quantum analogies of classical states, i.e., $|0\rangle$, $|1\rangle$, or a superposition of those states $|\varphi\rangle$, defined as

$$|\varphi\rangle = \alpha\,|0\rangle + \beta\,|1\rangle\,, \tag{1}$$

where $\alpha, \beta \in \mathbb{C}$ and $|\alpha|^2 + |\beta|^2 = 1$ [15]. Values $|\alpha|^2$ and $|\beta|^2$ represent the probability of measuring the qubit in state $|\varphi\rangle$ as $|0\rangle$ or $|1\rangle$, respectively, but until the measurement, both states are present and real [15]. Due to values α and β being complex numbers, and their combined magnitude being unitary, a qubit can be represented as a vector on a unitary complex sphere, called the Bloch sphere, shown in Fig. 1a and any unitary qubit operation can be treated as a rotation of a unit vector representing state $|\varphi\rangle$, on such sphere [15].

Quantum computing is based on the evolution in time of quantum systems that resemble operations performed on qubits by means of *quantum gates* [5]. The gates work not like their classical analogies but rather modify the current position of $|\varphi\rangle$ from (1). The result is obtained by initializing all qubits to desired states, application of quantum gates in predefined sequence forming a *quantum circuit* and finally measuring the desired qubits for their internal states. All operations are performed under the control of classical control systems and classical computers, that help to initialize the qubits and preprocess the data into its quantum representation [15].

The most basic quantum gate is the Pauli-X gate, i.e., a quantum analogy of NOT gate, that rotates the qubit state by π around its x axis and is represented by the Pauli-X matrix:

$$\mathrm{X} = \begin{bmatrix} 0 & 1 \\ 1 & 0 \end{bmatrix}. \tag{2}$$

Another notable example is the Hadamard gate [15], performing a rotation by π around $[1, 0, 1]/\sqrt{2}$ axis:

$$\mathrm{H} = \frac{1}{\sqrt{2}} \begin{bmatrix} 1 & 1 \\ 1 & -1 \end{bmatrix}, \tag{3}$$

which is often used to initialize the qubits into superpositions. The Hadamard gate, combined with a two-qubit controlled-NOT gate (CNOT) [15] that flips the target qubit state, only if the control qubit is in $|1\rangle$, as defined in

$$\mathrm{CNOT} = \begin{bmatrix} 1 & 0 & 0 & 0 \\ 0 & 1 & 0 & 0 \\ 0 & 0 & 0 & 1 \\ 0 & 0 & 1 & 0 \end{bmatrix}, \tag{4}$$

can be used to entangle two qubits. To put two qubits q_1 and q_2 into one of the entangled Bell states $|\beta_{q_1 q_2}\rangle$, q_1 is put into a superposition using the Hadamard gate. Next this qubit is used to control the flip of the q_2 in the CNOT gate, as shown in Fig. 1b. This results in the state of q_2 being directly dependant on state of q_1 and hence entangling it with q_1. The determination of a state of one qubit will result in the immediate determination of a state of the other.

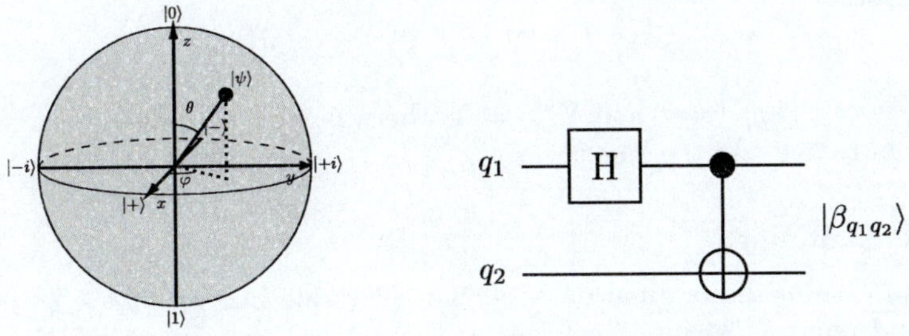

(a) The Bloch sphere

(b) The quantum circuit to introduce a Bell state

Fig. 1. Graphical description of quantum data processing basics

The quantum neural network used during the experiments relates to the model proposed in [6], combined with classical layers for merge and output. Thus the network forms a hybrid quantum-classical model discussed in [2]. The presented model is a modification of the quantum neural network proposed in [2] because it allows classifying all ten MNIST digits, and it is an extension of the authors' exploratory experiments reported in [17].

3 Experimental Framework

3.1 MNIST Data Set Split

In the study, out of the 60 000 images used to tune the network, 1/6 was dedicated for validation purposes. Let us denote the classification data set as the collection of N pairs of inputs $\boldsymbol{x}_i$ and outputs y_i, $(i = 1, 2 \ldots, N)$ that is exclusively divided into training (l), validation (v) and testing (t) subsets:

$$\left\{\boldsymbol{x}_i^{(l)}, y_i^{(l)}\right\}_{i=1}^{N^{(l)}}, \quad \left\{\boldsymbol{x}_i^{(v)}, y_i^{(v)}\right\}_{i=1}^{N^{(v)}}, \quad \left\{\boldsymbol{x}_i^{(t)}, y_i^{(t)}\right\}_{i=1}^{N^{(t)}}. \tag{5}$$

In all experiments $N^{(l)} + N^{(v)} = 60\ 000$ and $N^{(t)} = 10\ 000$. Let us also denote the general subset

$$\left\{\boldsymbol{x}_i^{(*)}, y_i^{(*)}\right\}_{i=1}^{N^{(*)}}, \tag{6}$$

that expresses data pairs without considering their relation to particular subset, but if necessary, general mark $(*)$ is replaced with suitable one $(l$, v or $t)$.

3.2 Class Label Determination

For a multi-class classification (opposed to the reference approach where only two digits: 3 and 6, are classified [6]), all ten digits are considered. According to (6) each data case $\boldsymbol{x}_i^{(*)}$ is related to its true class label

$$y_i^{(*)} \in \{1, 2, \ldots, c, \ldots, C\}, \tag{7}$$

where $C = 10$. Class label (7) can be also transformed to vector in one-hot encoding form

$$\mathbf{Y}_i^{(*)} = \left[Y_{i,1}^{(*)}, Y_{i,2}^{(*)}, \ldots, Y_{i,c}^{(*)}, \ldots, Y_{i,C}^{(*)}\right], \tag{8}$$

where $Y_{i,c}^{(*)} = 1$ if $y_i^{(*)} = c$ and $Y_{i,c}^{(*)} = 0$ otherwise, hence only one element of (8) is equal to one and the rest is zero.

3.3 Sub-sampling

Each data case in the original data set is a grayscale image of 28×28 pixels. This would require 784 input neurons in the network models. Simulation of a quantum circuit that processes such data dimension is hardly applicable on a classical machine. To reduce the complexity of simulating the neural network's quantum circuit and to allow operating on an ordinary machine, the images are scaled to the resolution of 8×8 pixels using the dedicated resize method from [1]. The resulting images are split into four parts of size 4×4 pixels, giving the final reduction of memory required to simulate the circuit using a modified version of quantum image representation proposed by [2] from 2^{784} to roughly $4 \cdot 2^{16}$ classical bits, giving the final mapping of

$$B^{28} \times B^{28} \rightarrow [E^4 \times E^4]^4, \tag{9}$$

where $B = \{0, 1, ..., 255\}$ and $E = [0, 1]$.

3.4 Quantum Data Representation

The image fragments with the size of 4 × 4 are represented using a method similar to [2], where the image pixel is correlated directly with a single qubit. The grayscale value determines the rotation of a qubit corresponding to the pixel position. The rotation is performed using the generalization of (2), being the $\mathrm{R_x}$ gate, allowing for rotation around the x axis by arbitrary angle θ, as defined in

$$\mathrm{R_x}(\theta) = e^{-i\theta \mathrm{X}/2} = \begin{bmatrix} \cos\frac{\theta}{2} & -i\sin\frac{\theta}{2} \\ -i\sin\frac{\theta}{2} & \cos\frac{\theta}{2} \end{bmatrix}, \quad (10)$$

where $\theta = \pi x$ is the rotation angle obtained by mapping the normalized pixel grayscale value onto $[0, \pi]$. Hence the original values between 0 and 255 (from 0.0 to 1.0 after normalization) are represented by $|0\rangle$ and $|1\rangle$ state, respectively. The example is shown in Fig. 2c.

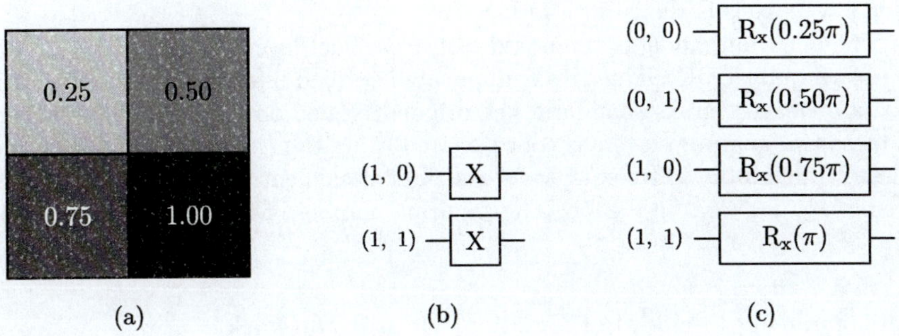

Fig. 2. Quantum encoding of example image (a) with X gates (b) [2] and $\mathrm{R_x}$ gates (c) [17]

3.5 Building a Model of a Quantum Neural Network

The quantum layer is used as an analogy of classical dense layer and is based on the idea of Parametrized Quantum Circuit (PQC). The layer responsible for insight is a quantum circuit with trainable parameters. The PQC uses parametrized CNOT gates defined as

$$\mathrm{CNOT}(z) = \begin{bmatrix} 1 & 0 & 0 & 0 \\ 0 & 1 & 0 & 0 \\ 0 & 0 & gc & -igs \\ 0 & 0 & -igs & gc \end{bmatrix}, \quad (11)$$

where $c = \cos(\frac{\pi}{2}z)$, $s = \sin(\frac{\pi}{2}z)$, $g = e^{\frac{i\pi z}{2}}$ [3], with z being the layer trainable parameters, i.e., the network learns what should be the scaling of all controlled negation operations. The PQCs are composed of 16 data qubits, that control

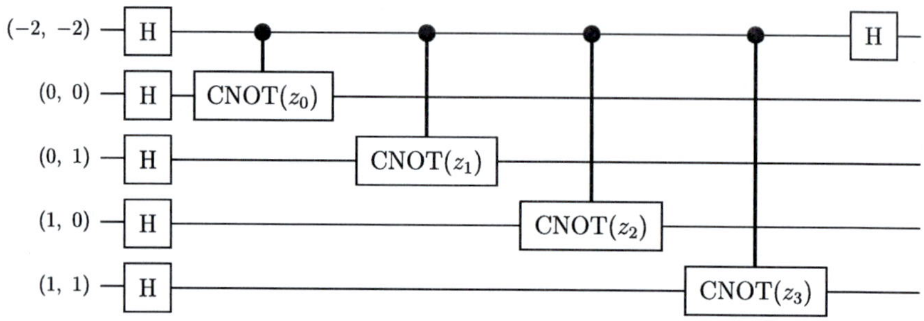

Fig. 3. Layout of a simplified version of PQC used for experiments, where $(-2, -2)$ qubit is the readout qubit and the other qubits are used for image pixels representation [17].

the single readout qubit. The simplified PQC, that would be used if the images were 2×2 pixels is shown in Fig. 3.

The quantum model is composed of two parallel layers, both constructed of four 4×4 qubit PQCs. The data is duplicated and fed into the layers in the form of 4×4 image chunks that form the original, scaled-down image. The output of these two quantum layers is concatenated using the classical merge layer and passed through the softmax layer for the final class membership certainty values prediction, as shown in Fig. 4b. The resulting number of trainable parameters is 218.

3.6 Building a Model of Classical Neural Network

The classical neural network model, compared to the prototype quantum model from the TensorFlow Quantum library [2], has a standard structure. Using the TensorFlow library [1], a network of the form of multi-layer perceptron with one hidden layer was built, ensuring that the number of parameters is similar to the quantum model. Thus, there is an input layer transforming the image to the vector shape, the hidden layer with four neurons, and the output layer with ten neurons. The resulting number of trainable parameters is 310, which is a number higher than that of the parameters of the quantum model, but this is related to the fact that the quantum images required to be split and that PQCs are not interconnected, leading to the reduction in the number of parameters in quantum model. However, the model layouts are the same.

3.7 Training, Validation, and Testing

Both models, quantum and classical neural networks (see Fig. 4), are designed to be $p = 64$ input, and $C = 10$ outputs classifiers. Hence the quality of the result of training, validation, and testing can be made analogous. The dedicated functions from the TensorFlow libraries [1,2] are used for training with validation. Predicted output vector of studied models can be of the following form:

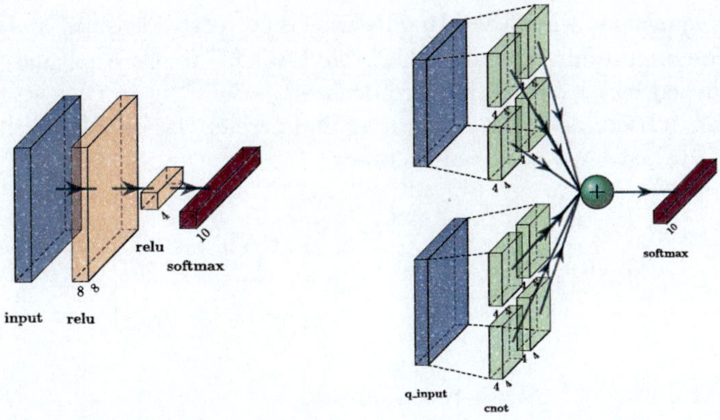

(a) (b)

Fig. 4. Overviews of the classical (a) and quantum neural network (b) model architectures

$$\widehat{\boldsymbol{Y}}_i^{(*)} = \left[\widehat{Y}_{i,1}^{(*)}, \widehat{Y}_{i,2}^{(*)}, \ldots, \widehat{Y}_{i,c}^{(*)}, \ldots, \widehat{Y}_{i,C}^{(*)}\right], \tag{12}$$

where $\widehat{Y}_{i,c}^{(*)}$ is the value of the cth output neuron of the network that indicate membership of the ith data case to the cth class. The output result is defined as

$$\hat{y}_i^{(*)} = \underset{1 \le c \le C}{\operatorname{argmax}} \left[\widehat{\boldsymbol{Y}}_i^{(*)}\right] \tag{13}$$

for the purpose of its comparison with (7). Implementation of the neural networks required some training and validation presumptions, e.g., about loss calculation and choosing the optimization method, hence the categorical cross-entropy (CCE) was chosen as loss function for the multi-class classifier training and validation. It is calculated for each true class vector $\boldsymbol{Y}_i^{(*)}$ (8) and predicted output $\widehat{\boldsymbol{Y}}_i^{(*)}$ (12), separately, and averaged for the subset:

$$\mathrm{CCE}^{(*)} = \frac{-1}{N^{(*)}} \sum_{i=1}^{N^{(*)}} \left[\sum_{c=1}^{C} Y_{i,c}^{(*)} \cdot \log \widehat{Y}_{i,c}^{(*)}\right]. \tag{14}$$

The Adam algorithm, which is a stochastic gradient descent method based on adaptive estimation of first- and second-order moments serves for a networks' parameters optimization [1]. Obviously, the network testing involved feeding it with $\boldsymbol{x}_i^{(t)}$ to predict $\hat{y}_i^{(t)}$ and finally, compare these with $y_i^{(t)}$, $(i = 1, \ldots, N^{(t)})$. To measure general performance of studied models, accuracy is calculated as

$$\mathrm{ACC}^{(*)} = \frac{1}{N^{(*)}} \sum_{i=1}^{N^{(*)}} \mathrm{I}\left(\hat{y}_i^{(*)} = y_i^{(*)}\right), \tag{15}$$

where I(·) equals one when the i-th data case is correctly classified and zero otherwise. Presented results relate to $\mathrm{ACC}^{(l)}$ and $\mathrm{ACC}^{(v)}$ that are calculated in each epoch (indexed by k), hence they are denoted as $\mathrm{ACC}_k^{(l)}$, $\mathrm{ACC}_k^{(v)}$, respectively. The testing subsets does not contain equinumerous classes, hence the second method of evaluation is balanced accuracy

$$\mathrm{BA}^{(t)} = \frac{1}{C}\sum_{c=1}^{C}\left(\frac{\sum_{i=1}^{N^{(t)}} \mathrm{I}\left(y_i^{(t)} = c \wedge \hat{y}_i^{(t)} = c\right)}{\sum_{i=1}^{N^{(t)}} \mathrm{I}\left(y_i^{(t)} = c\right)}\right), \tag{16}$$

which is the average of C class-precision values.

3.8 Results

The simulation of the quantum network is aimed at checking its classification possibilities in comparison with the classical network using a similar architecture of similar size. Naturally, the simulation of a quantum network is highly time-consuming because it requires the numerical equivalent of quantum computations, so the training and validation time cannot be compared. Nevertheless, the change of quantum and classical network performance from epoch to epoch allows their essential comparison. In order to carry out the calculations in a reasonable time, it was chosen to subject both networks to training and validation for ten epochs.

The aforementioned training and validation time was enough to check whether the training of the quantum network was successful and to compare its work with the classical network. Figures 6 and 5 present how $\mathrm{ACC}_k^{(l)}$, $\mathrm{ACC}_k^{(v)}$ and $\mathrm{CCE}_k^{(l)}$, $\mathrm{CCE}_k^{(v)}$ changed during training and validation of both models. As the result of each of five experiments $\mathrm{ACC}_k^{(t)}$ and $\mathrm{BA}_k^{(t)}$ are calculated.

According to both Fig. 5 and 6 one can see that the quantum neural network model (whether considering a change of accuracy and loss for training and validation steps) has a faster convergence, i.e., it achieves a stable result quicker. Greater stability can also be seen by analyzing the results of individual five experiments (transparent lines). Both models tend to minimize the loss function evenly, which cannot be said about the accuracy. For the latter, the classical model has a greater values dispersion for a single epoch, and they do not change as smoothly as for the quantum model. Particularly poor convergence stability of individual experiments can be seen in validation (right diagram in Fig. 6). The observation is also confirmed in the testing stage of the models. The advantage of the quantum model is proven by the accuracy and balanced accuracy values for each of the five experiments presented in Table 1.

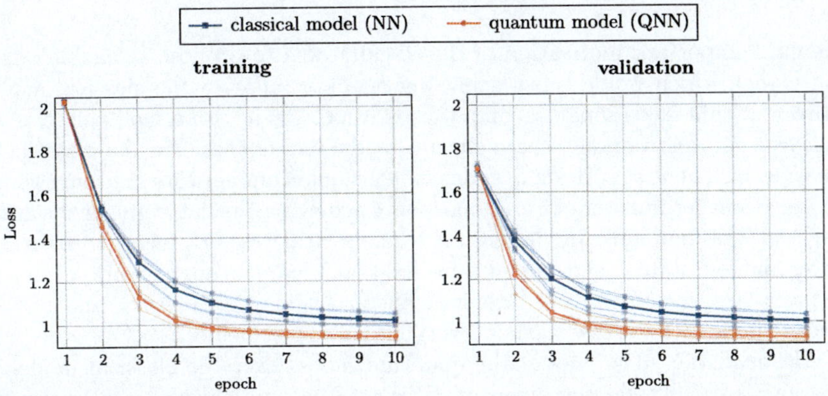

Fig. 5. Changes of loss function (14) for classical and quantum models in ten epochs. Transparent lines relates to single experiment and solid lines are mean results.

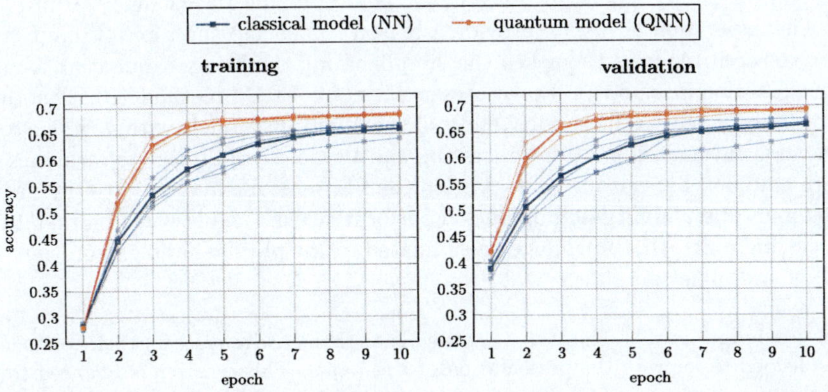

Fig. 6. Changes of accuracy (15) for classical and quantum models in ten epochs. Transparent lines relates to single experiment and solid lines are mean results.

Table 1. Comparison of $ACC_k^{(t)}$ (15) and $BA_k^{(t)}$ (16) of each of five hold-out experiments for classical and quantum model

Measure	Model	$k = 1$	$k = 2$	$k = 3$	$k = 4$	$k = 5$	Average
$ACC_k^{(t)}$	Classical	67.71%	67.04%	67.32%	68.43%	67.06%	67.51%
	Quantum	70.67%	70.59%	69.72%	70.78%	70.83%	70.52%
$BA_k^{(t)}$	Classical	67.24%	66.56%	66.88%	67.89%	66.52%	67.02%
	Quantum	70.15%	70.10%	69.20%	70.32%	70.31%	70.02%

4 Discussion and Conclusions

The most important motivation of the research was to propose a reliable quantum model, which would allow achieving results similar to the classical neural network within a reasonable computation time. As it turns out, this can be achieved, and additionally, the quantum model can outperform the classical in two aspects. The first is the convergence of the quantum network learning, which is a much smaller number of iterations before achieving the flattening of the accuracy and loss functions. In this aspect, the classical network seems to be much slower and less stable. The second aspect is that the quantum network, although having a similar architecture, achieved higher training, validation, and testing results for each of the five repeated verification experiments.

Although the advantage of the quantum model over the classical model in this work is still of a simulation nature, it is essential that the proposed algorithm converts numerical data into their quantum representation, i.e., it prepares image data to a form that can be directly sent to and processed by the quantum architecture. Thus, proposing algorithms for processing numbers into quantum representations and trying to solve real classification problems, such as image recognition, is of great importance in the field of quantum machine learning.

The experimental results provide a reason to develop the topic through further research. Among them lies the simplification of the used quantum image representation method or the representation by FRQI method [10]. Another interesting aspect is the serviceability of the quantum models with, e.g., custom layers for automated data preprocessing and the research on the performance of more complex hybrid quantum-classical neural network models, together with the topics of scalability and operation performance of the networks on quantum processing units, that introduce noise related to the physical implementations of qubits and quantum gates.

Acknowledgements. The authors would like to thank the referees for their comments that helped to improve the presentation of this paper. This research is financed from the statutory activities of the Faculty of Automatic Control, Electronics and Computer Science of the Silesian University of Technology.

References

1. Abadi, M., Agarwal, A., Barham, P., et al.: TensorFlow: large-scale machine learning on heterogeneous systems (2015). https://www.tensorflow.org/
2. Broughton, M., et al.: Tensorflow quantum: A software framework for quantum machine learning. arXiv:2003.02989 (2020)
3. Cirq Developers, Cirq, Zenodo (2020). https://doi.org/10.5281/zenodo.4064322
4. Deutsch, D.E.: Quantum theory, the Church-Turing principle and the universal quantum computer. Proc. R. Soc. London. A. Math. Phys. Sci. **400**(1818), 97–117 (1985). https://doi.org/10.1098/rspa.1985.0070
5. Deutsch, D.E., Barenco, A., Ekert, A.: Universality in quantum computation. Proc. R. Soc. London. Ser. A Math. Phys. Sci. **449**(1937), 669–677 (1995). https://doi.org/10.1098/rspa.1995.0065

6. Farhi, E., Neven, H.: Classification with quantum neural networks on near term processors. arXiv:1802.06002 (2018)
7. Fujisawa, T., Hayashi, T., Jung, S.W., Jeong, Y.H., Hirayama, Y.: Single-electron charge qubit in a double quantum dot. In: Ruggiero, B., Delsing, P., Granata, C., Pashkin, Y., Silvestrini, P. (eds.) Quantum Computing in Solid State Systems. Springer, New York, NY (2006). https://doi.org/10.1007/0-387-31143-2_34
8. Gupta, S., Zia, R.: Quantum neural networks. J. Comput. Syst. Sci. **63**(3), 355–383 (2001). https://doi.org/10.1006/jcss.2001.1769
9. Josephson, B.D.: The discovery of tunnelling supercurrents. Proc. IEEE **62**(6), 838–841 (1974). https://doi.org/10.1109/PROC.1974.9524
10. Le, P.Q., Dong, F., Hirota, K.: A flexible representation of quantum images for polynomial preparation, image compression, and processing operations. Quantum Inf. Process. **10**(1), 63–84 (2011). https://doi.org/10.1007/s11128-010-0177-y
11. LeCun, Y., Cortes, C., Burges, C.: The MNIST database of handwritten digits. Courant Institute, NYU, Google Labs, New York, Microsoft Research, Redmond (2010). http://yann.lecun.com/exdb/mnist/
12. Lucatto, B., Koda, D.S., Bechstedt, F., Marques, M., Teles, L.K.: Charge qubit in van der Waals heterostructures. Phys. Rev. B **100**(12), 121406 (2019). https://doi.org/10.1103/PhysRevB.100.121406
13. Morton, J.J., et al.: Solid-state quantum memory using the ^{31}P nuclear spin. Nature **455**(7216), 1085–1088 (2008). https://doi.org/10.1038/nature07295
14. Mosakowski, J., Owen, E., Ferrus, T., Williams, D., Dean, M., Barnes, C.: An optimal single-electron charge qubit for solid-state double quantum dots. arXiv:1603.05112 (2016)
15. Nielsen, M.A., Chuang, I.: Quantum Computation and Quantum Information. Cambridge University Press, Cambridge (2000). https://doi.org/10.1017/CBO9780511976667
16. Pla, J.J., et al.: A single-atom electron spin qubit in silicon. Nature **489**(7417), 541–545 (2012). https://doi.org/10.1038/nature11449
17. Potempa, R.: Simulation of quantum neural network with evaluation of its performance. Silesian University of Technology, Gliwice, Poland (2021)
18. Preskill, J.: Quantum computing in the NISQ era and beyond. Quantum **2**, 79 (2018). https://doi.org/10.22331/q-2018-08-06-79
19. Riedl, S., Lettner, M., Vo, C., Baur, S., Rempe, G., Dürr, S.: Bose-Einstein condensate as a quantum memory for a photonic polarization qubit. Phys. Rev. A **85**(2), 022318 (2012). https://doi.org/10.1103/PhysRevA.85.022318
20. Zhou, Z.-Q., Lin, W.-B., Yang, M., Li, C.-F., Guo, G.-C.: Realization of reliable solid state quantum memory for photonic polarization qubit. Phys. Rev. Lett. **108**(19), 190505 (2012). https://doi.org/10.1103/PhysRevLett.108.190505

Can Color Visual Cryptography Be Truly Random?

Leszek J. Chmielewski[1(✉)], Mariusz Nieniewski[2], and Arkadiusz Orłowski[1]

[1] Institute of Information Technology, Warsaw University of Life Sciences – SGGW, Nowoursynowska 159, 02-775 Warsaw, Poland
{leszek_chmielewski,arkadiusz_orlowski}@sggw.edu.pl
[2] Faculty of Mathematics and Informatics, University of Lodz, Banacha 22, 90-238 Łódź, Poland
mariusz.nieniewski@wmii.uni.lodz.pl

Abstract. The concept of color visual cryptography with truly random shares is proposed. The randomness was confirmed with the battery of 15 NIST randomness tests. The two shares are generated as two equal purely random sequence of R, G, B and black pixels. Coding is performed by swapping the pixels in the second share in small subregions called segments. The criterion for swapping is that after superimposing the shares, some pixels remain color and the remaining ones are covered so that they appear as black. The color pixels form a color image in the way of dithering in the RGB space. The secret is in principle decrypted with the unarmed human eye; however, it can be easily further reconstructed numerically to attain the quality very close to that of the dithered source image. It was observed that the coding can be optimized by varying the dimensions of the segments and the palettes in the dithering.

Keywords: Visual cryptography · Color · Random shares · Pixel swapping · Randomness tests · NIST

1 Introduction

Cryptography has been the subject of interest at least for several centuries. One of its branches is steganography – the action of hiding a secret message in another, overt message. The overt message is intended to hide the secret message, but also distracts the third parties from the fact that something is hidden. Transmitting a message which is unreadable evokes interest of cipher breakers.

Since the works of Naor and Shamir [6,7] visual cryptography has been continually developed. Their *classic* method of coding a black-and-white image in two or more other such images, has given a beginning for numerous versions of methods of visually coding grey-scale and color images in other images; many references can be found in [4,11]. Among them, let us name the concept of hiding a color image in other images, called *meaningful shares* [5,13]. The shares

M. Choraś et al. (Eds.): CORES 2021, IP&C 2021, ACS 2021, LNNS 255, pp. 72–86, 2022.
https://doi.org/10.1007/978-3-030-81523-3_7

Fig. 1. All possible 2 × 2 tiles and their indexes (grey separators used for display only).

usually have a specific structure, which besides being useful in the coding process, can also draw attention of an inquisitive observer. Although such coding is unbreakable, it is still susceptible to the most simple attack – the cutoff of the transmission channel. Therefore, hiding not only the secret, but also the fact that a transmission takes place, is of great value.

In this paper we shall present a concept of coding a color image in pure noise. Our world is full of noise, so noise does not attract too much attention. The visual coding of black-and-white images with purely random shares was proposed in [9], and the randomness of the shares was demonstrated in [10]. The technique was extended to color images in [8], where color was coded randomly, but the structure of the shares was far from random. The visual appearance of randomness of the shares was improved in [2]. In [3] more ideas to improve the randomness of the shares were explored. For the first time, the randomness was extensively tested, and it has been shown that the true randomness was not attained. In this way, it has been demonstrated what level of randomness could be achieved by purely mixing the pixels within the local regions of the image.

In this paper we use a different paradigm. We experiment with the coding on pure noise. We still use the concepts of coding the color in the dithered image similar to that of using tiles and segments, presented in [2,8], but we depart from using an organized segment containing determined numbers of red (R), green (G), blue (B) and black (K) pixels as a starting point. Instead, we use shares with pure RGBK noise.

The contents of this paper is as follows. In Sect. 2 the state-of-the-art concerning color visual cryptography is summarized. The truly random coding is introduced in Sect. 3. Examples of results for natural images are shown in Sect. 4. Problems of image quality are highlighted in Sect. 5. The randomness testing of shares is presented in Sect. 6. Section 7 concludes the paper.

2 Previous Methods

In the classic coding for black-and-white images, the *secret* image is represented with two images called the *shares* in which each pixel of the *secret* is represented by a 2 × 2-pixel *tile*. The tile for a pixel belonging to *share 1* is drawn at random from among the tiles belonging to pairs 4–13, 6–11, 7–10, where the tiles and their indexes are shown in Fig. 1. If the pixel in the *secret* is white, in the *share 2* the same tile as in the first one is taken. If this pixel is black, in the *share 2* the other tile from the pair is taken. A white pixel in the *secret* is represented by

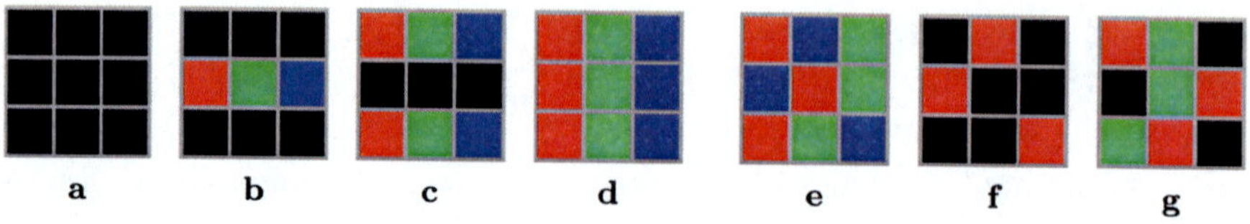

Fig. 2. Examples of a 3×3 segment which encodes a single color pixel (grey separators used for display only). (**a**) Zero brightness – black pixel; (**b**) brightness 1/3 – dark grey; (**c**) brightness 2/3 – bright grey; (**d**) full brightness – white pixel. (**e**), (**f**), (**g**) Segments with pixels randomly mixed within the segment: white, red and yellow.

a tile with two white pixels per four; a black one – by a tile with all black pixels. This state, which we shall call *classic*, has been described in [6,7].

If the tiles are selected not only from the listed pairs, but from all the 16 possible ones, the secret would still be represented, but with errors in white pixels. For example, if tile 16 (15, 2, 1) is drawn, then four (three, one, zero) pixels instead of two will be white. With this approach, not only the shares are uncorrelated with the secret, but they are also truly random. This *random* coding in BW images has been proposed and tested in [9,10].

In visual cryptography as described above, the pixel of the secret can be covered (black) or uncovered (white). With the use of the dithering technique, a color image can be formed of bright color pixels – red, green or blue – and black ones (denoted by K, as in CMYK), if the RGB color space is used. Let us represent each pixel of the secret with a 3×3 *segment*. Examples of segments are shown in Fig. 2a–d. Each of the colors R, G and B can be represented at four levels of brightness: black, 1/3 brightness, 2/3 brightness, and fully bright. This forms a palette of $4^3 = 64$ colors for the dithering. The pixels are organized in color columns, like in an old color TV. Each pixel of the segment can be coded with a tile, as in the *classic* or *random* coding, as described in [8].

The organization of pixels in R, G, B columns is far from randomness. Mixing them, like in Figs. 2e–g, would not decrease the readability of the images; on the contrary, it would reduce their apparent granularity. Mixing (permuting within a segment) can be done by tiles, or by pixels after the coding with tiles [2].

Let us code the yellow segment of Fig. 2g in two shares (full yellow is full red: 3R, full green: 3G). In each share, each pixel of the source segment is represented with a 2×2 tile, like in Fig. 3a. Now, let us form the two shares with the *classic* method. For simplicity, from all the tiles of Fig. 1 we shall use only tiles 7 and 10. Each *black* pixel in the source is represented in one share (Fig. 3b1) by tile 7 and in the other share (Fig. 3b2) by tile 10. Each *color* pixel is represented in both shares by tile 10, giving a half-color half-black tile. After superimposing the shares, the image of Fig. 3c is received. The coding with the *classic* method stops here. The pairs of tiles 4–13, 6–11, and 7–10 can be used at random. The same can be done with the *random* method with all the tiles used.

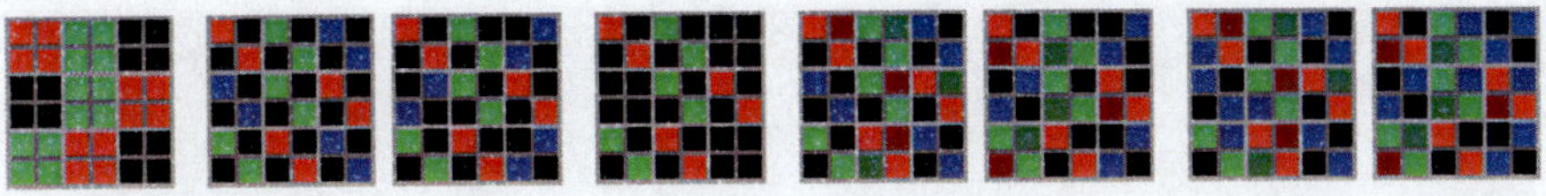

Fig. 3. Coding a yellow pixel of Fig. 2g represented by one segment (borders between tiles shown in light grey, and between pixels in tiles – in dark grey). (**a**) Source segment with 9 tiles; (**b1**), (**b2**) two shares coded classically; (**c**) decoded segment. Further, changed pixels are marked with darker color. (**d1**), (**d2**) Two shares with changes to R, G made in K pixels corresponding to K in the other segment. (**e1**), (**e2**) Two shares, further changes – K to B – can be made only in pixels which correspond to color pixels in the other segment.

Note that in Figs. 3b1, b2 in each share there are six red, green and blue pixels, and 18 black pixels. If true randomness is expected, these numbers should be close to equal. Let us try to change the nine of 18 black pixels into color: 3 to R, 3 to G and 3 to B, which would balance the numbers into nine pixels of R, G, B and K. Consider the shares b1, b2. Some of the pixels are black in both. Changing one of them into an arbitrary color wold not change the decoded image, as this color would be covered with black from another share. In this way, we can change three black pixels in one share having black counterparts in another share into red (marked in Fig. 3d1, d2 in darker red, only for better visualization) and into green (marked with darker green). However, here the process should stop as there are no more pixels with black in both shares. To overcome this, let us notice that if the color filters of which the shares are made are ideal (or close to ideal), then they pass the light in only a narrow color band. Superimposing such filters of different colors will cause no light pass through. Therefore, we can turn into blue such three pixels which have in the other share the counterpart in any color except blue. In Figs. 3e1, e2 this has been done in pixels marked with darker blue. Close examination reveals that these shares give the same image of Fig. 3c as the shares b1, b2.

Similar operations can be performed for a segment in any other color. If some changes of K into color still have to be made and there are no pixels in which this would not influence the decoded image, there is no other way than to draw the necessary number of pixels to be changed at random. This introduces additional errors into the decoded image.

There are four reasons of reduced quality of the decoded image. Firstly, colors are represented with separate R, G and B pixels in 3×3 segments. Secondly, colors are dithered into four levels only. Thirdly, by representing each pixel with a 2×2 tile with half of the pixels black, the brightness is halved. Finally, errors are introduced by using truly random tiles and by balancing the numbers of pixels in R, G, B and K.

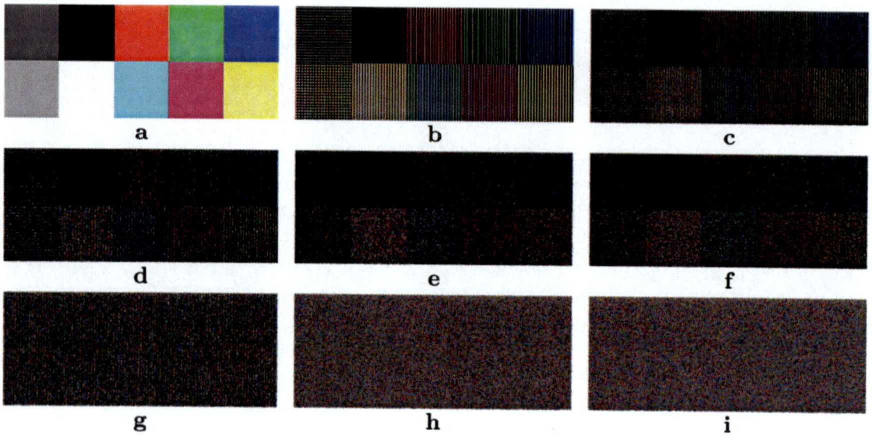

Fig. 4. Illustration of coding and decoding for a simple test image. (**a**) Secret, 100×40; (**b**) image **a** dithered and decomposed into color stripes, 300×120; all further images are 600×240: (**c**) coded classically; (**d**) coded randomly, no mixing [2]; (**e**) coded with *random coordinated* method [3]; (**f**) coded with *truly random* method proposed here. (**g**), (**h**), (**i**) Share 2 of **d**, **e**, **f**, respectively. The same example secret image **a** and the results except images **f**, **i** were used in [2,3,8].

These concepts have been explored in [3], where the just described method was denoted as the *coordinated* method (because changes of pixels were coordinated in shares within one segment). This method appeared to be the best from a number of methods tested. It seems that the limits of introducing the randomness into coding a segment were reached in this way. However, despite that some tests of randomness made with the NIST randomness test suite [1] performed on shares from some benchmark images succeeded, still some other failed (see Sect. 6 for more information on the tests).

The results from the methods presented above, and the one which will be proposed below, are shown in Fig. 4, where a simple test image was used. This useful image was employed also in [2,3,8]. The results and the received quality of the decoded image are well visible. In the images c-e the results received with the previously described methods are shown. In the image f the result calculated with the *truly random* method proposed in the next Section is presented.

3 Truly Random Coding

From the results described in [3] it seems to follow that an entirely new concept has to be introduced to obtain true randomness in the shares used to code the secret image. This concept originated from the idea to use the truly random image as the first share. As randomness it is understood that the pixels in R, G, B and K are placed in it with the use of a random number generator with four possible values. It is no surprise that for such an image (treated as a string of bits) all the randomness tests from the NIST test suite pass successfully (results will be presented further). The second share is generated by only swapping the positions of pixels with respect to those in the first share; values are left intact. The changes are made locally, within the regions of the image corresponding to the 6×6 segments used in the previously described methods. The changes are made in such a way that in the superimposed shares the numbers of R, G and B pixels should be equal (or as close as possible) to the numbers calculated by the dithering procedure described above.

As the result of dithering into a 64-color palette, in each segment the numbers of pixels should be: 0, 2, 4 or 6 R pixels, and the same for G and B. The remaining pixels should be black. So, the required numbers of R, G, B pixels should not be moved in the second share, and the remaining ones should be moved so that each of them covers a pixel in a different color (black at the best). In a segment, on the average there are 9 R, 9 G, 9 B, and 9 K pixels, so this should be possible in the majority of segments. Errors would appear in the segments where, randomly, there are less color pixels than needed. Experiments indicate that at least six pixels in each color, R, G, B, can be found in at least 75% of segments. This enables showing the full white, while not all the pixels of the secret are white.

It can be noticed that color can be obtained in the image decoded from random shares using other concepts than that of segments in RGB with dithering in a 64-color palette, as described above. This particular concept was used for the sake of continuity with the previous research, and because the probability of finding the necessary colors in each segment was sufficiently high. In the *truly random* coding the conditions related to the segments (Fig. 2) do not hold any more, so one could design a coding, for example, with segments 5×5, with 3 pixels corresponding to full color, which would enable showing the white in 90% of pixels, but would give decoded images with 0.72 brightness of those described above ($3 \times 3/25 = 0.36$ instead of $3 \times 6/36 = 0.5$). The criterion for a good coding would be that the probability of representing any color in any pixel is sufficiently high, that the palette for dithering the source image provides for sufficient quality, and that the raw decoded image is sufficiently bright. In principle, the segments do not have to be square, but care must be taken not to destroy the isotropy of the image representation.

The algorithm used to generate the segment in the second share for the given segment in the random first share, and for a given number of necessary R, G and B pixels in this segment in the reconstructed image is presented as Algorithm 1. In this description some secondary details are omitted. The algorithm is applied to segments independently, so parallel computing can be used in a natural way.

The algorithm was designed to keep all the operations as random as possible, rather than to perform the coding efficiently. This was done to check, first of all, whether it is possible to maintain the random nature of the second share to the maximum possible extent. Therefore, it can be expected that a more efficient algorithm can be proposed. For example, pairs of pixels with unsuccessful swap attempts can be excluded from further trials. The case of impossibility of further improvement can also be detected better. Now, attempts of swaps are made

Algorithm 1: Truly random coding algorithm in one segment.

```
Data: seg1: random segment in share1,
      ind[3colors]: required no. of pixels in R, G, B: 0, 2, 4, 6, from dithering.
Result: seg2: segment in share2, with all non-fixed pixels covered, if possible.
seg2 = seg1  // Initialize seg2 so that no pixel is covered.
state[6rows, 6cols] = free  // Table state of pixels. Values: free, fixed.
num[3colors] = count them in seg1  // Tab. num of colors: n. of pixs in color.
num[3colors] = min(ind[3colors], num[3colors])  // No more pixs available.
for (3colors, num[color] pixels in random locations [row, col]) do
    num[row, col] = fixed  // These pixs will never be covered so stay color.
while (any free pixels exist) and (admissible no. of trials not exceeded) do
    draw pix1[row1, col1] randomly until state[row1, col1] = free
    draw pix2[row2, col2] randomly until state[row2, col2] = free
    if (swapping (pix1, pix2) in seg2 increases number of covered pixels) then
        // (increasing number of covered pixels) ≡ (1 or 2 uncovered pix(s)
        // become covered) and (no already covered pix becomes uncovered).
        swap (pix1, pix2) in seg2
        state[row1, col1] = fixed
        state[row2, col2] = fixed
```

totally at random, in order not to spoil the randomness of the second share in any way. At this stage the efficiency was not our goal.

4 Examples with Natural Images

The examples of decoding two natural images, for which the results of randomness tests for the shares will be shown further, have been shown in Figs. 5, 6. The decoded images can be compared with those decoded from the *classic* coding, in which there are no errors resulting from any randomness. Visually the images coded with the *truly random* method are not inferior to those without randomness. Moreover, in the randomly coded images there is no organized structure of color columns, which makes them appear more smooth (the same can be received by mixing the pixels also in the classical coding). The comparison with the results of decoding from the *random coordinated* method would indicate results even more profitable for the *truly random* method (compare the results for a simpler image in Fig. 4d, e, f).

The possibility of decoding the secret without using any equipment is the basic advantage of visual cryptography. However, let us check how much information contained in the decoded image alone can be restored, should the numerical methods be used. This is not what is intended in everyday use, but it could reveal what kind of and how many errors are introduced.

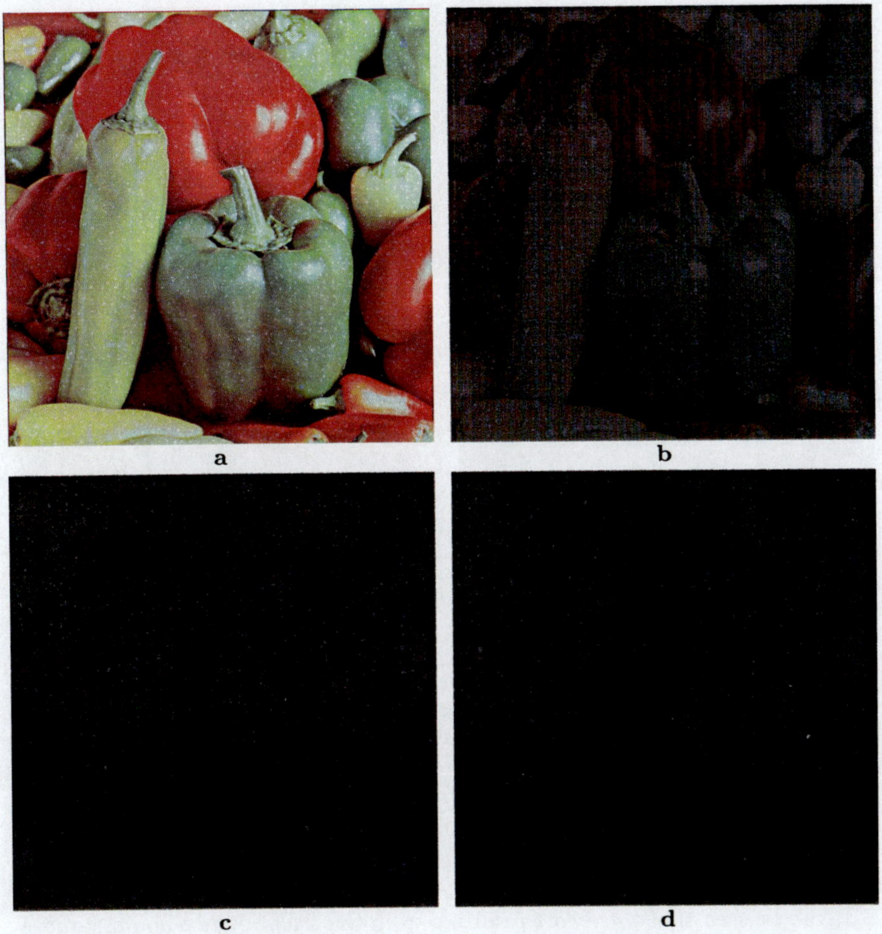

Fig. 5. Illustration of decoding for a natural image **peppers**. (**a**) Secret, 512 × 512; (**b**) image **a** dithered and decomposed into color stripes for coding, 1536 × 1536; further images are 3072 × 3072: (**c**) coded classically; (**d**) coded truly randomly.

Please recall that in the dithering, the intensities 0, 1/3, 2/3, 1 are assigned to the R, G and B components, and in the case of no errors, 0, 2, 4, or 6 pixels in R, G and B, respectively, are present in the decoded image. This makes it possible to restore the image by projecting the range $\langle 0, 6\rangle$ into the range of color intensities $\langle 0, 255\rangle$. The image obtained in this way should be close to the source image after dithering in the 64-color palette. The result of restoration for the image of Fig. 4a is shown in Fig. 7a, in comparison to the dithered image and the restored image from the *random coordinated* method (subfigures b, c; for

Fig. 6. Illustration of decoding for a natural image `parrots`. (**a**) Secret, 384 × 256; (**b**) image **a** dithered and decomposed into color stripes for coding, 1152 × 768; further images are 2304 × 1536: (**c**) coded classically; (**d**) coded truly randomly.

this image the dithering introduces no distortion; loss of quality seen in Fig. 4b comes from decomposition into color stripes). Results for the images of Figs. 5 and 6 are shown in Fig. 7d–i.

It can be seen that the errors resulting from the *truly random* coding are relatively rare, less frequent than those present in the *random coordinated* method. It would be easy to eliminate them with a suitable numerical filter.

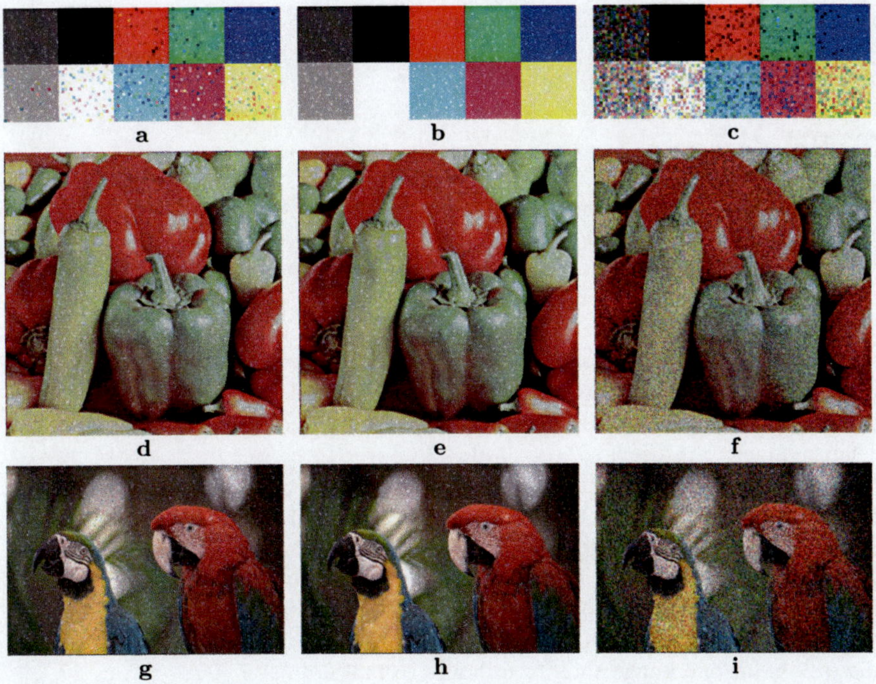

Fig. 7. Illustration of numerical restoration from the *truly random* method, in relation to the dithered image (without coding into color stripes). Restored images from the *random coordinated* method shown for comparison. (**a**) Restored image of Fig. 4f – *truly random* method; (**b**) dithered image of Fig. 4a (same as original in this case); (**c**) restored image of Fig. 4e – decoded from *random coordinated* method. (**d**) Restored image of Fig. 5d – decoded from *truly random* method; (**e**) dithered image of Fig. 5a; (**f**) restored image of Fig. 5 decoded from *random coordinated* method. (**g**) Restored image of Fig. 6d – decoded from *truly random* method; (**h**) dithered image of Fig. 6a; (**i**) restored image of Fig. 6 decoded from *random coordinated* method.

5 A Note on Image Quality

Reduced quality of a decoded image with respect to the original in RGB color visual cryptography results from several factors (cf. remarks in [3]). An image is decomposed into color channels and only some pixels of the shares are intended to be seen while others are covered. An image is dithered and the palette is reduced. Errors are introduced by the random elements in the algorithm. As it can be seen in the reconstructed image, the information loss with respect to the dithered image is not large, but if no processing is admissible then the loss in brightness is inevitable. It seems necessary to study the possibilities of improving the quality of decoded images. To this end, the suitable image quality measures would be useful; however, in such commonly used measures like SNR or SSIM it is assumed that dimensions and color representations in the compared images are the same. Pixel expansion and dithering in only black and pure basic colors introduces difficulties. The brevity of this presentation makes it necessary to leave these problems for future studies.

6 Randomness Tests

The randomness of the shares was tested with the Statistical Test Suite of the National Institute of Standards and Technology (NIST tests) [1]. A pixel of a share can take four values, so it was expressed as two bits: 00 for K, 01 for R, 10 for G and 11 for B (assignment is arbitrary and does not influence the randomness). Two shares were tested. Pixels were saved in two sequences: by rows and by columns. Hence, for one secret image there were four data sets to test. Each image was coded 100 times and the four p-values from the 15 NIST tests of four data sets were recorded. For the parameters of the NIST software the default values were used; so, $\alpha = 0.01$.

Results for the tests which have subtests were shown together. This concerns `CumulativeSums`, `NonOverlappingTemplate`, `RandomExcursions`, and `RandomExcursionsVariant`. Hence, 188 tests with their subtests were presented as just 15 histograms. The results from 100 random realizations of coding, four sets, 15 tests, were displayed in one page for each image, to enable capturing the whole result at a glance, according to the concept of *small multiples* [12]. Calculating the results for one image took from hours to half a day.

As written above, some tests succeeded but some other failed for the shares formed with the previously used *coordinated* method. The tests which usually failed were `Runs`, `LongestRun`, `OverlappingTemplates` and `ApproximateEntropy`. The tests which usually succeeded were `Frequency`, `BlockFrequency`, `CumulativeSums`, `Rank`, `RandomExcursions`, `RandomExcursionsVariant` and `LinearComplexity`. The remaining tests: `FFT`, `NonOverlappingTemplate`, `Serial` and `Universal` gave ambiguous results [3].

The results of the randomness tests for the *truly random* method introduced in this paper are presented in the form of histograms of p-values in Figs. 8 and 9 for two images of Figs. 5 and 6, respectively. In the further text the p-values will be referred to simply as p. The important results are the counts of the test failures, $p \leq \alpha$. They are indicated graphically as *low* and shown in red (two intensities). The counts for the test successes, indicated as *good*, are shown in grey (two intensities). Some tests have preconditions (`RandomExcursions`, `RandomExcursionsVariant`, and `Universal`); the failures to meet them are shown in blue at the right of the graphs and are indicated with a question mark as *not applicable*. The data for share 1 are shown with full symbols, and those for share 2 with empty symbols. The data for pixels read by columns, vertically, are shown with bars, and by rows, horizontally, with circles.

The data for $p \leq \alpha$ which indicate failures are important, so the p-axis is nonlinear. The transformation $p \rightarrow p^a$ is such that the point 0.01 takes the place of the former point 0.1. The width of the bins is $\alpha/4$ everywhere.

The results for these and other benchmark images have been summarized in Table 1 where only the numbers of failures and successes of the tests are given.

Table 1. Abstracted results of statistical randomness tests shown as pairs (or triplets, where applicable) of numbers of realizations, shown with color text: (*low* *good* *n/a*).

Image	size	Frequency	Block Frequency	Cumulative Sums	Runs	LongestRun	Rank	FFT	Non Overlapping Template	Overlapping Template	Universal	Approximate Entropy	Random Excursions	Random Excursions Variant	Serial	Linear Complexity
test100	100× 40	8 392	5 395	23 777	3 397	0 400	2 398	4 396	643 58557	4 396	0 0 400	3 397	18 1094 2088	18 2484 4698	7 793	0 400
baboon	125×120	0 400	10 390	2 798	8 392	9 391	5 395	6 394	625 58575	7 393	6 394 0	1 399	25 2087 1088	31 4721 2448	10 790	5 395
parrots	384×256	8 392	7 393	21 779	4 396	2 398	2 398	3 397	634 58566	6 394	7 393 0	6 394	26 2518 656	57 5667 1476	7 793	8 392
peppers	512×512	0 400	2 398	0 800	5 395	4 396	7 393	4 396	612 58588	17 383	6 494 0	5 395	25 2839 336	71 6373 756	9 791	3 397
Lena	512×512	8 392	3 397	16 784	5 395	6 394	7 393	8 392	575 58625	14 386	2 398 0	2 398	31 2801 368	67 6305 828	5 795	2 398
boats	787×576	8 392	2 398	17 783	3 397	7 393	6 394	8 392	613 58587	30 370	2 398 0	1 399	36 2892 272	59 6529 612	6 794	7 393

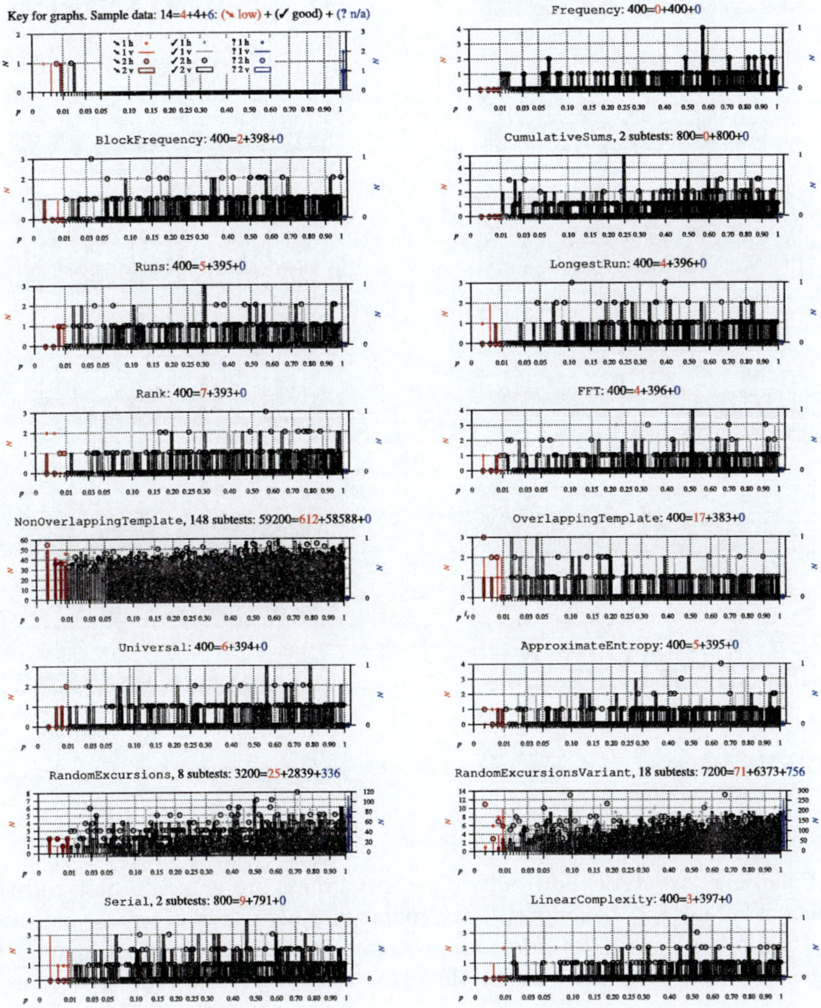

Fig. 8. Histograms of p for color image peppers: *truly random* method. Key shown in the upper left sub-image.

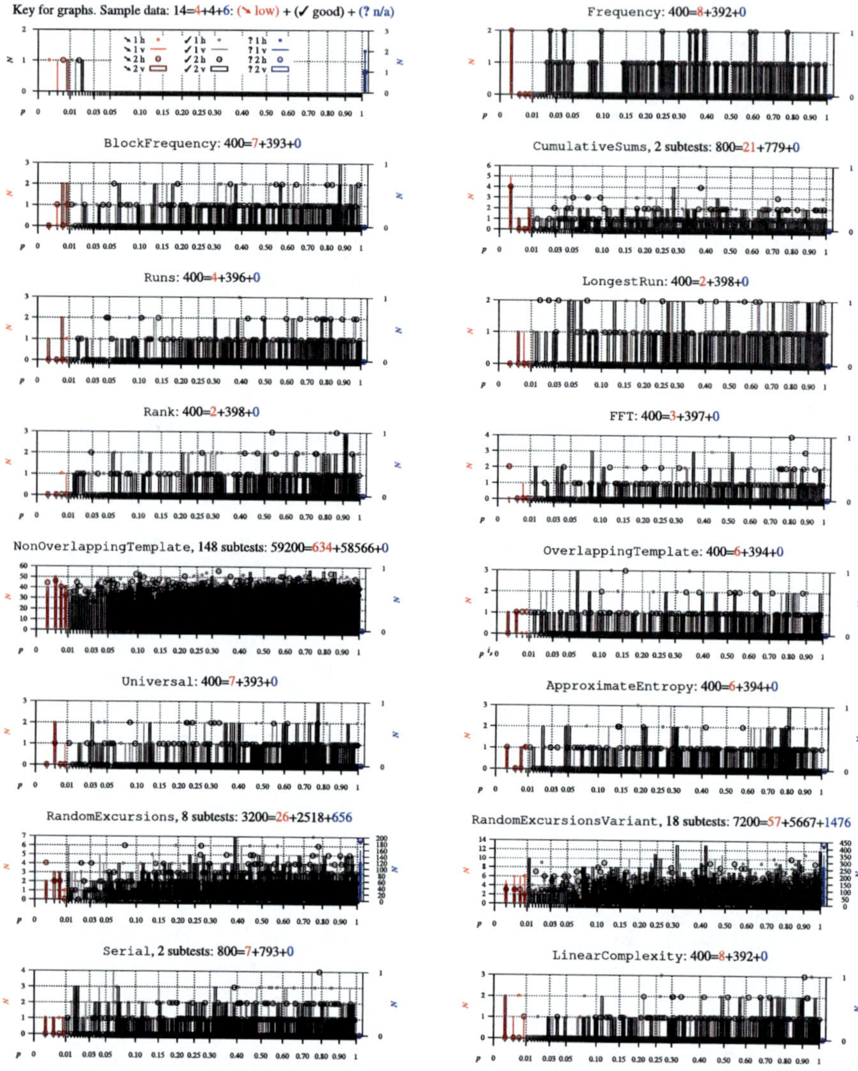

Fig. 9. Histograms of p for color image `parrots`: *truly random* method. Key shown in the upper left sub-image.

Finally, let us state the results. The histograms are generally flat and the counts of failures $p \leq \alpha$ conform to the counts of other values of p. This indicates that the shares can be considered as random, irrespective whether share 1 or share 2 is considered, and whether the pixels are read by rows or by columns.

7 Conclusion

It seems that the possibilities of supporting the randomness of the shares in color visual cryptographic coding based upon dithering, random coding of shares, balancing the number of color and black pixels and mixing the pixels has attained its limit without reaching the full randomness. This was shown in our previous publications. Therefore, a new concept of *truly random* coding was presented. The first share is generated in a fully random way and the second share is formed from the first share by merely changing the positions of some of its pixels (swapping), locally within segments. The criterion for swapping the pixels is that after superimposing the shares, the numbers of pixels required by the dithering process remain color, why the other pixels stay or become black, due to covering by a pixel from the other share which is a black pixel or a pixel in a color different from that of the pixel in the first share.

It has been demonstrated that the swapping, which is performed in a random manner, does not alter the random nature of the second share (while the first share remains random due to its random generation). This was checked with the battery of NIST randomness tests for the series of 100 realizations of coding, for the examples of one artificial and five natural images.

Besides that the decoding can be performed by a bare human eye, as it is in the visual cryptographic methods, the quality of the decoded image can be easily reconstructed up to the level of quality close to that of the source secret image after the dithering stage.

The proposed coding makes it possible to apply various dithering schemes and can be optimized for decoding quality. Such optimization based on experiments with a larger set of images will be the subject of further studies.

References

1. Bassham, L.E., Rukhin, A.L., Soto, J., et al.: A statistical test suite for random and pseudorandom number generators for cryptographic applications. Technical report, National Institute of Standards and Technology, Gaithersburg, MD, USA (2010). http://www.nist.gov/manuscript-publication-search.cfm?pub_id=906762. Series: Special Publication (NIST SP), Rep. No. 800–22 Rev 1a
2. Chmielewski, L.J., Gawdzik, G., Orłowski, A.: Towards color visual cryptography with completely random shares. In: Proceedings of the Conference on PP-RAI 2019 - Konf. Polskiego Porozumienia na rzecz Rozwoju Sztucznej Inteligencji, pp. 150–155. Wrocław University of Science and Technology, Wrocław, Poland (2019). http://pp-rai.pwr.edu.pl/PPRAI19_proceedings.pdf
3. Chmielewski, L.J., Nieniewski, M., Orłowski, A.: Testing the randomness of shares in color visual cryptography. Pattern Anal. Appl. (2021). https://doi.org/10.1007/s10044-021-00999-5
4. Cimato, S., Yang, C.N.: Visual Cryptography and Secret Image Sharing (Digital Imaging and Computer Vision), 1st edn. CRC Press Inc., Boca Raton, FL, USA (2012). https://doi.org/10.1201/b11068
5. Dhiman, K., Kasana, S.S.: Extended visual cryptography techniques for true color images. Comput. Electr. Eng. **70**, 647–658 (2018). https://doi.org/10.1016/j.compeleceng.2017.09.017

6. Naor, M., Shamir, A.: Visual cryptography. In: De Santis, A. (ed.) Advances in Cryptology – EUROCRYPT 1994. Proceedings of the Workshop on the Theory and Application of Cryptographic Techniques. Lecture Notes in Computer Science, vol. 950, pp. 1–12. Springer, Perugia, Italy (1995). https://doi.org/10.1007/BFb0053419
7. Naor, M., Shamir, A.: Visual cryptography II: improving the contrast via the cover base. In: Lomas, M. (ed.) Security Protocols. Proceedings of the International Workshop on Security Protocols, Lecture Notes in Computer Science, vol. 1189, pp. 197–202. Springer, Cambridge, United Kingdom (1997). https://doi.org/10.1007/3D-540-62494-5_18
8. Orłowski, A., Chmielewski, L.J.: Color visual cryptography with completely randomly coded colors. In: Vento, M., Percannella, G. (eds.) Proceedings of the International Conference on Computer Analysis of Images and Patterns CAIP 2019. Lecture Notes in Computer Science, vol. 11678, pp. 589–599. Springer, Salerno, Italy (2019). https://doi.org/10.1007/978-3-030-29888-3_48
9. Orłowski, A., Chmielewski, L.J.: Generalized visual cryptography scheme with completely random shares. In: Petkov, N., Strisciuglio, N., Travieso, C.M. (eds.) Proceedings of the 2nd International Conference on Applications of Intelligent Systems APPIS 2019, pp. 1–6. Association for Computing Machinery, Las Palmas de Gran Canaria, Spain (2019). https://doi.org/10.1145/3309772.3309805
10. Orłowski, A., Chmielewski, L.J.: Randomness of shares versus quality of secret reconstruction in black-and-white visual cryptography. In: Rutkowski, L., et al. (eds.) Proceedings of the International Conference on Artificial Intelligence and Soft Computing ICAISC 2019, vol. 11509, pp. 58–69. Springer, Zakopane, Poland (2019). https://doi.org/10.1007/978-3-030-20915-5_6
11. Stinson, D.R., Paterson, M.: Cryptography: Theory and Practice, 4th edn. Chapman and Hall/CRC Press, Boca Raton (2018). https://doi.org/10.1201/9781315282497
12. Tufte, E.R.: The Visual Display of Quantitative Information. Graphic Press, Cheshire, CT, USA (1983)
13. Wang, D., Yi, F., Li, X.: On general construction for extended visual cryptography schemes. Pattern Recognit. **42**(11), 3071–3082 (2009). https://doi.org/10.1016/j.patcog.2009.02.015

Description-Based Ranking of Visual Instances: *Feasibility Study for Keypoints*

Andrzej Śluzek(✉)

Institute of Information Technology, Warsaw University of Life Sciences-SGGW, Warsaw, Poland
andrzej_sluzek@sggw.edu.pl

Abstract. The paper introduces (and investigates) a novel methodology for improving performances of detection of visual objects (keypoints are selected as objects for the feasibility study). Using any (i.e. either hand-crafted or deep-learned) detector in conjunction with the selected descriptor, we rank the extracted keypoints using stability of their descriptors. Stability is defined by the magnitude of partial derivatives of the descriptor over parameters determining deformations of the original images. In practice, discrete approximations of the derivatives are used. We have preliminarily verified that by using the keypoints top-ranked by the proposed criterion (i.e. those with least-sensitive descriptors) instead of keypoints top-ranked by the standard prominence criteria, performances of matching (we primarily focus on *precision*) can be substantially improved. The results have been obtained using a popular benchmark set of images.

Keywords: Visual objects · Keypoints · Detectors · Descriptors · Ranking · Stability · Matching

1 Introduction

In spite of (and partially thanks to) rapid development of deep learning methods for machine vision (in particular CNN-based), detection of significant items in visual data remains an important and stimulating topic. The types of those items could be very diversified, starting from simple corners (e.g. [5,8,23]), through general-application keypoints (e.g. [3,13–15,29]), visual saliencies (e.g. [7,9,26]), to detectable objects with assigned semantics (e.g. [10,20,21,30]).

Evaluation criteria for reliability and quality of detection results are, nevertheless, an important supplement to the detectors themselves. In case of semantic object detection, the subjective estimates (i.e. evaluation by the human observer) are usually sufficient, even though they are often supported by more objective confidence estimates (e.g. [18] or [21] for CNN-based detectors).

However, in case of more generic items (i.e. mainly keypoints/key regions), performances of detectors are usually evaluated by their repeatability and stability under changing viewing/geometric conditions, e.g. [11,17]. This is frequently

M. Choraś et al. (Eds.): CORES 2021, IP&C 2021, ACS 2021, LNNS 255, pp. 87–96, 2022.
https://doi.org/10.1007/978-3-030-81523-3_8

done in conjunction with evaluation of the corresponding descriptors, e.g. [6,16], where the matching results are compared against the ground-truth correspondences. For such evaluations, we need a substantial database of related images to obtain various statistical estimates, including *precision*, *recall*, etc.

The main purpose of this paper is to introduce and preliminarily investigate an alternative methodology for performance evaluation of *detection results*. For convenience, we focus on keypoint detection, but the methodology is suitable for detection of other types of instances, as long as the detection results are represented by regular or irregular shapes (e.g. rectangular bounding boxes, circular/elliptic approximations, patches, etc.), i.e. all categories mentioned above can be included. It should be assumed, however, that the detected items can be characterized by some *numerical descriptor*.

Given a *single* image (no alternative views of the scene are available for comparison) we localize visual objects (i.e. keypoints in the included feasibility study) using the provided detector, and describe them by the corresponding descriptor. Subsequently, a number of parameters is selected to characterize *expected* sources of image distortion in the discussed scenarios. Those parameters may represent geometric distortions (shifts, scaling, rotations, camera non-linearities, etc.) as well as photometric ones (e.g. linear intensity transformations, gamma correction, blurring, additive noise, etc.). In a way, we perform the *data augmentation* by creating a number of slightly corrupted variations of the original image. Then, partial derivatives of the *descriptors* over those parameters are approximated for all original keypoints, i.e. keypoints are not re-detected in the corrupted images.

Keypoints are ranked based on the minimum absolute values of the partial derivatives. In other words, the top keypoints are those for which the descriptor values are maximally stable under the assumed types of image distortion. It should be noted that the rankings can change (sometimes significantly) if an alternative keypoint descriptor is used.

This approach is very different from standard methods of ranking detected keypoints (or other visual objects), where they are ranked only by their *prominence* (e.g. magnitude of operators in hand-crafted detectors, or confidence values in CNN detectors) and no descriptors are involved.

Mathematical fundamentals and implementation details of this methodology are given in Sect. 2. Section 3 contains limited-scale experimental results, while the concluding Sect. 4 discussed the intended applications and future works.

In some sense, the paper presents the reverse of the method discussed in [25], where the *keypoint detection* was driven by the selected descriptor stability over the fluctuations of the keypoint geometry. In is this paper, we *rank a given set of keypoints* based on how their descriptors are stable under minor fluctuations of image content.

2 Formal Model

Detectors of keypoints (interest points, keypatches, local features, etc.) are operators extracting prominent locations (with their relevant neighborhoods) in given images. The neighborhoods most frequently have regular shapes (e.g. circles, ellipses or rectangles) or are approximated by such shapes (e.g. elliptic approximations of MSER [14] and SIMSER [24] key regions). Popular detectors are either hand-crafted (e.g. [4,13–15]) or, more recent ones, learned (e.g. [19,27,28]).

In this work, we do not distinguish between these two categories, i.e. regardless the type of detector *Det* we just take the collection of keypoints extracted by *Det* from an*Det* image *Im*. Those keypoints are represented by a selected descriptor *Desc*, which can also be either hand-designed (e.g. [4,13]) or learned (e.g. [22]).

Altogether, we consider the values of *Desc* for the whole collection $\mathbf{C} = \{C_1, ..., C_M\}$ of extracted keypoints as the input data for the process of ranking $\mathbf{C}$ keypoints in the context of *Desc* descriptor. The process is explained in the following subsections.

2.1 Parametrization of Image Distortions

A number of parameters $\{p_1, ..., p_N\}$ are introduced to embed distortions into the original image. Parameters can be of different nature. Some of them represent geometric transformations, e.g. horizontal and vertical shifts, rotations, scaling (in one or two dimensions), non-linear camera geometry, etc.

The parameters representing viewing conditions can include intensity shift, gamma correction, contrast scaling, etc. We can also include parameters to describe other effects, like additive noise, image smoothing or sharpening, etc.

From the practical perspective, such parameterization models actual inaccuracies (e.g. in camera setups) expected in the process of image acquisition.

Altogether, the descriptor of a given keypoint C_i, i.e. $Desc(C_i)$, can be considered the output of an N-parameter function for some nominal (but not necessarily explicitly known) parameters:

$$Desc\{C_i(p_1(0), ..., p_N(0))\}\,. \tag{1}$$

If the image is modified (we assume rather minor levels of those corruptions) by the distortion corresponding to p_k parameter, we can again calculate the descriptor value for the same keypoint C_i (which does not change, but the image fragment overlapped by the keypoint area would be slightly different). In this way, we obtain values of the same function for slightly different arguments:

$$Desc\{C_i(p_1(0), ..., p_k(0 + \Delta k), ..., p_N(0))\}\,, \tag{2}$$

and, eventually, an approximate of the partial derivative $\partial Desc/\partial p_k$ can be obtained as

$$\frac{\partial Desc(C_i)}{\partial p_k} \approx Desc\{C_i(..., p_k(0 + \Delta k), ...)\} - Desc\{C_i(..., p_k(0), ...)\}\,, \tag{3}$$

2.2 Keypoint Ranking

Equation 3 estimates sensitivity of the keypoint descriptor to small fluctuations of the images content (geometry, illumination, noise, etc.). In practice, it is more suitable to use the magnitude of such changes relatively to the magnitude of descriptors.

Therefore, we propose N growth-rate functions $q_k(p_k)$, where $k = 1, ..., N$, defined by the ratio of the partial derivative norm over the descriptor magnitude:

$$q_k(C_i) = \frac{\left\| \frac{\partial Desc(C_i)}{\partial p_k} \right\|}{\| Desc\{C_i(p_1, ..., p_N)\}\|}, \tag{4}$$

where $\|\cdot\|$ is a norm (usually Euclidean) of the descriptor vector (or its partial derivative).

This definition has some resemblance to MSER keypoint detection (where a single growth-rate function is defined by sensitivity of the region's area to the threshold level) or SIMSER keypoint detection (where both threshold levels and image scales are used as the parameters). The smaller the growth-rate function, the less sensitive the descriptor is to the corresponding image distortion.

Eventually, the overall sensitivity of the keypoint descriptors to the assumed image deformations would be straightforwardly defined by the following function:

$$Q(C_i) = \sum_{k=1}^{N} q_k(C_i). \tag{5}$$

It can be referred to as *sensitivity rate* (the larger its value, the more sensitive keypoint descriptor is to the image distortion represented by the parameters $\{p_1, ..., p_N\}$).

Based in Eq. 5, any collection of M keypoints (or other visual instances) can be ranked in the context of selected descriptor *Desc*. We just list keypoints according to the increasing values of their *sensitivity rate*.

$$Q(C_{i_1}) \leq ... \leq Q(C_{i_k}) \leq .... \leq Q(C_{i_M}). \tag{6}$$

By selecting keypoints from the top of such a list, we prospectively improve performances of algorithms in various keypoint-based tasks (including visual data retrieval, vision-based tracking, image stitching, etc.) where minor distortions of acquired data are unavoidable.

3 Experimental Verification

The exemplary feasibility study is based on a popular MSER keypoint detector, [14]. It has been selected because the returned keypoints are represented by ellipses, so that a wide range of image distortions can be tested, including rotations (which are irrelevant in case of circular keypoints).

3.1 Test Scenario

In the exemplary scenario we assume a number of image distortions, which (jointly or individually) represent most of the standard deformations expected in typical real-world applications of machine vision, where both visibility conditions and geometric/hardware setups may vary, even if nominally constant.

In each case, the physical values of increments Δk for the nominal parameters (see Eq. 2) are experimentally selected in such a way that:

a. They are small enough to approximately maintain the linearity of descriptor variations (since we use them as approximations of partial derivatives),
b. they are large enough to noticeably affect the descriptor values in typical images, and
c. they produce hardly noticeable visual effects in the image contents.

Altogether, we consider (somehow arbitrarily) the following distortions:

1. Image rotation (with $2deg$ increment);
2. 2D image enlarging (with 1.1 scale);
3. Horizontal image stretching (with 1.05 factor);
4. Vertical image stretching (with 1.05 factor);
5. Horizontal shift (by 2% of the lower dimension of image resolution);
6. Vertical shift (by 2% of the lower dimension of image resolution);
7. Gamma correction (with 1.3 factor).

We deliberately excluded images blurring from the applied distortions since typical keypoint descriptors have some smoothing mechanisms embedded.

Figure 1 shows an exemplary image and the corresponding images corrupted by the above-defined distortions.

3.2 Keypoint Ranking Experiments

To better understand characteristics and properties of the proposed technique, it has been applied to a number of images, including images from an extremely popular benchmark dataset[1]. This dataset provides homographies between images (i.e. ground truth matches between keypoints). As mentioned above, MSER detector has been used as the baseline keypoint detector.

Keypoint descriptions are obtained using four popular descriptors, namely SURF [4], BRISK [12], FREAK [1] and KAZE [2].

The experiments have been coded in Matlab, using some standard tools for handling image regions and evaluating their descriptors (e.g. `MSERregions` and `extractFeatures`).

First, MSER keypoints (in a form of oriented ellipses) are found in the original image. Then, using the same ellipses as keypoints, we calculate descriptors not only in the original image but also in the corrupted images, as specified Subsect. 3.1. The differences between the-same-ellipse descriptors, in the original

[1] http://www.robots.ox.ac.uk/~vgg/research/affine/.

Fig. 1. The original image (top left) and images distorted by operations defined in Subsect. 3.1.

image and its corrupted versions, are used to approximate the keypoint *growth rate functions* (see Eq. 4) and, subsequently, its *stability rate* (see Eq. 5).

Keypoints are then ranked by the ascending order of their stability rates (see Eq. 6).

The first significant observation is that the keypoint rankings are not determined just by the size of keypoints (which would be a naive but justifiable expectation, because the visual contents of larger ellipses change relatively less). Instead, rankings strongly depend both on the keypoint content and, more importantly, on the selected descriptor.

Figure 2 displays 30 top-ranked (according to Eq. 6) MSER keypoints identified by using four types of keypoint descriptors in an exemplary image. Highly diversified collections of keypoints can be noticed.

3.3 Keypoint Matching Experiments

In these experiments, we evaluated whether using only highly-ranked keypoints can improve performances of matching processes. In typical natural images, the number of detected keypoints is usually very large so that it makes sense to use only some of them.

Normally, keypoints can be ranked during the detection process so that only the most *prominent* ones (i.e. maximizing the detection criteria) are retained. The proposed technique actively involves the descriptor so that the retained keypoints are those which are most stable in the context of applied descriptor (regardless the prominence of keypoints, as estimated by the detector).

Thus we performed the following test.

In selected images of the benchmark dataset, keypoint are extracted by MSER detector (or any other detector) and described by selected descriptors. From that, we match *the-same-class* images in two different ways:

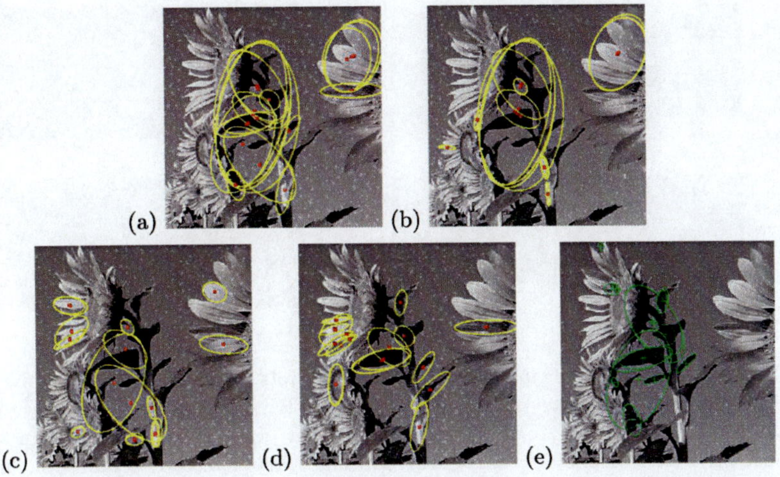

Fig. 2. 30 top-ranked (i.e. most stable) MSER keypoints in the context of four keypoint descriptors: (a) SURF, (b) BRISK, (c) FREAK and (d) KAZE. For the reference, 30 original top keypoints (i.e. with the highest magnitude of the detector, and without taking into account any descriptor values) are shown in (e).

a. Keypoint are ranked according to their prominence determined by the detector. The top half of the keypoints are used for image matching.
b. Keypoints are ranked according to Eq. 6, and the top half of so-ranked keypoints are used for image matching.

Exemplary comparisons between Options (a) and (b) are given in Figs. 3 and 4. The examples are selected to show cases of varying difficulties (in terms of image matching) and somehow surprising cases. Standard options of `matchFeatures` Matlab functions are used.

By discarding half of available keypoints from each image, we unpredictably affect *recall* of keypoint matching (some of the *ground-truth* matches may disappear, if one or both involved keypoints are discarded). Thus, our primary focus is to compare *precision* of Options (a) and (b).

The average results from all tested images (and from all four descriptors SURF, BRISK, FREAK and KAZE) indicate that *precision* of keypoint matching has been increased from **72%** to **88%**. We do not split the result into performances of individual descriptors (this is not our objective to rank them). Nevertheless, we have noticed that the best improvements happen to SURF descriptor.

In our opinion, this is the expected result because SURF is the most "sophisticated" descriptor and, thus, can better reflect image variations (or uniformities) under minor distortions.

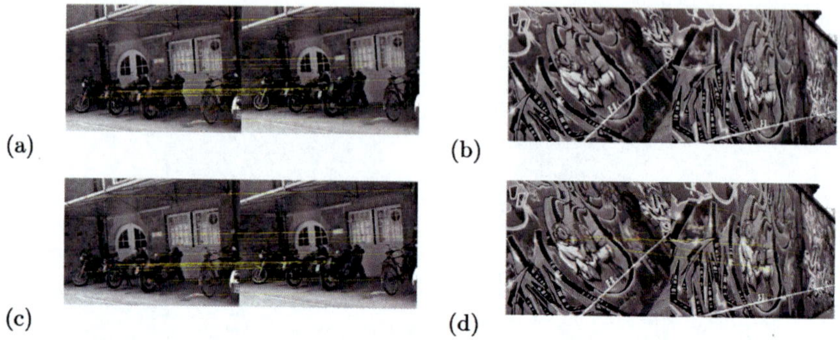

Fig. 3. Results of matching most prominent keypoints (a,b) and matching the top-ranked (descriptor-based) keypoints (c,d). In (a,c), FREAK descriptor is used, in (b,d) the descriptor is KAZE. Note the surprising effect in (c) where no matches are found for most prominent keypoints, while the top-ranked keypoints (d) produce a number of correct(!) matches.

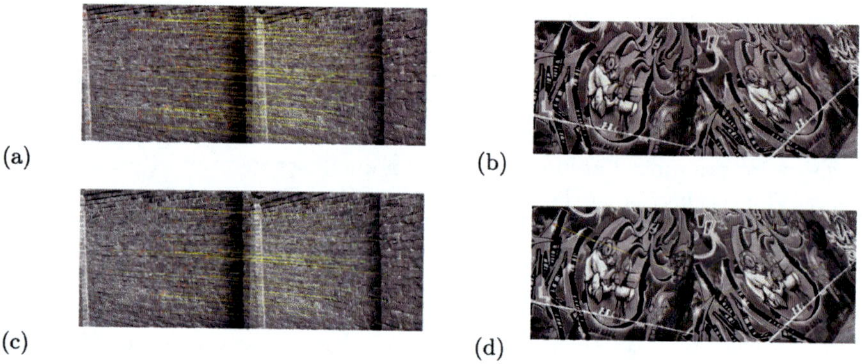

Fig. 4. Results of matching most prominent keypoints (a,b) and matching the top-ranked (descriptor-based) keypoints (c,d). In (a,c), SURF descriptor is used, in (b,d) the descriptor is BRISK. Note different, but equally correct results in (b) and (d).

4 Summary

The paper presents a small, but prospectively useful improvement for algorithms handling visual items (keypoints are used as example). In general, the method consists in ranking *detected items* by stability of their *descriptors.*

Stability of keypoint descriptors is defined by discrete approximations of their partial derivatives over parameters defining minor distortions of the image contents. Only the top-ranked (i.e. the most stable in the above sense) keypoints are retained for further operations, e.g. for matching images.

In the study, we compared *precision* of keypoint matching, where only half of the extracted keypoints were retained. In the first case, we took the most prominent (in the sense of their detector) keypoints, while in the second case the most stable (in the context of their descriptor)keypoints were used.

Substantial improvements have been obtained in the second case, which we consider the *proof-of-concept* for the proposed approach.

Continuation of this works involving more types of keypoint detectors (including DL-based), other categories of visual objects, and other tasks (e.g. tracking applications) is planned for the near future.

References

1. Alahi, A., Ortiz, R., Vandergheynst, P.: Freak: fast retina keypoint. In: 2012 IEEE Conference on Computer Vision and Pattern Recognition (CVPR), pp. 510–517 (2012)
2. Alcantarilla, P.F., Bartoli, A., Davison, A.J.: Kaze features. In: Computer Vision—ECCV 2012 , pp. 214–227 (2012)
3. Barroso-Laguna, A., Riba, E., Ponsa, D., Mikolajczyk, K.: Key.net: keypoint detection by handcrafted and learned cnn filters. In: Proceedings of the IEEE/CVF International Conference on Computer Vision (ICCV) (2019)
4. Bay, H., Ess, A., Tuytelaars, T., Van Gool, L.: Speeded-up robust features (surf). Comput. Vis. Image Underst. **110**(3), 346–359 (2008)
5. Clady, X., Ieng, S.H., Benosman, R.: Asynchronous event-based corner detection and matching. Neural Netw. **66**, 91–106 (2015)
6. Dai, Z., Huang, X., Chen, W., He, L., Zhang, H.: A comparison of CNN-based and hand-crafted keypoint descriptors. In: 2019 International Conference on Robotics and Automation (ICRA), pp. 2399–2404 (2019)
7. Duan, L., Wu, C., Miao, J., Qing, L., Fu, Y.: Visual saliency detection by spatially weighted dissimilarity. In: CVPR 2011, pp. 473–480 (2011)
8. Harris, C., Stephens, M.: A combined corner and edge detector. In: Alvey Vision Conference, pp. 147–152 (1988)
9. Itti, L., Koch, C., Niebur, E.: A model of saliency-based visual attention for rapid scene analysis. IEEE Trans. Pattern Anal. Mach. Intell. **20**(11), 1254–1259 (1998)
10. Jiao, L., et al.: A survey of deep learning-based object detection. IEEE Access **7**, 128837–128868 (2019)
11. Komorowski, J., Czarnota, K., Trzcinski, T., Dabala, L., Lynen, S.: Interest point detectors stability evaluation on apolloscape dataset. In: Proceedings of the European Conference on Computer Vision (ECCV) Workshops (2018)
12. Leutenegger, S., Chli, M., Siegwart, R.Y.: Brisk: binary robust invariant scalable keypoints. In: 2011 International Conference on Computer Vision (ICCV), pp. 2548–2555 (2011)
13. Lowe, D.G.: Distinctive image features from scale-invariant keypoints. Int. J. Comput. Vis. **60**(2), 91–110 (2004)
14. Matas, J., Chum, O., Urban, M., Pajdla, T.: Robust wide baseline stereo from maximally stable extremal regions. In: British Machine Vision Conference, pp. 384–393 (2002)
15. Mikolajczyk, K., Schmid, C.: Scale and affine invariant interest point detectors. Int. J. Comput. Vis. **60**, 63–86 (2004)
16. Mikolajczyk, K., Schmid, C.: A performance evaluation of local descriptors. IEEE Trans. PAMI **27**, 1615–1630 (2005)

17. Mikolajczyk, K., et al.: A comparison of affine region detectors. Int. J. Comput. Vis. **65**, 43–72 (2005)
18. Nascimento, J.C., Marques, J.S.: Performance evaluation of object detection algorithms for video surveillance. IEEE Trans. Multimed. **8**(4), 761–774 (2006)
19. Ono, Y., Trulls, E., Fua, P., Kwang, M.Y.: LF-NET: learning local features from images. In: 32nd International Conference on Neural Information Processing (NIPS 2018), pp. 6237–6247 (2018)
20. Prasad, D.K.: Survey of the problem of object detection in real images. Int. J. Image Process. **6**(6), 441–466 (2012)
21. Redmon, J., Farhadi, A.: Yolo9000: better, faster, stronger. In: 2017 IEEE Conference on Computer Vision and Pattern Recognition (CVPR), pp. 7263–7271 (2017)
22. Revaud, J., et al.: R2D2: repeatable and reliable detector and descriptor (2019). http://arxiv.org/abs/1906.06195
23. Rosten, E., Drummond, T.: Machine learning for high-speed corner detection. In: Computer Vision—ECCV 2006, pp. 430–443 (2006)
24. Śluzek, A.: Descriptor-driven keypoint detection. In: 2018 Digital Image Computing: Techniques and Applications (DICTA), pp. 1–8 (2018)
25. Śluzek, A.: Improving performances of mser features in matching and retrieval tasks. LNCS (ECCV 2016 Workshops) 9915, pp. 759–770 (2016)
26. Ullah, I., et al.: A brief survey of visual saliency detection. Multimed. Tools Appl. **79**, 34605–34645 (2020)
27. Yi, K.M., Fortuny, E.T., Lepetit, V., Fua, P.: Lift: learned invariant feature transform. In: Proceedings of the 14th European Conference on ECCV 2016), vol. 2, pp. 467–483 (2016)
28. Zhang, X., Yu, F.X., Karaman, S., Chang, S.: Learning discriminative and transformation covariant local feature detectors. In: 2017 IEEE Conference on Computer Vision and Pattern Recognition (CVPR), pp. 4923–4931 (2017)
29. Zheng, L., Yang, Y., Tian, Q.: Sift meets CNN: a decade survey of instance retrieval. IEEE Trans. PAMI **40**(5), 1224–1244 (2018)
30. Zou, Z., Shi, Z., Guo, Y., Ye, J.: Object detection in 20 years: a survey (2019). http://arxiv.org/abs/1905.05055

Fuzzy System for Lip Print Identification

Przemysław Kudłacik(✉) and Tomasz Orczyk

Faculty of Science and Technology, University of Silesia in Katowice, Będzińska 39, 41-200 Sosnowiec, Poland
{przemyslaw.kudlacik,tomasz.orczyk}@us.edu.pl

Abstract. The paper presents the design and performance study of a fuzzy system for lip print identification. The problem applies to the general field of image analysis. It is a particular environment where uncertainty plays a very important role. The solution assumes a classical rule-based fuzzy system that automates the fuzzification of premises and facts based on appropriately selected ranges of learning data. Comparative analysis of several solutions has shown that the proposed approach obtains the best results.

Keywords: Lip print · Classification · Fuzzy system

1 Introduction

Person identification based on the analysis of the lip prints is a relatively new area as compared to the analysis of fingerprints or handwritten signatures. In recent years, researchers have shown a significant increase of interest in this subject. Many articles have been published, showing the usefulness of lip prints or lip photos [4,8,10,19] in person identification. Alongside, the new lip print features that could be used to recognize or authenticate people were also researched. Examples of such features are: the lip contour shape [24], area and outline dimensions [11,18,23,26]. These features can be analyzed using various methods: PCA based method [11], Zernike moments [5], Radon transform [6], SIFT or SURF algorithms [1], and hidden Markov models [23]. Accuracy levels in above mentioned papers were around 90% [1,11,23]. In [3,12] lip print furrows have been analyzed. In the first paper accuracy of the system reached 96% on the basis of 20 high quality lip prints, whereas in the second one 95% on the basis of lip prints registered from 24 individuals. The pattern matching method together with the "brute force" algorithm were used to verify people in work [20]. The obtained verification efficacy was 85% based on 200 complete, good quality lip imprints. In the work [21], the identification efficacy was 58% based on an analysis of 300 lip prints. The proposed approach is based on the fast normalized cross-correlation method, which shows that methods based only on image processing algorithms do not allow to obtain a satisfactory results. The method of identifying lip prints using the DTW algorithm was proposed in [15]. The effectiveness of the method was 78%, on a collection of 120 complete lip prints.

M. Choraś et al. (Eds.): CORES 2021, IP&C 2021, ACS 2021, LNNS 255, pp. 97–106, 2022.
https://doi.org/10.1007/978-3-030-81523-3_9

The identification efficacy of 77% was achieved at work [25]. In this approach the methods of fingerprint bifurcation analysis was adapted to the analysis of lip prints. It should be noticed that these investigations were performed on the lip prints from databases containing data obtained by high resolution cameras in laboratories, where the optimal lighting conditions were ensured. In contrast, in our method, more real life quality lip prints were analyzed.

Rule-based fuzzy systems are characterized by great flexibility, as they do not impose methods of generating rules and allow to freely adjust parameters defining generalization (fuzzification) through membership functions of generated fuzzy sets. Fuzzy systems are repeatedly applied to image analysis [13]. In the context of lip print identification, they allow for very flexible expansion of the knowledge base by adding new rules when needed, or removing those that are no longer necessary for some reason (e.g., removing samples of a person from the database). Such flexibility is very difficult, or even impossible, to achieve in artificial neural networks, which generally require a re-learning process after placing a new vector in the knowledge base. For this reason, this research focused on building and testing a fuzzy system for lip print identification.

2 Lip Print

The uniqueness of lip prints makes them an especially effective trait in criminology. Whenever appropriate lip prints are secured on the crime scene, they may be used as an evidence. This new form of identification is accepted and introduced worldwide as an evidence in lawsuits [4,10].

The variety of surfaces from which lip prints are taken, as well as the flexibility of the lips may cause some fragments of the lip print to be invisible or incomplete. Because currently the lip print analyses are carried out manually by forensic technicians, this is a tedious and time-consuming operation.

In a real life quality of lip print samples is far from perfect, even when we are talking about laboratory taken reference samples. This may have a negative impact on person identification accuracy. Examples of real life examples of lip prints are presented in Fig. 1: Fig. 1a is the example of lip print, where most of lip print furrows are clearly visible, Fig. 1b is an example, where most of the lip furrows on the upper lip are invisible, and the incomplete lip print is shown in Fig. 1c.

For the purpose of automated person identification, based on a lip prints, the digital form of the lip prints are pre-processed in order to enhance the visibility of the furrows on each lip. The detailed description of the lip print acquisition, pre-processing, and feature extraction, and region quality assessment is beyond the scope of this article, and has been presented in [27], here only highlights of these stages will be mentioned.

The pre-processing consists of three stages:

- Image (Fig. 2a) is subjected to the Gaussian smoothing procedure. The result of using this method is shown in Fig. 2b.

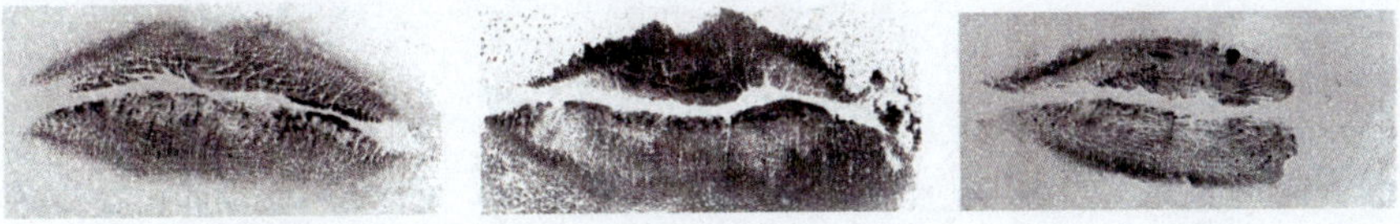

(a) (b) (c)

Fig. 1. Real life examples of lip prints taken in a laboratory environment: (a) high quality, (b) blurred (c) incomplete lip imprint.

- The Otsu image thresholding algorithm is applied [14] to binarize the analyzed image (Fig. 2c).
- Image is cleaned from artifacts that may occur during the binarization process. In the cleaning process, the area of all clusters of black pixels in the lip print is calculated. Then, clusters that size is less than 0.1% of the sum of all black pixels in the lip print are removed. The cleaned image is depicted in Fig. 2d.

(a) (b) (c) (d)

Fig. 2. Lip furrows enhancement: (a) original image, (b) smoothed image, (c) binarized image, (d) cleaned image.

After image is binarized and cleaned, the feature extraction can be done. For the purpose of feature extraction the lip print image is divided into S segments $reg_1, \ldots, reg_S$ of equal width. The relevance of applying local analysis and the impact of the number of segments on the effectiveness of identification was examined in the experimental part. An example of lip print divided into $S = 20$ segments is shown in Fig. 3.

The following features are then extracted from each of the reg_S segments:

- v_1 - number of furrows,
- v_2 - average length of all furrows,
- v_3 - standard deviation of all furrows lengths,
- v_4 - average slope to the OX axis of all furrows,
- v_5 - standard deviation of the slope to the OX axis of all furrows,
- v_6 - average value of the center coordinates x^c of all furrows,
- v_7 - average value of the center coordinates y^c of all furrows.

Fuzzy systems show great effectiveness when analyzing missing or uncertain data [13]. Properly constructed rules allow correct pattern matching of images that differ in some areas.

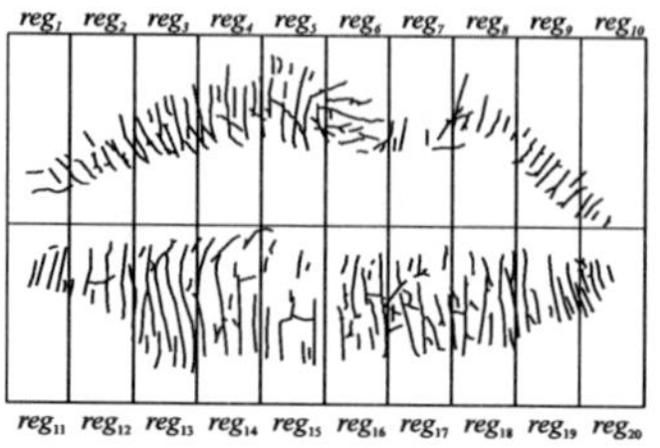

Fig. 3. Division of the lip imprint into S=20 segments.

3 System Structure

The base elements defining the structure of the system are its number of inputs, number of outputs and the set of premises used in the rule base. On this basis, rules can be defined by combining facts (inputs) with premises and by building conclusions, which already have a direct impact on the outputs of the fuzzy system.

In the designed system, the number of system inputs is closely matched with the number of features in the input vectors. As described earlier, the number of features strictly depends on the number of segments in the analyzed lip print image. Thus, assuming 7 features per segment and S segments, the system has a total of $7 \times S$ features, which is exactly related to the number $7 \times S$ of system inputs. Let the inputs of the system be denoted by $i \in [1, 7 \times S]$ ("i" from the word "input").

Similarly, we can determine the number of outputs of the system, which in this case depends strictly on the number of classes, or simply the number of people sampled. Thus, assuming 100 sampled individuals in the available test base, the system will have the same number of outputs, where each output is assigned to a different class/person. Let the outputs of the system be denoted by $p \in [1, 100]$ ("p" from the word "person"). This parameter will also be used in rule construction to indicate the class/person number.

3.1 Rule Base of the Proposed System

Due to the small number of samples in the database taken from each person, which is a typical feature of biometric systems, the system assumes the construction of one fuzzy rule for each feature vector in the database. Assuming the natural generalization feature of fuzzy systems, the rules will in theory allow matching the most similar test sample to a given person. In cases where there are many samples representing a class, it is possible to aggregate or average them in advance using, for example, clustering algorithms. In the current database, where there are only 6 samples for each person, this is not necessary. Thus, the database is balanced and the size of the fuzzy system is relatively small.

The following is a general rule notation of the designed fuzzy system:

Algorithm 1: General rule notation of the designed fuzzy system

---

if X_1 *is* A_{1jp} *and* X_2 *is* A_{2jp} *and* ... *and* X_i *is* A_{ijp} **then**
  Y_p *is* B_p ;
end

---

where sets X represent the inputs of the system, sets A the premises, while Y and B represent the outputs and conclusions respectively. As can be seen, in addition to the indices i and p, describing the number of the input (feature) and the number of the person, respectively, there is an index j in the rule, which corresponds to the number of the sample of the person. Thus, the set A_{ijp} is a premise built for the i-th input (feature), based on the j-th sample of the p-th person. The details of building the premise are described later in the paper. In Algorithm 1, it can also be seen that all premises in the rules are connected by an *and* conjunction.

Considering 6 feature vectors for each person, the used lip-print database will therefore allow to generate up to 600 rules. Each rule at the best matching of the test sample generates a triangular fuzzy set B'p of height 1.0 on the output p, i.e. assigned to the person (class). The exact form of this set does not matter at all, since the classification result will be the output number with the highest activation level, which is directly related to the best fit of a particular rule.

Figure 4 shows the structure diagram of the proposed fuzzy classification system. It is consistent with the general rule presented above. It can be seen how the outputs of the system are affected by each rule built for a particular sample of individuals.

It can be said that each rule makes a "vote" for a particular person (class) with the level of activation mapped to its assigned output. After aggregating the results from all the rules, the classification result of a given sample is influenced by the highest value obtained on a specific output p, i.e. indirectly by the person, and directly by the rule that fits best in the designed knowledge base.

3.2 Configuration of Premises and Facts

A key component of the system is the appropriate configuration of uncertainty. In the designed system, uncertainty is modeled by several parameters defining fuzzy sets defining premises and facts, i.e., system inputs during classification.

The structure described earlier does not explain these concepts precisely so they must be specified.

At first, it should be emphasized that all membership functions in the system are of Gaussian type, defined by the following equation

$$\mu(x) = exp\left(-\frac{(x-b)^2}{2c^2}\right) \tag{1}$$

where the parameter b determines the peak center, while c determines the width, or fuzzification level. In preliminary studies, trapezoidal and triangular functions

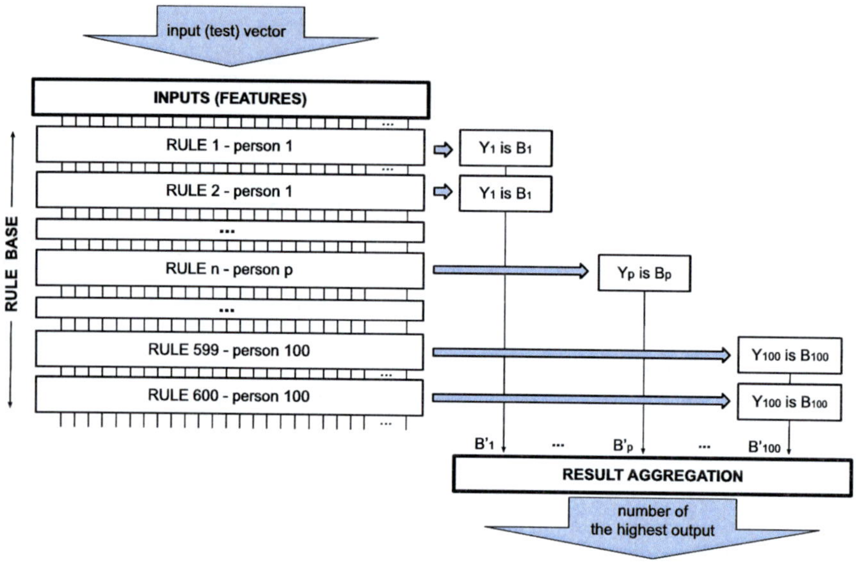

Fig. 4. System structure

were also tested, but it was the Gaussian ones that gave the best results. The only exception is the conclusion membership functions, which were assigned a triangular form due to their less complicated structure. As mentioned earlier, the type of the membership function of the conclusion did not matter, since the result is only affected by the highest value, which is the very peak of the membership function.

A range of feature values, obtained from the learning set, was used to determine the baseline value of the fuzzification parameter c, which is the width of the Gaussian function. Let the range be denoted as Z_i , where i denotes the feature number, and thus $i \in [1, 7 \times S]$. The range of feature values alone would be too large, so for final fuzzification purposes it was divided by a fixed parameter. This parameter was chosen experimentally. In the case of the facts it was equal to 20, while in the case of the premises it took the value of 3. Thus, the final values of the parameters c for the purposes of determining the Gaussian fuzzification of the membership functions of the facts (inputs of the system) took the form

$$c_i^F = \frac{Z_i}{20}, \; i \in [1, 7 \times S] \tag{2}$$

On the other hand, the fuzzification values for each premise associated with the i-th input (feature) took the form

$$c_i^P = \frac{Z_i}{3}, \; i \in [1, 7 \times S] \tag{3}$$

Thus, the baseline fuzzification for facts (inputs) is $\frac{1}{20}$ range of feature values, while for premises it is $\frac{1}{3}$ range of these values.

Experimental tests to select the values were performed for a limited number of tests, only to verify that changing any of the parameters affects the results. It should also be noted that these parameters are dependent on each other. When increasing the fuzziness of the input data, the fuzziness of the premises should be decreased and vice versa.

3.3 Results

The designed fuzzy system has been tested on a described database of lip prints using the leave-one-out method, which is used in classification problems for a relatively small number of vectors in a learning set. It was the case in this situation. Thus, in each iteration of the test, one consecutive vector was selected and treated as the test set. The remaining vectors were used as the learning set.

A comparative analysis was performed for 7 different approaches. In addition to the designed system, tests were also performed for SFkNN [16], PNN [22], kNN [9], SVM [7], FRI [2], and C4.5 [17] classifiers.

The first group of experiments was performed in order to identify the best solutions for more accurate testing. The initial choice was to divide the lip prints into 12 and 32 segments. They considered a relatively small number of segments and a higher number of segments. This study allowed first to indicate how the systems' performance behaves with respect to increasing the number of segments (splits) and secondly, to identify the approaches with the best results.

Accuracy of all tested approaches for given two numbers of segments are presented in Table 1. Proposed fuzzy system is denoted by FRBS (Fuzzy Rule-Based System).

Analyzing the results, it can be seen that the best results were achieved for the following approaches: FRBS, SFkNN and PNN. FRI gave promising results for low number of segments, however, ultimately did not increase accuracy like the mentioned solutions. Thus, a more thorough study was performed for the three listed systems to analyze the behavior of these different approaches in relation to changes in the number of segments in a broader spectrum. The obtained results of this tests are presented in Table 2. The same results are also presented graphically in Fig. 5.

In all three cases, one can clearly see an increase in classification accuracy relative to an increase in the number of segments. The best results in each case were obtained for the maximum examined number of segments, except that for the last three tested values the changes were no longer so significant.

Table 1. Accuracy(%) results for examined systems

Segments	FRBS	SFkNN	PNN	KNN	SVM	FRI	C4.5
12	32.5	49.3	**64.3**	0.0	5.6	62.1	16.7
32	**92.5**	91.2	82.0	11.1	11.1	65.9	9.0

Table 2. Accuracy(%) according to number of segments in lip prints for the best examined systems

Segments	FRBS	SFkNN	PNN
4	**20.5**	7.8	3.0
8	27.2	**33.5**	8.3
12	32.5	49.3	**64.3**
16	40.3	58.8	**71.0**
20	59.0	70.0	**73.0**
24	75.5	**81.2**	75.0
28	87.7	**87.8**	77.2
32	**92.5**	91.2	82.0
36	**96.0**	94.3	83.0
40	**97.8**	94.8	84.0

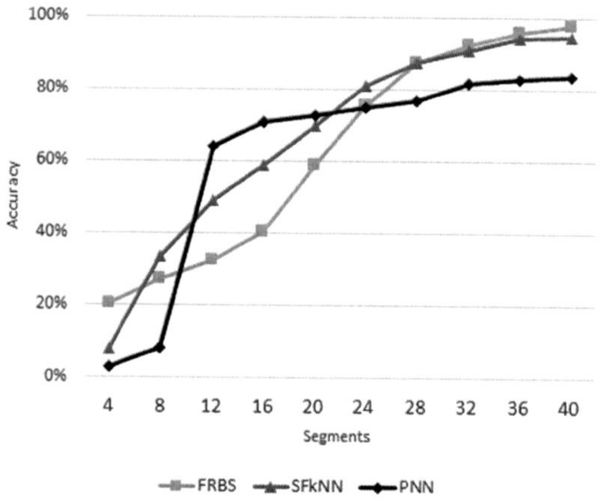

Fig. 5. Accuracy change according to number of segments in lip prints for the best examined systems

4 Conclusions

In this paper, a rule-based fuzzy system for person identification based on lip prints is proposed. The designed approach is analyzed with several other solutions for a variable number of lip print segments.

Finally, and most significantly, the proposed fuzzy system obtained the best results. However, in general, the results clearly indicate the problematic nature of the approach when dividing into fewer segments. This is a characteristic of all investigated approaches.

The problem may be related to the uneven quality distribution of the lip print areas. Some are repeatable in all samples and some are not. Subsequent research will focus on this very problem. The authors hope that including segment/area quality in the analysis will significantly improve the classification results.

Acknowledgements. We would like to thank Piotr Porwik, Rafał Doroz, and Krzysztof Wróbel for sharing with us their pre-pocessed lip print data base [27].

References

1. Bakshi, S., Raman, R., Sa, P.K.: Lip pattern recognition based on local feature extraction. In: 2011 Annual IEEE India Conference (INDICON), India, pp. 1–4 (2011)
2. Berthold, M.R.: Mixed fuzzy rule formation. Int. J. Approximate Reasoning **32**(2–3), 67–84 (2003)
3. Bhattacharjee, S., Arunkumar, S., Bandyopadhyay, S.K.: Personal identification from lip-print features using a statistical model. Int. J. Comput. Appl. **55**(13), 30–34 (2012)
4. Bécue, A., Eldridge, H., Champod, C.: Interpol review of fingermarks and other body impressions 2016–2019. Forensic Science International: Synergy. Article in press (2020)
5. Choras, M.: The lip as a biometric. Pattern Anal. Appl. **13**(1), 105–112 (2010)
6. Choras, R.: Lip-prints Feature Extraction and Recognition. Springer, Berlin Heidelberg, Berlin, Heidelberg (2011)
7. Cortes, C., Vapnik, V.: Support-vector networks. Mach. Learn. **20**(3), 273–297 (1995)
8. Das, S., Muhammad, K., Bakshi, S., Mukherjee, I., Sa, P.K., Sangaiah, A.K., Bruno, A.: Lip biometric template security framework using spatial steganography. Pattern Recogn. Lett. **126**, 102–110 (2019)
9. Fix, E.: Discriminatory analysis: nonparametric discrimination, consistency properties, vol. 1. USAF school of Aviation Medicine (1985)
10. Fonseca, G.M., Ortíz-Contreras, J., Ramírez-Lagos, C., López-Lázaro, S.: Lip print identification: current perspectives. J. Forensic Leg. Med. **65**, 32–38 (2019)
11. Gomez, E., Travieso, C.M., Briceno, J.C., Ferrer, M.A.: Biometric identification system by lip shape. In: Proceedings. 36th Annual 2002 International Carnahan Conference on Security Technology, Atlantic City, NJ, USA, pp. 39–42 (2002)
12. Kim, J.O., Baik, K.S., Chung, C.H.: On a lip print recognition by the pattern kernel with multi-resolution architecture. In: Liu, J., Cheung, Y., Yin, H. (eds.) IDEAL 2003. LNCS, vol. 2690, pp. 561–568. Springer, Heidelberg (2003). https://doi.org/10.1007/978-3-540-45080-1_77
13. Kudłacik, P., Porwik, P., Wesołowski, T.: Fuzzy approach for intrusion detection based on user's commands. Soft. Comput. **20**, 2705–2719 (2015)
14. Otsu, N.: A threshold selection method from gray-level histograms. IEEE Trans. Syst. Man Cybern. **9**(1), 62–66 (1979)
15. Porwik, P., Orczyk, T.: DTW and voting-based lip print recognition system. In: Cortesi, A., Chaki, N., Saeed, K., Wierzchoń, S. (eds.) CISIM 2012. LNCS, vol. 7564, pp. 191–202. Springer, Heidelberg (2012). https://doi.org/10.1007/978-3-642-33260-9_16
16. Porwik, P., Orczyk, T., Lewandowski, M., Cholewa, M.: Feature projection k-NN classifier model for imbalanced and incomplete medical data. Biocybernetics Biomed. Eng. **36**(4), 644–656 (2016)

17. Quinlan, J.R.: C4. 5: Programs for Machine Learning. Elsevier (2014)
18. Raman, R., Sa, P.K., Majhi, B., Bakshi, S.: Fusion of shape and texture features for lip biometry in mobile devices. Security, Institution of Engineering and Technology (2017)
19. Saini, M., Kapoor, A.K.: Biometrics in forensic identification: applications and challenges. J. Forensic Med. **1**(2), 1000,108 (2016)
20. Sharma, P., Deo, S., Venkateshan, S., Vaish, A.: Lip print recognition for security systems: an up-coming biometric solution. In: Tsihrintzis, G.A., Virvou, M., Jain, L.C., Howlett, R.J. (eds.) Intelligent Interactive Multimedia Systems and Services, pp. 347–359. Springer, Berlin Heidelberg, Berlin, Heidelberg (2011)
21. Smacki, L.: Latent lip print identification using fast normalized cross-correlation. In: 2013 International Conference on Biometrics and Kansei Engineering (ICBAKE), pp. 189–192 (2013)
22. Specht, D.F.: Probabilistic neural networks. Neural Networks **3**(1), 109–118 (1990). https://doi.org/10.1016/0893-6080(90)90049-Q, URL https://www.sciencedirect.com/science/article/pii/089360809090049Q
23. Travieso, C.M., Zhang, J., Miller, P., Alonso, J.B.: Using a discrete hidden Markov model kernel for lip-based biometric identification. Image Vis. Comput. **32**(12), 1080–1089 (2014)
24. Travieso, C.M., Zhang, J., Miller, P., Alonso, J.B., Ferrer, M.A.: Bimodal biometric verification based on face and lips. Neurocomputing **74**(14–15), 2407–2410 (2011)
25. Wrobel, K., Doroz, R., Palys, M.: Lip print recognition method using bifurcations analysis. In: 7th Asian Conference on Intelligent Information and Database Systems (ACIIDS). Proceedings, Part II, pp. 72–81 (2015)
26. Wrobel, K., Doroz, R., Porwik, P., Naruniec, J., Kowalski, M.: Using a probabilistic neural network for lip-based biometric verification. Eng. Appl. Artif. Intell. **64**, 112–127 (2017)
27. Wrobel, K., Porwik, P., Doroz, R.: Effective lip prints preprocessing and matching methods. In: Burduk, R., Jackowski, K., Kurzyński, M., Woźniak, M., Żołnierek, A. (eds.) Proceedings of the 9th International Conference on Computer Recognition Systems CORES 2015. AISC, vol. 403, pp. 347–357. Springer, Cham (2016). https://doi.org/10.1007/978-3-319-26227-7_33

Assessment of Correlations Between Age and Textural Features of CT Images of Thoracic Vertebrae

Weronika Węgrzyn[1], Monika Pierzchała[1], Paulina Bałon[1], Robert Paweł Banyś[2], and Adam Piórkowski[1(✉)]

[1] Department of Biocybernetics and Biomedical Engineering, AGH University of Science and Technology, Mickiewicza 30 Av., 30–059 Krakow, Poland
pioro@agh.edu.pl

[2] Department of Radiology and Diagnostic Imaging, John Paul II Hospital, Pradnicka St 80, 31-202 Krakow, Poland

Abstract. The human spine undergoes degenerative changes with age. One of the symptoms of aging is osteoporosis, which is manifested by loss of bone mass and reduced density of trabeculae. Previous studies have shown that some image features are correlated with age. In this research, image features' values were obtained using textural analysis. CCTA images of the thoracic vertebra of patients of different ages were analyzed using qMaZda software. The correlation between features' values and the age of the person was determined using Spearman and Pearson coefficients. The research was carried out on two groups: a group consisting of male and female patients, and a group consisting of only women. In the entire research group, the image attributes derived from GRLM and GLCM showed the highest correlation with age. In the female group, GLCM features were the most correlated with age; moreover, the values of the correlation's strength were noticeably higher than the highest values obtained in the group of both sexes. The influence of sex and image parameters on the relationship between age and textural features is discussed.

Keywords: Texture analysis · Age · Thoracic vertebrae · GLCM · GRLM

1 Introduction

As with all body tissues, the human spine is deformed and exposed to damage as a result of the ageing process. During childhood and early adulthood, the vertebral body's cortical shell is dense and thin. The end-plates of vertebral bodies are covered with cartilage and contain vascular perforations [14]. With ageing, the density of transversely oriented trabeculae decreases under the influence of extensive activity or a lack of it, hormonal changes, calcium deficits, nutrition

M. Choraś et al. (Eds.): CORES 2021, IP&C 2021, ACS 2021, LNNS 255, pp. 107–115, 2022.
https://doi.org/10.1007/978-3-030-81523-3_10

and other lifestyle factors. The shape of vertebrae becomes wider, shorter and more oblate. This phenomenon leads to osteoporosis or osteopenia, which is recognized as the early stage of osteoporosis. These diseases lead to increased risk of spine fractures [7,9]. Osteoporosis and osteopenia usually affect elderly women more often than elderly men due to hormonal deficiencies, which can cause changes in bone tissue. Around one third of postmenopausal women suffer from osteoporosis [10]. Moreover, selective loss of trabeculae bone tissue has an impact on the shape of intervertebral discs and the elasticity of the spine [9].

Many age-related spine deformations can be detected by analysis of medical images. Recent research has shown that morphological and shape features of bones can be extracted from images produced with techniques such as radiography, ultrasonography or MRI [1,3]. Moreover it has been proven that some image attributes are more associated with age or sex than others. One study [8] described research on MRI images of thoracic and cervical intervertebral discs. Data were collected from patients who completed a lifestyle questionnaire. The parameters of intervertebral discs such as disc space narrowing, decrease in the signal intensity of the intervertebral discs (DSI) and posterior disc protrusion (PDP) were estimated. Almost half of the patients demonstrated a deformation in at least one thoracic intervertebral level. DSI was strongly correlated with age, whereas PDP was correlated not only with age but also with smoking and degenerative changes in the cervical spine, which were more frequent than those in the thoracic spine.

Another interesting study was carried out with use of texture analysis [7]. This method ensures access to image data that is extracted with mathematical techniques but cannot be evaluated by the human eye; however, these methods are not frequently used to calculate bone features because such a large amount of data is not easy to interpretate. The authors of the article noted that although texture analysis can be conducted on CT images, capturing them delivers high radiation doses to a patient's body. Because of the fact that the radius and the calcaneus have similar trabecular microarchitecture to vertebrae, these authors hypothesized that it is possible to assess age-related osteoporotic degenerations using radiographic scans of these bones. The measured parameters included gray-level run length. The research revealed a correlation between the run-length of the radius and the run-length of non-fractured vertebrae; however, there was no relationship between vertebrae and the calcaneus.

Analysis of texture is not only conducted on bone tissue images. One study [6] presented the influence of age and gender on structural brain asymmetry measured with texture analysis; cooccurrence matrices were calculated to define the spatial image structure. The obtained results confirmed the thesis that male brains are more asymmetrical than female ones. Correlation of brain asymmetry with age was not remarkable in this study [2].

Sexual dimorphism can also be studied by analyzing bones. One study presented the possibility of assessing age from the 12th thoracic vertebra among Egyptians [2]. However, sex-related parameters can vary between human races and populations. The study also showed that the 12th thoracic vertebra provides better sex estimation than the 1st lumbar vertebra among Egyptians.

2 Materials and Methods

The research was carried out using contrast-enhanced CT images obtained with coronary computed tomography angiography using a Siemens clinical scanner. The scans contained whole or partial images of thoracic vertebrae Th4-Th11. The CCTA (Coronary Computed Tomography Angiography) images were taken from men and women aged 21–84 years. Some of datasets were rejected due to poor image quality; this was caused by motion artifacts, or the visible part of vertebrae was too small and would therefore not be useful for further analysis. Finally, 1,007 good-quality images of vertebrae were collected from the study group, which consisted of 71 women and 95 men (Fig. 1). Most patients were 39–77 years old. The files were stored in DICOM format with a color depth of 12 bits. All images were anonymized but age, sex, and pixel-spacing information were preserved. The pixel spacing of the analyzed scans was in the range 0.291–0.482 (Fig. 2), but most of the images had a pixel spacing of 0.376. On average, pixel spacing was lower in the group of women than in the group of women and men. Similarly, the mean ROI surface area for the female group was lower than for the group of both sexes: for women it was 3540 pix^2; and for men and women together it was 3616 pix^2.

First step of data preparation was selection of representative scans of every visible thoracic vertebrae in patients' datasets. In the selection process, the following factors were taken into consideration: the clear border between the inner trabecular part of a vertebra and its contour as constituted by compact bone; as few artifacts as possible; and the nearest distance to the middle of vertebrae's height in order to avoid the difference between ROI areas that can occur when vertebrae are measured too close to the intervertebral disc below or above. The best quality of scans was almost always achieved in the systolic or diastolic frame of the heart cycle.

Subsequently, regions of interest were defined on the previously specified representative slices. The contours and filled ROIs were depicted manually in qMaZda software and exported to PNG files (Fig. 3). The trabecular bone of vertebrae was selected as the ROI area. Afterwards the textural features of the scans were extracted in qMaZda [12,13]. The obtained feature set included GLCM (Gray-Level Co-occurrence Matrix, [4]), GRLM (Gray-Level Run Length Matrix) and ROI surface area. Statistical analysis of the obtained data was conducted; between each attribute and the age of the patients, the Spearman and Pearson's r coefficients were calculated, which made it possible to determine which attributes represent evidence of ageing.

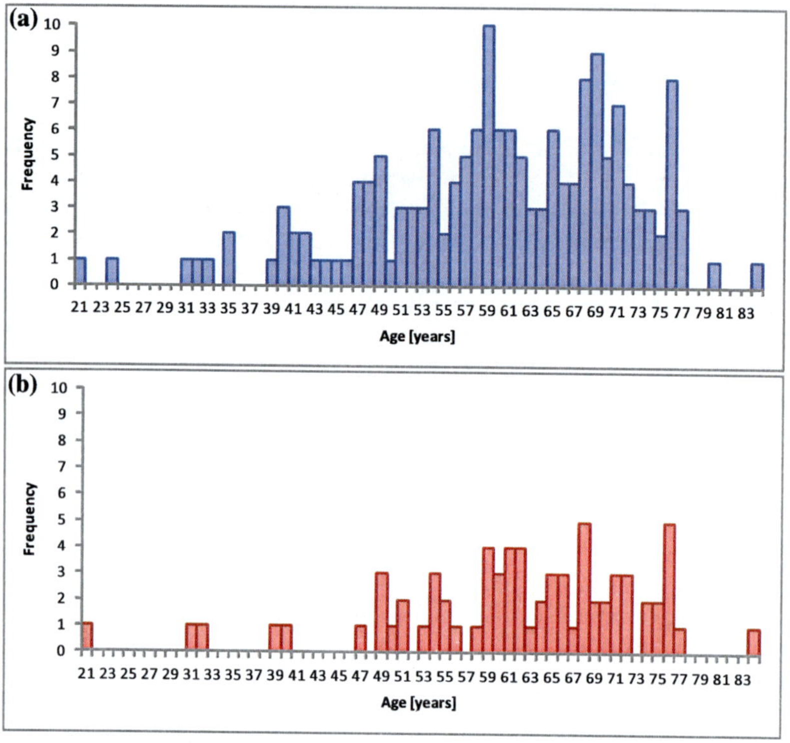

Fig. 1. Histogram of patients' ages: male and female (a), only female (b).

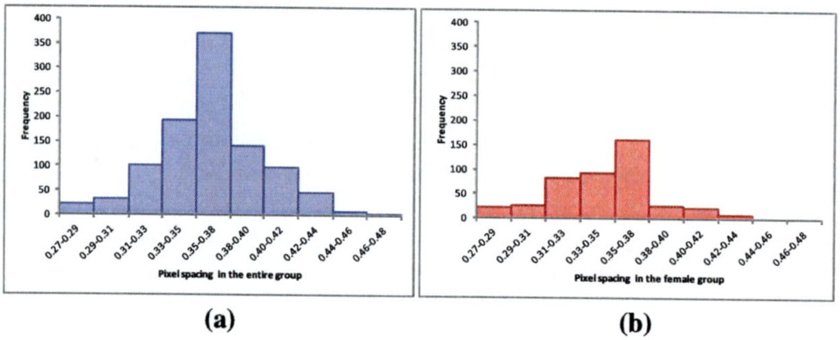

Fig. 2. Pixel spacing histogram for male and female patients (a), and for female patients (b).

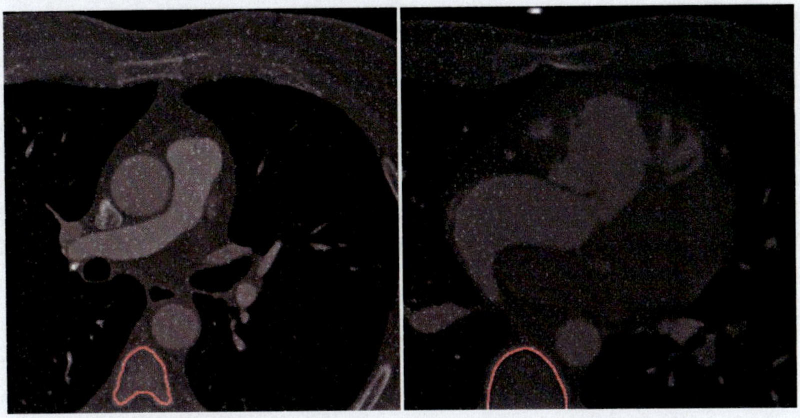

(a) (b)

Fig. 3. ROI contours plotted on exemplary representative cross-sections: 57-year-old man (a) and 21-year-old woman (b).

3 Experimental Results

Firstly, the correlations between the features calculated in qMaZda and the age of the patients were studied on the whole study group. The Spearman coefficient was calculated for every attribute from the set to determine how strongly each feature correlated with a patient's age. The features that showed the biggest negative and positive correlations were all derived from GLCM or GRLM and reached a modulus above 0.4 (Table 1). The *YD5* abbreviation means that the Y component of the YUV color model of the image was analyzed, no normalization was applied, and the image was quantized with 5 bits per pixel [5, 11]. The two features that correlated strongly revealed the most obvious trend among the study group: with increasing age of patients, the value of *YD5GrlmVMGLevNonUni* (Fig. 4) decreased and *YD5GlcmV1DifEntrp* increased statistically. The correlations determined by Pearson coefficient were less significant.

Table 1. The highest correlations of the texture parameters for the whole study group: men and women.

Texture feature	Spearman's coefficient value
YD5GrlmVMGLevNonUni	−0.410
YD5GrlmNMGLevNonUni	−0.4075
YD5GlcmV1Contrast	0.406
YD5GlcmV1DifVarnc	0.406
YD5GlcmV1DifEntrp	0.406

An interesting dependency was observed when the correlation between texture features was calculated only between images of women's spines and the age of the women. In that study group, a significant increase of the negative and positive correlations between age and the features derived from GLCM was noticed. As previously, the results were based on the Spearman and Pearson coefficients, but the strongest relationships were found with the former (Table 2). The biggest increase among the five highest results occurred in the case of *YD5GlcmZ2SumAverg*: the difference between the value of the Spearman's coefficient for the entire set and for the set of women was 0.09.

Table 2. The highest correlations of the texture parameters for the female study group.

Texture feature	Spearman's coefficient value
YD5GlcmV1Contrast	0.4958670
YD5GlcmV1DifVarnc	0.4958670
YD5GlcmV1DifEntrp	0.4958670
YD5GlcmV1InvDfMom	−0.4958670
YD5GlcmZ2SumAverg	−0.4956205

Moreover, in more than half of the cases, the correlation was stronger in the female group for both Spearman and Pearson coefficients.

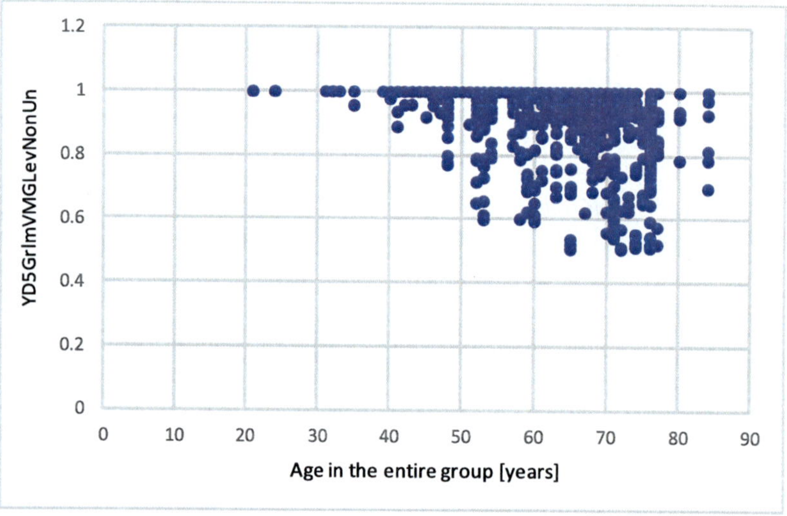

Fig. 4. Change of the YD5GrlmVMGLevNonUni parameter depending on patients' age.

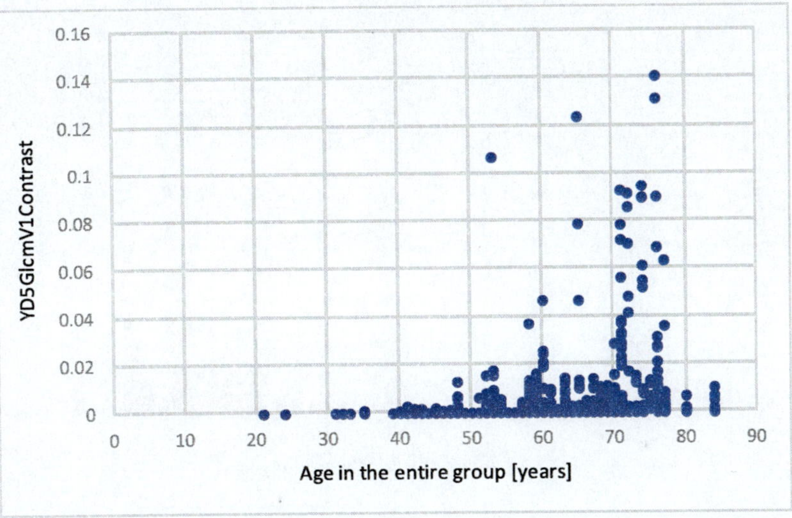

Fig. 5. Change of the YD5GlcmV1Contrast parameter depending on patients' age.

4 Discussion

The study found that some image features obtained by texture analysis were more correlated with age than others. The correlation between the given features and the age of the entire study group showed that the attributes most correlated with age are those derived from GLCM and GRLM. The situation in the group of women was slightly different. The significant increase in the correlation between age and GLCM characteristics in the female group allows us to hypothesize that GLCM features are correlated with both the age and sex of the patient. Considering the fact that osteoporotic changes are more frequent in women, the causes for the high correlation between these parameters and the increased occurrence of osteoporosis symptoms in the images of older women can be found. The five parameters that correlated most strongly with age in both females and both sexes include *YD5GlcmV1Contrast* (Fig. 5, 6), *YD5GlcmV1DifVarnc*, and *YD5GlcmV1DifEntrp* (Fig. 7); however, for the female group they were about 0.09 higher. All the strongest relationships in both groups were determined by Spearman's coefficient.

Some dependencies could be observed for both the study conducted on both sexes and the study conducted only on women. For some older individuals, it can be noticed in the presented charts that the features that were most correlated with age were much higher (the trend increased for *YD5GlcmV1DifEntrp*, *YD5GlcmV1Contrast*) or lower (the trend decreased for *YD5GrlmNMGLevNonUni*) compared to the average value of this parameter. For *YD5GrlmNMGLevNonUni*, the differences between the values were the smallest. One may wonder whether these are random dependencies or whether

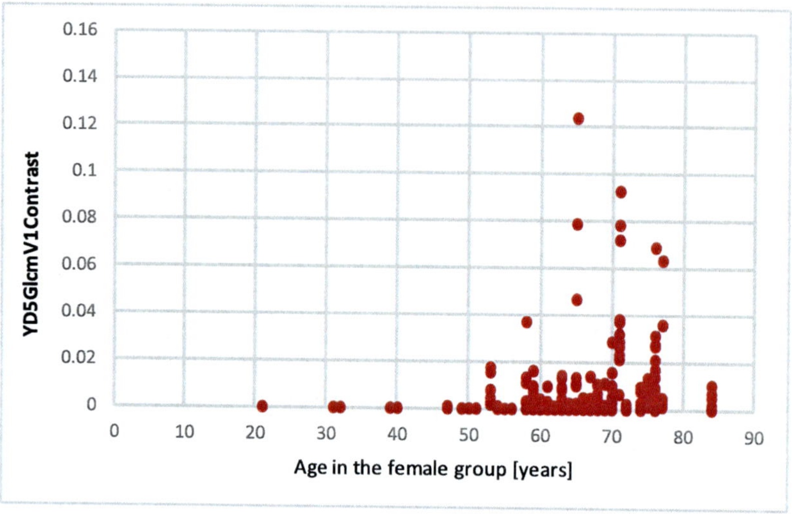

Fig. 6. Change in the YD5GlcmV1Contrast parameter, depending on patients' age.

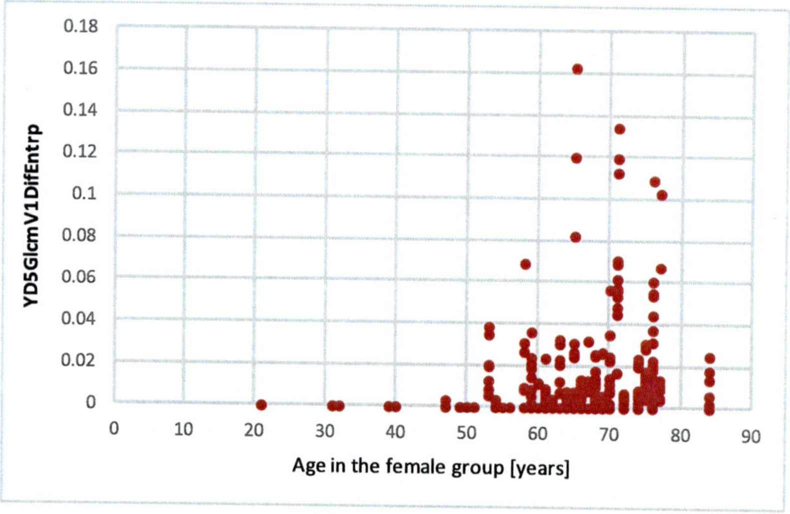

Fig. 7. Change in the YD5GlcmV1DifEntrp parameter, depending on patients' age.

these data came from people with increased changes in image textures and thus larger degenerative changes in the spine.

An important aspect of the research is that the pixel spacing was mostly lower in the female spine images than in the male spine images. A higher value of this parameter makes the scans less sensitive to detail, therefore bone mass losses could be more visible in images with a lower pixel spacing value. The question is how much this parameter affects the final results.

Acknowledgement. This work was financed by the AGH – University of Science and Technology, Faculty of EAIIB, KBIB no 16.16.120.773.

References

1. Donaldson, M., McGrother, C., Clayton, D., Clarke, M., Osborne, D.: Calcaneal ultrasound attenuation in an elderly population: measurement position and relationships with body size and past fractures. Osteoporos. Int. **10**(4), 316–324 (1999)
2. El Dine, F.M.B., El Shafei, M.M.: Sex determination using anthropometric measurements from multi-slice computed tomography of the 12th thoracic and the first lumbar vertebrae among adult Egyptians. Egypt. J. Forensic Sci. **5**(3), 82–89 (2015)
3. Gaidel, A., Khramov, A.: Application of texture analysis for automated osteoporosis diagnostics by plain hip radiography. Pattern Recognit. Image Anal. **25**(2), 301–305 (2015)
4. Haralick, R.M., Shanmugam, K., Dinstein, I.H.: Textural features for image classification. IEEE Trans. Syst. Man Cybern. **6**, 610–621 (1973)
5. Kociołek, M., Strzelecki, M., Obuchowicz, R.: Does image normalization and intensity resolution impact texture classification? Comput. Med. Imaging Graph. **81**, 101,716 (2020)
6. Kovalev, V.A., Kruggel, F., von Cramon, D.Y.: Gender and age effects in structural brain asymmetry as measured by MRI texture analysis. Neuroimage **19**(3), 895–905 (2003)
7. Mallard, F., Bouvard, B., Mercier, P., Bizot, P., Cronier, P., Chappard, D.: Trabecular microarchitecture in established osteoporosis: relationship between vertebrae, distal radius and calcaneus by X-ray imaging texture analysis. Orthop. Traumatol.: Surg. Res. **99**(1), 52–59 (2013)
8. Matsumoto, M., et al.: Age-related changes of thoracic and cervical intervertebral discs in asymptomatic subjects. Spine **35**(14), 1359–1364 (2010)
9. Mazur, P., Obuchowicz, R., Piórkowski, A.: The influence of age on morphometric and textural vertebrae features in lateral cervical spine radiographs. In: Information Technology in Biomedicine, AISC, vol. 1186, pp. 71–80. Springer (2021)
10. Sadeghi, H., Ashraf, A., Zeynali, N., Ebrahimi, B., Jehu, D.A.: Balance and functional mobility predict low bone mineral density among postmenopausal women undergoing recent menopause with osteoporosis, osteopenia, and normal bone mineral density: a cross-sectional study. Geriatr. Nurs. **42**(1), 33–36 (2021)
11. Strzelecki, M., Kociolek, M., Materka, A.: On the influence of image features wordlength reduction on texture classification. In: International Conference on Information Technologies in Biomedicine, pp. 15–26 (2018)
12. Szczypiński, P.M., Klepaczko, A., Kociołek, M.: QMaZda - software tools for image analysis and pattern recognition. In: 2017 Signal Processing: Algorithms, Architectures, Arrangements, and Applications (SPA), pp. 217–221. IEEE (2017)
13. Szczypiński, P.M., Strzelecki, M., Materka, A., Klepaczko, A.: MaZda - a software package for image texture analysis. Comput. Methods Program. Biomed. **94**, 66–76 (2009)
14. Twomey, L., Taylor, J.: The lumbar spine: structure, function, age changes and physiotherapy. Aust. J. Physiother. **40**, 19–30 (1994)

Preprocessing of Document Images Based on the GGD and GMM for Binarization of Degraded Ancient Papyri Images

Hubert Michalak, Robert Krupiński, Piotr Lech, and Krzysztof Okarma(✉)

Department of Signal Processing and Multimedia Engineering, Faculty of Electrical Engineering, West Pomeranian University of Technology in Szczecin, Sikorskiego 37, 70-313 Szczecin, Poland
{michalak.hubert,rkrupinski,piotr.lech,okarma}@zut.edu.pl

Abstract. Thresholding of document images is one of the most relevant operations that influence the final results of their further analysis. Although many image binarization methods have been proposed during recent several years, starting from global thresholding, through local and adaptive methods, to more sophisticated multi-stage algorithms and the use of deep convolutional neural networks, proper thresholding of degraded historical document images is still an open challenge. Due to the release of the recent challenging DIBCO 2019 dataset, containing two categories of images with the particularly demanding papyri dataset (track B), most of the earlier proposed methods turned out to be significantly less effective for them. Analysing the competition's results of the DIBCO 2019, apart from the variety of submitted trendy methods based on the use of deep learning, the great importance of image preprocessing may be observed for such demanding images. Hence, in this paper, we present the applicability analysis of the previously proposed preprocessing methods based on the use of the GGD and GMM in combination with various image binarization algorithms for this purpose. The obtained results are comparable with the state-of-the-art methods, including the top algorithms submitted to the recent DIBCO series contest.

Keywords: Document images · Gaussian mixture models · GGD · Image binarization

1 Introduction

Binary image analysis is frequently used in many application areas related to the necessity of a fast shape analysis as well as limited by the amount of the analysed image data. Such applications include e.g. Optical Character Recognition (OCR), Optical Mark Recognition (OMR), recognition of QR codes, morphological image processing as well as the analysis of silhouettes and object tracking

M. Choraś et al. (Eds.): CORES 2021, IP&C 2021, ACS 2021, LNNS 255, pp. 116–124, 2022.
https://doi.org/10.1007/978-3-030-81523-3_11

in automation and robotics, medical image analysis, marking inspection in the semiconductor industry, etc. Since in most cases natural images are used as the input data, their appropriate binarization undoubtedly plays an important role. Nevertheless, probably the most widely recognized application is the analysis of document images, hence newly developed image binarization methods, strongly influencing its results, are usually verified using some databases containing degraded document images.

Some of the most widely known datasets used for this purpose have been released by the organizers of the Document Image Binarization COmpetitions (DIBCO), held during the recent editions of the International Conference on Document Analysis and Recognition (ICDAR) and the International Conference on Frontiers in Handwriting Recognition (ICFHR). To increase the diversity of the test images used in the contests, the most recent dataset, namely DIBCO 2019, contains 20 images organized in two categories (tracks), where track A contains 10 historical machine-printed and handwritten document images with the most representative types of appearing degradations, including a combination of handwritten and machine-printed text in a single document, differently than in previous datasets.

Nevertheless, the most demanding part of the DIBCO 2019 dataset is the track B that concerns a papyri dataset containing 10 different images of ancient papyri in various states of preservation written at various antique centuries using different kinds of papyrus quality, ink and handwriting styles [19]. As the images from the DIBCO papyri dataset do not have homogeneous properties (lighting, resolution or digitization material) and they are affected both by preservation and restoration issues (holes, erased ink, dust, misplaced fibers, etc.), they are considered as the most challenging collection without any similarities with previous DIBCO datasets. To illustrate their originality and the complexity of the binarization task, different than in previous contests, the original 10 images have been presented in Fig. 1, preserving their original scale in relation to each other.

2 Related Works

Some overviews of popular binarization methods and their properties can be found in available survey papers or books, e.g. [21]. In some cases, more or less time-consuming operations are additionally required, e.g. training of deep neural networks, median filtering, morphological processing or background removal. Since our motivation is the verification of the usefulness of selected preprocessing operations in combination with some of the classical methods, we have not focused on the use of the neural networks and some other multi-stage methods in this paper, also due to recently raised time-quality efficiency challenges [12].

Comparing the binarization results with the binary ground-truth (GT) images provided in the DIBCO 2019 dataset, some of the locally adaptive methods, e.g. based on the Niblack's algorithm [16], have led to relatively good results in comparison to some other binarization methods, including the classical global Otsu's method [18]. As stated in the conclusions formulated by the organizers of

Fig. 1. The original 10 images from DIBCO 2019 dataset - track B (based on [19]).

the competition [19], "*the use of an explicit postprocessing stage which in certain cases is coupled with the use of a preprocessing stage, proving that those stages have a major impact on the success of the binarization process*".

To verify the potential advantages of the appropriate preprocessing stage, applied before the use of the previously known image binarization methods, some experiments have been conducted, assuming the use of the first preprocessing step according to the methods proposed in the papers [10,11]. A "rough" thresholding has been postulated as the first step for the initial elimination of background information based on the calculation of a global threshold using some of the properties of the histogram approximation using the Generalized Gaussian Distribution (GGD) and Gaussian Mixture Model (GMM).

3 Image Preprocessing Based on the GGD and GMM

A general assumption for the use of the proposed preprocessing images is based on the similarity of histograms of natural document images converted to greyscale with those obtained for binary images corrupted by Gaussian noise. Such similarity can also be observed for two Gaussian distributions and the histograms of the GT images with Gaussian noise [10]. Therefore, the histograms of such images may be modelled using the Generalized Gaussian Distribution (GGD) which also covers some other popular distributions, e.g. Gaussian and

Laplacian distributions, a uniform distribution, as well as an impulse function and some other special cases [8,9]. Hence, the parameters of the GGD, estimated using various methods [23], may be used for characterization of the modelled image histogram.

The probability density function of the GGD is given as [4]:

$$f(x) = \frac{\lambda \cdot p}{2 \cdot \Gamma\left(\frac{1}{p}\right)} e^{-[\lambda \cdot |x|]^p}, \tag{1}$$

where p is the shape parameter, $\Gamma(z) = \int_0^\infty t^{z-1} e^{-t} dt, z > 0$, the parameter λ is based on the standard deviation σ of the distribution and $\lambda(p, \sigma) = \frac{1}{\sigma}\left[\frac{\Gamma(\frac{3}{p})}{\Gamma(\frac{1}{p})}\right]^{\frac{1}{2}}$. For the parameter $p = 1$ the Laplacian distribution is obtained and $p = 2$ represents the Gaussian distribution. For $p \to 0$ an impulse function is obtained whereas for $p \to \infty$ the function $f(x)$ becomes a uniform distribution.

An extension of this idea is the use of the Gaussian Mixture Model (GMM) built from Gaussian distribution components with two general parameters: locations μ and standard deviations σ together with a vector defining the mixing proportions. Assuming two classes for the binary images, two Gaussian distribution components have been assumed that represented the background and the foreground text. Hence, it is assumed that the initial threshold is located between the locations μ of these two distributions and the parameters may be calculated using the iterative Expectation-Maximization (EM) algorithm based on two general steps – the first estimates the expected value for each observation and the second step optimizes the parameters of the probability distributions using the maximum likelihood [10].

Therefore, the initial step, making it possible to remove the partial background information, may be conducted by choosing a threshold located between two peaks of the approximated histogram (the use of two Gaussians in the GMM model is referred to as the GMM2), defined by their location parameters μ, as well as the location parameter of a single Gaussian distribution or the GGD, as proposed earlier [11]. After the normalization of the intensity range, such images may be considered as the input for further binarization using classical methods.

Nevertheless, in practical applications an appropriate choice of the initial threshold can be made using various approaches based on the parameters μ and σ of a single Gaussian distribution, GGD and GMM2 [10], significantly influencing final thresholding results, particularly for the most challenging images. In the conducted experiments, the following variants of thresholds X_{thr} have been used:

- location parameter μ_{GGD} of the GGD (as in the paper [11]),
- location parameter μ_G of the single Gaussian distribution,
- location parameter μ_G with subtracted Gaussian standard deviation σ_G,
- intersection point of two GMM2 curves (thr),
- higher from two location parameters of two GMM2 distributions μ_{GMMmax},
- weighted average of two GMM2 locations: $\mu_{GMM01} \cdot w_{01} + \mu_{GMM02} \cdot w_{02}$,
- weighted average of two GMM2 locations with subtracted standard deviations: $(\mu_{GMM01} - \sigma_{01}) \cdot w_{01} + (\mu_{GMM02} - \sigma_{02}) \cdot w_{02}$,

– minimum values from all above thresholds.

where the normalized weights, so that $w_{01} + w_{02} = 1$, have been determined during the GMM calculations conducted using the fast approximated method based on the standardized moment [7].

4 Discussion of Experimental Results

The experiments have been made for both categories of the DIBCO 2019 dataset, containing 10 images in each track, using the initial thresholds determined using the methods described above. Since the final results also strongly depend on the algorithms used in the second stage, several known local and adaptive thresholding methods have been used as well to determine the influence of the preprocessing and the possibly best combination of the two processing stages. All the colour images from the dataset have been converted to greyscale using the most popular ITU-R Recommendation BT.601.

The obtained results have been compared with those provided by the direct application of the binarization methods without the first preprocessing stage using commonly used methodologies [17], including F-Measure and accuracy metrics based on the number of true positive (TP) foreground pixels as well as true negatives (TN) and false positives/negatives (FP and FN respectively). The accuracy values may be calculated as

$$ACC = \frac{TP + TN}{TP + TN + FP + FN}, \tag{2}$$

whereas the F-Measure is defined as

$$FM = 2 \cdot \frac{PR \cdot RC}{PR + RC} = \frac{2 \cdot TP}{2 \cdot TP + FP + FN}, \tag{3}$$

where PR and RC denote precision (ratio of true positives to the sum of all positives) and recall (true positives to the sum of true positives and false negatives), respectively.

Additionally, the PSNR and Distance Reciprocal Distortion (DRD) [13] values have been calculated. The DRD metric may be determined by the analysis of the changed pixels' surroundings. For the n-th changed pixel it can be calculated locally using the formula:

$$DRD_n = \sum_{u=-2}^{2} \sum_{v=-2}^{2} |GT_n(u,v) - BW_n(u,v)| \cdot W(u,v), \tag{4}$$

where GT denotes the reference "ground-truth" and BW stands for the result of binarization. The normalized weight matrix W used in the formula (4) is defined as:

$$W = \begin{pmatrix} 0.0256 & 0.0324 & 0.0362 & 0.0324 & 0.0256 \\ 0.0324 & 0.0512 & 0.0724 & 0.0512 & 0.0324 \\ 0.0362 & 0.0724 & 0 & 0.0724 & 0.0362 \\ 0.0324 & 0.0512 & 0.0724 & 0.0512 & 0.0324 \\ 0.0256 & 0.0324 & 0.0362 & 0.0324 & 0.0256 \end{pmatrix}. \tag{5}$$

Hence, the final DRD value is computed as:

$$DRD = \frac{1}{K} \cdot \sum_{n=1}^{N} DRD_n \quad (6)$$

where N denotes the number of the changed pixels in the image and K is the number of the non-uniform blocks of size 8×8 pixels in the GT image (blocks containing 64 black or 64 white pixels are excluded).

Better binarization results produce lower DRD values but higher F-Measure, accuracy and PSNR. Since all metrics lead to similar conclusions, only DRD and accuracy results have been presented in Tables 1 and 2, respectively. An illustration of the influence of the preprocessing on the obtained results for the sample image from track B is presented in Fig. 2.

Table 1. The average DRD values obtained for both DIBCO 2019 subsets.

DIBCO 2019 subset	Track A			Track B (papyri)		
Method	Direct	With μ_G	Proposed	Direct	With μ_G	Proposed
Otsu [18]	9.58	7.06	5.56	209.34	18.82	13.24
Niblack [16]	22.30	12.98	6.86	39.21	16.87	14.90
Sauvola [20]	29.10	9.63	6.56	62.42	13.36	13.06
Bradley (mean) [2]	25.53	11.63	6.33	86.51	13.49	**12.47**
Chou [3]	46.31	10.65	5.83	187.08	**12.30**	13.17
Wolf [22]	37.03	10.08	6.86	76.00	13.50	13.13
Bradley (Gaussian) [2]	39.96	12.69	6.53	140.52	13.76	**12.68**
Bernsen [1]	96.92	35.55	9.21	130.75	19.96	15.94
Gatos [5]	12.08	10.50	6.76	38.14	20.40	24.17
Meanthresh	13.63	9.96	6.44	29.93	14.78	13.77
NICK [6]	**6.09**	**6.50**	**5.31**	**26.23**	**13.17**	19.00
JUCS [14]	10.09	**7.05**	**5.56**	37.47	13.49	13.36
Entropy [15]	18.09	**6.81**	**5.54**	40.33	**12.74**	**12.48**

Significantly worse results for the track B may be easily noticed for all methods as shown in Tables 1 and 2. The best results for track A may be obtained directly using the NICK method [6], which is one of the modifications of the Niblack's adaptive algorithm, and the use of the additional preprocessing is useful mainly for the lower performance methods. Nevertheless, for the papyri dataset (track B) all methods produce much worse results, considering both the DRD values and accuracy. For this subset, the advantages of the background removal by the initial thresholding using the location parameter of the Gaussian distribution μ_G are clearly visible. Due to its combination with the previously proposed entropy-based method, it is possible to achieve the best results in terms

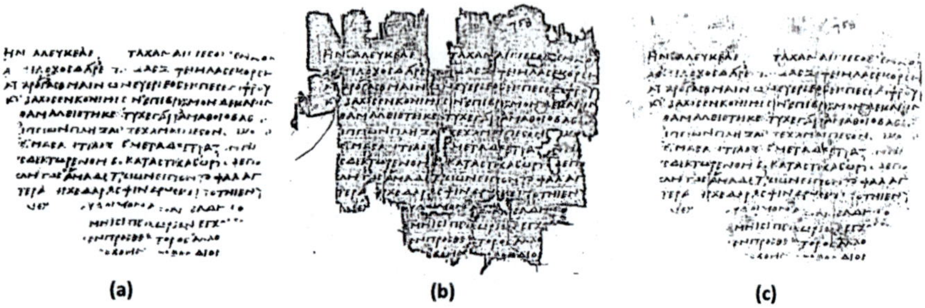

Fig. 2. Illustration of the results for the sample image from DIBCO 2019 Track B (papyri dataset): (a) - sample GT image, (b) - result of the Niblack's thresholding, (c) - result of the Niblack's binarization with the proposed preprocessing.

Table 2. The average accuracy values obtained for both DIBCO 2019 subsets.

DIBCO 2019 subset	Track A			Track B (papyri)		
Method	Direct	With μ_G	Proposed	Direct	With μ_G	Proposed
Otsu [18]	0.9475	0.9556	**0.9608**	0.4633	**0.9587**	0.9569
Niblack [16]	0.9149	0.9462	0.9579	0.8952	0.9494	0.9530
Sauvola [20]	0.8864	0.9424	0.9549	0.8380	0.9574	0.9576
Bradley (mean) [2]	0.9118	0.9486	0.9591	0.7863	0.9574	**0.9589**
Chou [3]	0.8856	0.9521	0.9607	0.5359	**0.9583**	0.9572
Wolf [22]	0.8668	0.9390	0.9531	0.8060	0.9575	0.9577
Bradley (Gaussian) [2]	0.8797	0.9450	0.9579	0.6600	0.9570	**0.9586**
Bernsen [1]	0.7060	0.8976	0.9476	0.6608	0.9447	0.9531
Gatos [5]	0.9394	0.9420	0.9461	0.8931	0.9329	0.9183
Meanthresh	0.9404	0.9517	0.9589	0.9185	0.9542	0.9557
NICK [6]	**0.9670**	**0.9655***	**0.9617**	**0.9259**	0.9392	0.9374
JUCS [14]	0.9547	**0.9594**	**0.9616**	0.9022	0.9559	0.9554
Entropy [15]	0.9306	**0.9568**	0.9607	0.8737	**0.9587**	**0.9582**

* obtained for maximum of two location parameters μ_{GMMmax} (better than for μ_G)

of the DRD and one of the best in terms of accuracy. Its application for the track A images produces results being among the top three methods as well.

To enhance the results obtained for lower performance binarization methods, as well as further improve the binarization accuracy and the DRD values for track A images, another initial threshold proposed in the paper may be successfully applied, namely the location of the first GMM curve with subtracted its standard deviation ($\mu_{GMM01} - \sigma_{01}$), leading to good results as shown in Tables 1 and 2. Its application for the papyri dataset improves mainly the results obtained by some lower performance methods, although the increase of performance for the best methods is slightly smaller than for the use of the μ_G for the initial background removal.

5 Conclusions and Future Work

The best preprocessing methods applied for the first stage of image binarization of the DIBCO 2019 dataset are different than previously considered in the papers [10,11]. Nevertheless, the use of the proposed approach makes it possible to improve the results significantly, particularly for the most challenging papyri images. Particularly encouraging results have been obtained for the combination of the considered preprocessing with previously proposed entropy-based thresholding.

To accelerate the calculations in the proposed procedure, our further research will concentrate on the development of fast histogram approximation methods based on the Monte Carlo method or polynomial approximation, as well as the consideration of the use of pixel's voting methods in combination with the proposed approach.

References

1. Bernsen, J.: Dynamic thresholding of grey-level images. In: Proceedings of 8th International Conference on Pattern Recognition (ICPR), pp. 1251–1255 (1986)
2. Bradley, D., Roth, G.: Adaptive thresholding using the integral image. J. Graph. Tools **12**(2), 13–21 (2007). https://doi.org/10.1080/2151237X.2007.10129236
3. Chou, C.H., Lin, W.H., Chang, F.: A binarization method with learning-built rules for document images produced by cameras. Pattern Recogn. **43**(4), 1518–1530 (2010). https://doi.org/10.1016/j.patcog.2009.10.016
4. Clarke, R.J.: Transform Coding of Images. Academic, New York (1985)
5. Gatos, B., Pratikakis, I., Perantonis, S.: Adaptive degraded document image binarization. Pattern Recogn. **39**(3), 317–327 (2006). https://doi.org/10.1016/j.patcog.2005.09.010
6. Khurshid, K., Siddiqi, I., Faure, C., Vincent, N.: Comparison of Niblack inspired binarization methods for ancient documents. In: Document Recognition and Retrieval XVI, vol. 7247, pp. 72470U-1–72470U-9. SPIE (2009). https://doi.org/10.1117/12.805827
7. Krupiński, R.: Approximated fast estimator for the shape parameter of generalized Gaussian distribution for a small sample size. Bull. Pol. Acad. Sci. Tech. Sci. **63**(2), 405–411 (2015). https://doi.org/10.1515/bpasts-2015-0046
8. Krupiński, R.: Reconstructed quantized coefficients modeled with generalized Gaussian distribution with exponent 1/3. Image Process. Commun. **21**(4), 5–12 (2016)
9. Krupiński, R.: Modeling quantized coefficients with generalized Gaussian distribution with exponent 1/m, m = 2, 3, ... In: Gruca, A., Czachórski, T., Harezlak, K., Kozielski, S., Piotrowska, A. (eds.) Man-Machine Interactions 5: 5th International Conference on Man-Machine Interactions, ICMMI 2017, pp. 228–237. Springer, Cham (2018). https://doi.org/10.1007/978-3-319-67792-7_23
10. Krupiński, R., Lech, P., Okarma, K.: Improved two-step binarization of degraded document images based on Gaussian mixture model. In: Krzhizhanovskaya, V., et al. (eds.) Computational Science - ICCS 2020. Lecture Notes in Computer Science, vol. 12141, pp. 467–480. Springer, Heidelberg (2020). https://doi.org/10.1007/978-3-030-50426-7_35

11. Krupiński, R., Lech, P., Tecław, M., Okarma, K.: Binarization of degraded document images with generalized Gaussian distribution. In: Rodrigues, J., et al. (eds.) Computational Science - ICCS 2019. Lecture Notes in Computer Science, vol. 11540, pp. 177–190. Springer, Heidelberg (2019). https://doi.org/10.1007/978-3-030-22750-0_14
12. Lins, R.D., Bernardino, R.B., de Jesus, D.M.: A quality and time assessment of binarization algorithms. In: Proceedings of the 15th IAPR International Conference on Document Analysis and Recognition, ICDAR 2019, Sydney, Australia, 20–25 September 2019, pp. 1444–1450. IEEE (2019). https://doi.org/10.1109/ICDAR.2019.00232
13. Lu, H., Kot, A.C., Shi, Y.Q.: Distance-reciprocal distortion measure for binary document images. IEEE Signal Process. Lett. **11**(2), 228–231 (2004). https://doi.org/10.1109/LSP.2003.821748
14. Michalak, H., Okarma, K.: Fast binarization of unevenly illuminated document images based on background estimation for optical character recognition purposes. J. Univ. Comput. Sci. **25**(6), 627–646 (2019). https://doi.org/10.3217/jucs-025-06-0627
15. Michalak, H., Okarma, K.: Improvement of image binarization methods using image preprocessing with local entropy filtering for alphanumerical character recognition purposes. Entropy **11**(2), 286 (2019). https://doi.org/10.3390/e21060562
16. Niblack, W.: An Introduction to Digital Image Processing. Prentice Hall, Englewood Cliffs (1986)
17. Ntirogiannis, K., Gatos, B., Pratikakis, I.: Performance evaluation methodology for historical document image binarization. IEEE Trans. Image Process. **22**(2), 595–609 (2013). https://doi.org/10.1109/TIP.2012.2219550
18. Otsu, N.: A threshold selection method from gray-level histograms. IEEE Trans. Syst. Man Cybern. **9**(1), 62–66 (1979). https://doi.org/10.1109/TSMC.1979.4310076
19. Pratikakis, I., Zagoris, K., Karagiannis, X., Tsochatzidis, L., Mondal, T., Marthot-Santaniello, I.: ICDAR 2019 competition on document image binarization (DIBCO 2019). In: 2019 International Conference on Document Analysis and Recognition (ICDAR), Sydney, NSW, Australia, pp. 1547–1556 (2019). https://doi.org/10.1109/ICDAR.2019.00249
20. Sauvola, J., Pietikäinen, M.: Adaptive document image binarization. Pattern Recogn. **33**(2), 225–236 (2000). https://doi.org/10.1016/S0031-3203(99)00055-2
21. Shrivastava, A., Srivastava, D.K.: A review on pixel-based binarization of gray images. In: Advances in Intelligent Systems and Computing, vol. 439, pp. 357–364. Springer, Singapore (2016). https://doi.org/10.1007/978-981-10-0755-2_38
22. Wolf, C., Jolion, J.M.: Extraction and recognition of artificial text in multimedia documents. Formal Pattern Anal. Appl. **6**(4), 309–326 (2004). https://doi.org/10.1007/s10044-003-0197-7
23. Yu, S., Zhang, A., Li, H.: A review of estimating the shape parameter of generalized Gaussian distribution. J. Comput. Inf. Syst. **21**(8), 9055–9064 (2012)

Application of Image Entropy Analysis for the Quality Assessment of Stitched Images

Krzysztof Okarma(✉) and Mateusz Kopytek

Department of Signal Processing and Multimedia Engineering, Faculty of Electrical Engineering, West Pomeranian University of Technology in Szczecin, Sikorskiego 37, 70-313 Szczecin, Poland
{okarma,km46880}@zut.edu.pl

Abstract. Image stitching is one of the relevant operations useful in virtual reality as well as in remote sensing applications which is used for the generation of panoramic images. The quality of the obtained panoramic images may be decreased due to the presence of various factors, e.g., geometric distortions, blur, ghosting, or colour distortions. Since the distortion types specific for the stitched images are different than those that may be found in the databases used for general-purpose image quality assessment, the development of new quality metrics, as well as the testing image databases, are necessary. One of the recently proposed methods for the evaluation of the stitched images is based on the comparison of 36 features of the constituent and finally stitched images, including shape parameters of the Generalized Gaussian Distribution (GGD) and the eigenvalues of a bivariate distribution obtained from the Gaussian Mixture Model (GMM). Although this method, known as Stitched Image Quality Evaluation (SIQE), provides a high correlation with subjective scores for the ISIQA dataset developed by its authors, its verification has been originally made assuming the use of only 20% of images for testing and 80% for training. Nevertheless, a noticeably smaller correlation with subjective quality scores obtained for the whole dataset may be meaningfully increased, employing an additional image entropy analysis proposed in the paper. The obtained results are encouraging and the proposed extension of the SIQE metric increases significantly both the Pearson's Linear Correlation Coefficient and Rank-Order Correlations with Mean Opinion Scores provided in the ISIQA database.

Keywords: Image quality assessment · Image stitching · Panoramic images · Image entropy

1 Introduction

The most typical methods for obtaining panoramic images utilize a number of images captured by cameras assuming that they contain some overlapping areas.

M. Choraś et al. (Eds.): CORES 2021, IP&C 2021, ACS 2021, LNNS 255, pp. 125–132, 2022.
https://doi.org/10.1007/978-3-030-81523-3_12

Hence, an appropriate choice of the distinctive image features in these regions, as well as their proper assignment, influence the overall quality of the final panoramas, similarly to some other post-processing operations such as e.g. interpolation and blending. Actually, image stitching operation may be divided into three major parts: image registration, calibration and post-processing (blending). Since there are many freeware and commercial software packages for image stitching, they might produce various quality panoramas, depending on the applied algorithms and the selected parameters during consecutive processing stages.

One of the most widely known issues in image stitching is the influence of non-identical lighting conditions in the constituent images leading to a visible seam, similarly to the impact of background changes. To deal with the problems caused by lens distortions and to ensure a sufficient number of detected keypoints for further feature matching, constituent images should have a reasonable overlap of at least 15–20%. Nevertheless, some distortions may still be visible even after the calibration, correction of vignetting issues, chromatic aberrations and final blending used to reduce the visibility of seams. Some other factors which should be considered, particularly for moving subjects, are the presence of ghosting artifacts, blur as well as visible colour and geometric distortions.

One of the dynamically developing application areas of image registration and image stitching methods is the fusion of multiple images obtained by drones equipped with cameras [3,13] as well as the other solutions regarding the Visual Simultaneous Localization and Mapping (VSLAM) in mobile robotics. Considering the vision-based control of mobile robots, the quality of the obtained images play an important role and in the case of detection of low quality stitched images, they might be rejected from further analysis.

Nevertheless, similarly as in the general-purpose image quality assessment (IQA), the development of reliable IQA metrics for stitched images is based on their correlation with subjective assessment. One of the recently proposed metrics, verified using the specially created database, known as the Indian Institute of Science Stitched Image Quality Assessment (ISIQA) dataset, has been developed by Madhusudana and Soundararajan [6]. The authors of this metric, namely Stitched Image Quality Evaluator (SIQE), have proposed the use of 36 features including 12 shape parameters of the Generalized Gaussian Distribution (GGD) and the 24 features based on the eigenvalues of a bivariate distribution obtained from the Gaussian Mixture Model (GMM).

Although SIQE achieves a high correlation with Mean Opinion Scores (MOS) for the ISIQA database, its authors have tested this metric only for the randomly chosen 20% of the dataset images. Regardless of the numerous repetitions of the random drawing of the training and test images, the correlation values for smaller datasets are usually higher, therefore in this paper, the additional verification has been made using the whole ISIQA database. Since the performance of the SIQE metric has decreased for the whole dataset, the extension of this metric with the use of image entropy features has been considered in the paper that leads to a noticeable increase of correlation of the proposed metric with subjective opinions expressed as the MOS values.

2 Brief Overview of Stitched Image Quality Assessment

Assuming the knowledge of the constituent images as well as the final stitched image, the use of the full-reference (FR) approach, being the most popular also in general-purpose IQA, seems to be possible for the evaluation of stitched images. Actually, such an approach would be equivalent to the indirect assessment of the stitching algorithm. Nevertheless, due to the unavailability of the perfect quality undistorted panorama, a comparison of features of constituent images and the stitched image cannot be considered as a "purely" FR approach. On the other hand, the use of the features of constituent source images testifies to the fact that it would not also be the typical no-reference (NR) method, since "blind" metrics rely only on the assessed image data. Finally, a specificity of distortions that may be found in image panoramas is another reason to develop new IQA methods for this purpose.

A problem of colour correction and colour balancing in the image and video stitching was considered by Xu and Mulligan [14], whereas Solh and AlRegib [11] proposed the quality assessment for multi-camera images sensitive to photometric distortions and geometric distortions. This method, known as Multi-view Image Quality Measure (MIQM), utilized luminance, contrast, spatial motion and edge-based structure components. Paalanen et al. analysed mainly the mosaicing performance [8] and Qureshi et al. [10] utilized the well-known Structural Similarity (SSIM) index [12] for the comparison of high-frequency information of the stitched and constituent images in overlapping regions.

Recently, some other methods were proposed, e.g. by Cheung et al. [2] who measured the geometric and structure errors with the use of local variance of optical flow field energy computed between the distorted and reference images combined with the intensity and chrominance gradient in highly-structured patches, respectively. A framework for the classification of colour correction methods for image stitching was presented by Bellavia and Colombo [1] with the use of improved Color Image Difference (iCID) measure [9] as well as Feature Similarity (FSIM) [15], originally applied for the general-purpose IQA. Another method based on colour contrast similarity and colour value difference was proposed by Niu et al. [7], however, this approach is also limited to the analysis of colour inconsistency issues. As stated by Hou et al. [4], the potential development of the CNN-based methods is limited by the availability of the ground truth data, hence the trained model may lead to overfitting.

An interesting more general approach to quality assessment of stitched images was presented in the article [6], being fundamental for the experiments discussed in this paper. The idea of the SIQE framework is based on the detection of features from the constituent and stitched images using steerable pyramids using marginal and bivariate statistics. Subband coefficients for marginal statistics were modelled using a Generalized Gaussian Distribution (GGD), utilizing its

Fig. 1. Sample constituent images from ISIQA dataset and the stitched image with the highest global differential entropy and visible geometrical and colour distortions.

Fig. 2. Sample constituent images from ISIQA dataset and the stitched image with the lowest global differential entropy and no visible distortions.

shape parameters as 12 features for a specified 100×100 pixels patch. To detect the presence of ghosting artifacts the analysis of bivariate statistics for neighbouring passbands was proposed with the use of the Gaussian Mixture Model (GMM). Since the increased correlation values may be found in such case, they may be reflected in 24-dimensional feature vectors obtained for constituent and stitched images. Finally, a vector containing 36 differences of features, calculated for the constituent and stitched images, is obtained, used as the input for the Support Vector Regression (SVR) function.

3 Proposed Extension and Its Experimental Verification

The analysis of the quality assessment results computed for 264 stitched images obtained for 26 scenes that are included in the ISIQA dataset, as well as for a similar number of additional stitched images generated using the freeware Hugin software with various parameters, has demonstrated the potential of further improvements of the SIQE metric. Nevertheless, due to the lack of subjective evaluations for the additionally generated images, further development of the metric has been made using only the images from the ISIQA database that contain the MOS values.

The motivation of the extension of the original SIQE metric is the influence of the presence of distortions typical for stitched images on the local and global image entropy. As expected, the presence of some distortions, such as ghosting or colour inconsistency as well as local geometric distortions as illustrated in Fig. 1, increases not only the global entropy of the stitched images but also the values of the local entropy. On the other hand, small entropy differences between the constituent and the stitched images are characteristic for high-quality images, as illustrated in Fig. 2, even if the entropy of the individual images is high due to many small visible details.

Assuming a similar approach as for the other 36 features originally applied in SIQE, based on the calculation of their differences between constituent and stitched images, the additional features have been added after the SVR step and combined with the SIQE values. Hence, two additional entropy-based features may be defined as:

$$ent_{global} = ent^{c}_{global} - ent^{s}_{global} \tag{1}$$

and

$$\overline{ent_{local}} = \overline{ent^{c}{}_{local}} - \overline{ent^{s}{}_{local}}, \tag{2}$$

where the global entropy values ent_{global} and the average local entropy values $\overline{ent_{local}}$ for constituent (c) and stitched (s) images are subtracted, respectively.

Finally, the proposed combined metrics in two variants, referred to as EntSIQE, based on the weighted sum and weighted product, have been considered:

$$\text{EntSIQE}_1 = \alpha \cdot \text{SIQE} + \beta \cdot ent_{global} + \gamma \cdot \overline{ent^{s}{}_{local}} \tag{3}$$

and

$$\text{EntSIQE}_2 = (\text{SIQE})^{\alpha} \cdot (ent_{global})^{\beta} \cdot \left(\overline{ent^{s}{}_{local}}\right)^{\gamma} \tag{4}$$

where the parameters α, β, and γ are subjected to optimization, maximizing the correlation with the MOS values provided in the ISIQA database. Nevertheless, the application of the ent_{global} features, calculated as the differences for the constituent and the stitched images, has led to worse results than those presented in Table 1.

Table 1. Correlations with subjective scores obtained for the ISIQA database.

Metric	Parameters			Correlation with MOS		
	α	β	γ	PLCC	SROCC	KROCC
SIQE	–	–	–	0.7488	0.7057	0.5308
EntSIQE$_{-1}$	0.3592	0.7938	2.3176	0.8012	0.7920	0.5971
EntSIQE$_{-2}$	0.8597	−0.0036	0.4579	0.8101	0.7945	0.5990

As the measures of correlation three commonly used approaches have been applied. The most popular Pearson's Linear Correlation Coefficient (PLCC) between the objective metric Q calculated for a set of images and their known MOS values may be defined as their covariance divided by the product of their standard deviations:

$$r = \frac{\text{cov}(Q, MOS)}{\sigma_Q \cdot \sigma_{MOS}}. \tag{5}$$

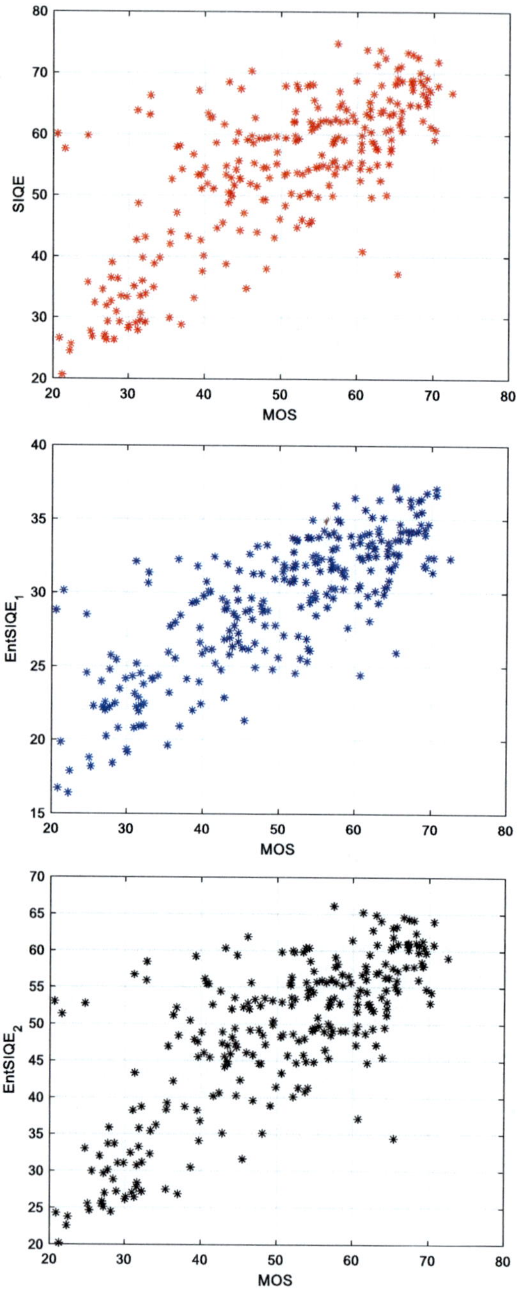

Fig. 3. Scatter plots for SIQE and two proposed variants of the EntSIQE metric.

Pearson's r coefficient generally reflects the prediction accuracy of the image quality. Due to the combination of features the additional use of the logistic function, often applied for nonlinear regression, has not led to a meaningful change of the correlation coefficient (differences below 0.003), also for the original SIQE.

The two other correlations used in experiments belong to the family of rank-order correlations. Spearman's Rank Order Correlation Coefficient (SROCC) is defined as:

$$\rho = 1 - \frac{6 \cdot \sum d_i^2}{n \cdot (n^2 - 1)}, \tag{6}$$

where d_i is the difference between the ranks of corresponding images in two sets sorted according to objective (Q) and subjective (MOS) quality scores and n is the number of images. An alternative is Kendall Rank Order Correlation Coefficient (KROCC) defined as:

$$\tau = 2 \cdot \frac{n_c - n_d}{n \cdot (n - 1)}, \tag{7}$$

where n_c and n_d are the number of concordant and discordant, respectively, that are considered as the pairs of images ordered in the same way and reversely.

The scatter plots illustrating the increase of the linearity of the correlation with subjective quality scores are shown in Fig. 3 where a more linear concentration of points representing individual stitched images in the ISIQA dataset may be noticed for the proposed EntSIQE metrics.

4 Conclusions and Future Work

The proposed extensions of the recently delivered Stitched Image Quality Evaluator (SIQE) make it possible to increase significantly the correlation of the extended objective quality metrics (EntSIQE) with the subjective evaluations due to the use of the additional entropy-based features. Obtained results are encouraging and our further research will be oriented towards the development of even better-performing quality metric utilizing some additional features as well as its validation for some other types of images, containing similar types of distortions. One of the interesting recent ideas might also be the quality assessment of parallax-tolerant image stitching methods [5]. Another direction of future work may be the development of some other models that might be applied for the combination of features.

References

1. Bellavia, F., Colombo, C.: Dissecting and reassembling color correction algorithms for image stitching. IEEE Trans. Image Process. **27**(2), 735–748 (2018). https://doi.org/10.1109/tip.2017.2757262

2. Cheung, G., Yang, L., Tan, Z., Huang, Z.: A content-aware metric for stitched panoramic image quality assessment. In: 2017 IEEE International Conference on Computer Vision Workshops (ICCVW). IEEE (2017). https://doi.org/10.1109/iccvw.2017.293
3. Duan, H., Liu, Y., Huang, H., Wang, Z., Zhao, H.: Image stitching algorithm for drones based on SURF-GHT. IOP Conf. Ser. Mater. Sci. Eng. **569**, 052025 (2019). https://doi.org/10.1088/1757-899x/569/5/052025
4. Hou, J., Lin, W., Zhao, B.: Content-dependency reduction with multi-task learning in blind stitched panoramic image quality assessment. In: 2020 IEEE International Conference on Image Processing (ICIP). IEEE (2020). https://doi.org/10.1109/icip40778.2020.9191241
5. Jung, K., Hong, J.: Quantitative assessment method of image stitching performance based on estimation of planar parallax. IEEE Access **9**, 6152–6163 (2021). https://doi.org/10.1109/access.2020.3048759
6. Madhusudana, P.C., Soundararajan, R.: Subjective and objective quality assessment of stitched images for virtual reality. IEEE Trans. Image Process. **28**(11), 5620–5635 (2019). https://doi.org/10.1109/tip.2019.2921858
7. Niu, Y., Zhang, H., Guo, W., Ji, R.: Image quality assessment for color correction based on color contrast similarity and color value difference. IEEE Trans. Circuits Syst. Video Technol. **28**(4), 849–862 (2018). https://doi.org/10.1109/tcsvt.2016.2634590
8. Paalanen, P., Kämäräinen, J.-K., Kälviäinen, H.: Image based quantitative mosaic evaluation with artificial video. In: Salberg, A.-B., Hardeberg, J.Y., Jenssen, R. (eds.) SCIA 2009. LNCS, vol. 5575, pp. 470–479. Springer, Heidelberg (2009). https://doi.org/10.1007/978-3-642-02230-2_48
9. Preiss, J., Fernandes, F., Urban, P.: Color-image quality assessment: from prediction to optimization. IEEE Trans. Image Process. **23**(3), 1366–1378 (2014). https://doi.org/10.1109/tip.2014.2302684
10. Qureshi, H., Khan, M., Hafiz, R., Cho, Y., Cha, J.: Quantitative quality assessment of stitched panoramic images. IET Image Proc. **6**(9), 1348–1358 (2012). https://doi.org/10.1049/iet-ipr.2011.0641
11. Solh, M., AlRegib, G.: MIQM: a novel multi-view images quality measure. In: 2009 International Workshop on Quality of Multimedia Experience. IEEE (2009). https://doi.org/10.1109/qomex.2009.5246953
12. Wang, Z., Bovik, A., Sheikh, H., Simoncelli, E.: Image quality assessment: from error visibility to structural similarity. IEEE Trans. Image Process. **13**(4), 600–612 (2004). https://doi.org/10.1109/tip.2003.819861
13. Xiong, P., Liu, X., Gao, C., Zhou, Z., Gao, C., Liu, Q.: A real-time stitching algorithm for UAV aerial images. In: Proceedings of the 2nd International Conference on Computer Science and Electronics Engineering (ICCSEE 2013). Atlantis Press (2013). https://doi.org/10.2991/iccsee.2013.405
14. Xu, W., Mulligan, J.: Performance evaluation of color correction approaches for automatic multi-view image and video stitching. In: 2010 IEEE Computer Society Conference on Computer Vision and Pattern Recognition (CVPR). IEEE (2010). https://doi.org/10.1109/cvpr.2010.5540202
15. Zhang, L., Zhang, L., Mou, X., Zhang, D.: FSIM: a feature similarity index for image quality assessment. IEEE Trans. Image Process. **20**(8), 2378–2386 (2011). https://doi.org/10.1109/tip.2011.2109730

Practical Applications of Computer Recognition Systems

Gyroscope-Based Remote Text Entry Interface

Adam Nowosielski[1](✉) and Patryk Krasa[2]

[1] Faculty of Computer Science and Information Technology, West Pomeranian University of Technology, Żołnierska 52, 71-210 Szczecin, Poland
anowosielski@wi.zut.edu.pl

[2] TietoEVRY, Product Development Services, Radio, Aleja Piastów 30, 71-064 Szczecin, Poland

Abstract. This paper analyzes the problem of remote text entry using a gyroscope. Its rotary movements are mapped to the screen coordinate system and employed for a virtual laser pointer behaviour. We analyze different placement for the unit and choose held in a hand smartphone with a buil-in gyroscope for further investigations. In the study, the analysis of different modes of text entry is examined, including simple point and click procedure on different layouts, and trajectory-based stroke or gesture typing.

Keywords: Gyroscope · Gesture-based user interface · Text entry · Virtual keyboard · Human Computer Interaction (HCI) · Natural User Interface (NUI)

1 Introduction

The classic human-computer interface consists of a keyboard, mouse and display. Newer solutions are based on a touch screen or contrary, are based on a touchless interaction [1]. In the latter, hand gestures or the movement of the entire body are most often used. They are recognized with held or wearable inertial sensors or with the aid of computer vision and pattern recognition approaches. Regardless the mode of interaction, the interface is always the intermediary between an electronic device and a human being and two most important tasks in this interaction include text input and pointer control.

In the paper we focus on the problem of text entry using drawn in the air gestures captured by a gyroscope. The problem of text entry is well known in the literature with many alternative proposals to the physical keyboard. One example is speech recognition [2]. The technology is researched since the 1950s and is considered mature. Beside dictation, speech recognition has a great potential in dialogue systems or voice assistants. Another proposal to writing is eyetracking. In principle, the eye tracker captures humans' gaze direction. Example of gaze-based text entry is reported in [3] where authors adopted their already proposed Dasher system [4]. The use of EEG signal for typing is in turn shown

M. Choraś et al. (Eds.): CORES 2021, IP&C 2021, ACS 2021, LNNS 255, pp. 135–144, 2022.
https://doi.org/10.1007/978-3-030-81523-3_13

in [5]. Authors proposed a brain-computer interface (BCI) for typing a letter or word using two intentions of left or right hand movement (motor imagery, MI) and a dwell time. Real, not imagery, hand movements in the air have also been considered for text entry. In [6] a special glove with reflective markers is used in order to track the position of hand and fingers. In [7] interface components are controlled by the Kinect-based hands motion tracking module. Both solutions offer word-gesture (swipe) technique on virtual onscreen keyboard. Another way to text input is by using head movements. In [8] authors analyzed the efficiency of six different virtual keyboards in head mounted devices context. In [9] head movements are captured in the thermal spectrum and the interface proposed offers hierarchical character selection procedure without any additional mechanisms for clicking events. [10] presents a tongue controlled oral interface. The paper reports a full alphabet text input using 10 inductive sensors placed on the palate of the oral cavity operated by the activation unit, glued or pierced to the tongue [10].

The above presented few examples show great variety in approaches to the text entry. Many of these proposals are dedicated to people with disabilities. Others are intended for a spontaneous or casual interaction (casual short writing). All the solutions mentioned differ in their complexity, universality and application. The cost of purchase is also different. In this context, we examine in the paper a proposal of text entry using a gyroscope. The topic is not new and some proposals have already been reported. However, it is worth to observe that now, thanks to recent advances in micro-electro-mechanical systems (MEMS), gyroscopes are combined with accelerometers, and also with magnetometers in the inertial measurement units (IMU). They are low-cost, small-sized, light-weighted and most importantly precise [11]. Many electronic devices (e.g. smartphones, smartwatches) are equipped with them. As a consequence, they find utility in human–computer interfaces. Some proposals use only accelerometer [12] which captures effectively linear gestures (e.g., up/down or left/right patterns), others employ gyroscope individually [11], best for detecting circular motions (rotation, hand waving, etc.). The combination of accelerometer and gyroscope enables recognition of both linear and rotational gestures and exemplary interfaces are reported in [13] and [14]. In [14] whole sentences are written by hand in the air based on inertial sensors attached to the back of the hand. It resembles, according to authors, a writing on an imaginary blackboard. By holding the phone like a pen, pretending the phone's corner to be the pen-tip, in the PhonePoint Pen solution [12] the user is able to write short messages in the air. The phone identifies the hand gestures as one of multiple strokes [12] using accelerometer. Similar solution is proposed in [13] where handwriting on any surface is recognized from the trajectory of the phone's 'writing' corner.

The rest of the paper is structured as follows. Section 3 presents concept of the interface. In Sect. 4 different means of text entry are presented. Experimental results are provided and discussed in Sect. 5. The paper ends with a summary.

2 Interface Concept

A three-axis electronic gyroscope measures rotation about its own three axes, i.e. angular velocity in rad/s. The principle of operation of most gyroscopes built into current smart devices (smartphone, smartwatches, etc.) is based on oscillator. A vibrating structure of micro-electro-mechanical system measures the Coriolis effect. For a user in a large-screen scenario (see example in Fig. 1a) a gyroscope can measure roll, pitch, and yaw movements (RPY) as depicted in Fig. 1b. Held in a hand smartphone can be utilized as a virtual laser pointer. Such concept has been analysed in [13] for a handwriting in the air.

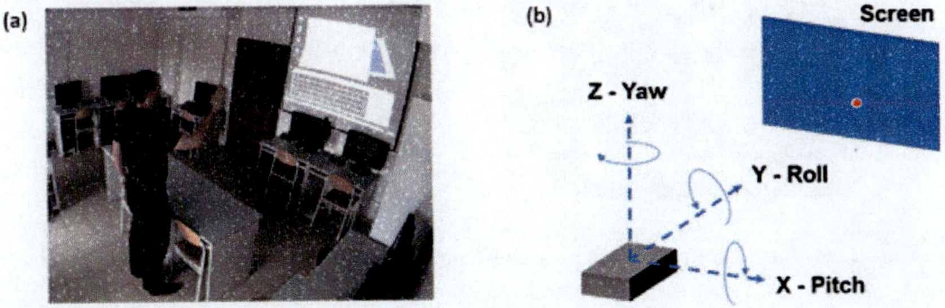

Fig. 1. Exemplary large-screen environment (a) and sketch of the gyroscope to the screen placement (b)

The RPY movements can be efficiently recognized on the basis of raw signal sample recordings or from a set of statistical parameters derived from signals [15]. In our implementation we treat a gyroscope readings as a digital signal. In order to transfer data in the form of angular velocities to data in the form of pointer movements on a two-dimensional surface, the following transformations are performed in our interface:

- the angular velocity of the X axis (Pitch) is converted into cursor movements along the Y axis of the two-dimensional screen surface (up/down movements),
- the angular velocity of the Z axis (Yaw) is converted into cursor movements along the X axis of the two-dimensional screen surface (left/right movements).

The signals are sampled at an interval of 10 ms and the mapped coordinates are refined by applying a moving average over the last five readings. We decided to employ a simple algorithm to reduce the computational complexity and minimize the delay. To compensate for differences in the distance between the user and the screen a scale factor parameter is employed. It is, in the present version, set manually and used to modify the readings of current deviation according to the appropriate axis.

It is worth to notice that such use of the gyroscope is not limited to the discussed smartphone, held in the hand. Another possibility is to use a gyroscope built into a smartwatch or a smartband placed on the wrist. In such solution the interface could be operated with natural hand gestures, similarly as in vision-based approaches but with a simpler setup and less computationally complex algorithms. The approach can also be used in the head operated interfaces. Figure 2 presents a prototype from our text entry by head movement project (under research). Similar concept but in the context of signal analysis (from both gyroscope and accelerometer) has been presented in [15]. It has been proven in [15] that signals from an IMU positioned on the user's forehead carry rich enough information to confidently recognize the head gestures.

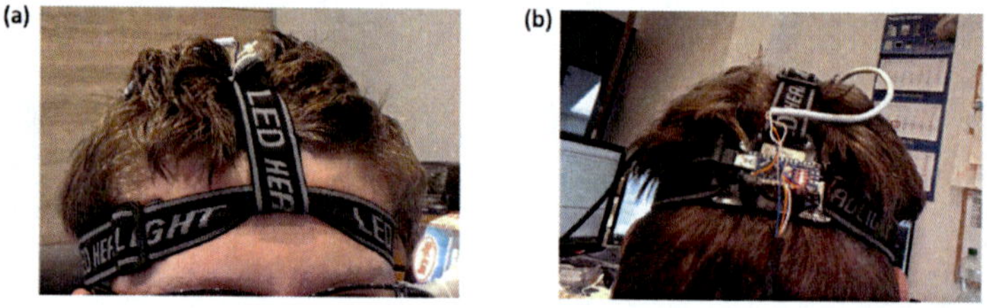

Fig. 2. Gyroscope unit placed on the user head, front (a) and back (b) view

Different placement of the sensor allows for various applications. As a wearable head mounted unit it can prove its value for people with disabilities who cannot control standard electronic equipment. Used as a hand held device can serve as a virtual laser pointer to gain control in large-display environments, intermittent or casual interaction, places where the standard keyboard is not a choice (e.g. public spaces, home cinema systems, television, presentation).

The motion readings captured by a gyroscope are translated into the movements of a pointer on a display. Typing in such an interface can be achieved through handwriting characters in the air (similarly as in [13] or through the virtual keyboard operated with a mouse pointer. Since the real-world application scenario causes that signals are measured continuously and may contain motions from real-life activities, typing is activated by pressing or pressing and holding (for swipe gestures) a button. In our study we analyzed different modes of text entry including simple point and click keyboards of different layouts, and trajectory based stroke or gesture typing. In the latter, the system analyzes a trajectory of (x, y) coordinates over time and gestures are recognized from sequence of directions.

3 Examined Modes of Text Entry

In this section, we briefly introduce keyboard layouts we considered in our examinations. We focused our attention on the most widespread QWERTY keyboard and also on other not so popular solutions. The QWERTY key arrangement has been designed in the 19th century and now more efficient and ergonomic layouts have been introduced. Despite the fact that many argue the QWERTY key arrangement is not suited for the current needs and interfaces, it still remains the standard. We examined the QWERTY layout in the key press mode and with a dwell time option (named QwertyTimer in our interface and handled with a dwell time of 750 ms).

In Fig. 3 the QWERTY layout is contrasted with Fitaly and Opti. Both Fitaly and Opti have been designed for devices operated with a stylus or a touch screen (smartphones, tablets). Their layouts, contrary to the QWERTY, take into account not only the frequency of individual letters occurrences, but also the frequency of digrams. Characters forming the most common digrams are close to each other. Performance is improved by minimizing movement distances.

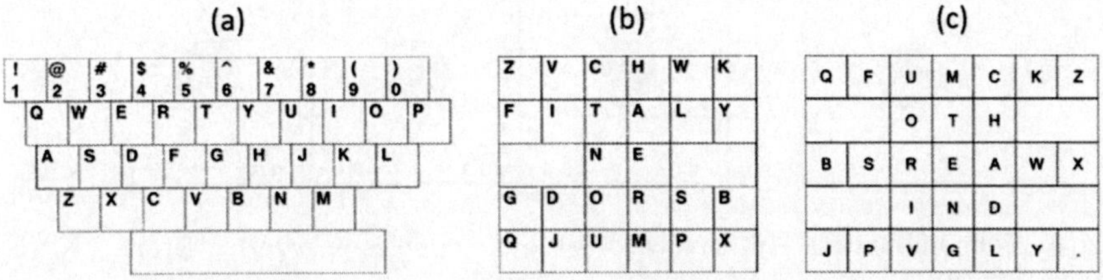

Fig. 3. QWERTY (a), Fitaly (b) and Opti (c) keyboard layouts

Another popular solution in keyboard design is the reduction of the number of buttons. The best example here is a phone numeric keypad where each numeric key is associated with several alphabet characters. A more contemporary solution based on this concept is the TapTap depicted in Fig. 4a. Access to any character is a two-step. First, user selects a button with grouped characters. Then, the interface changes, expanding letters from the selected group into separate buttons (see Fig. 4b). However, minimizing finger or stylus movements increases the number of clicks needed.

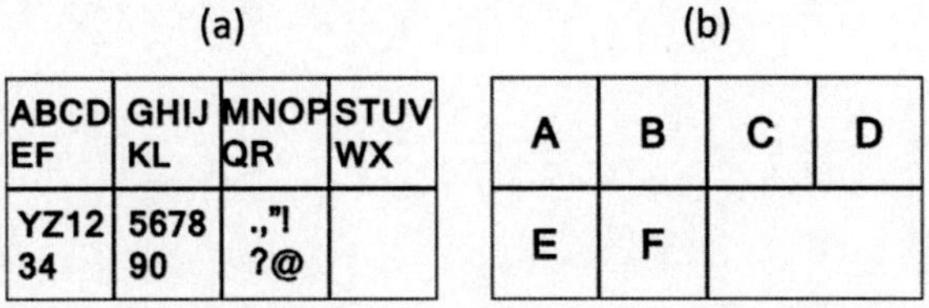

Fig. 4. TapTap keyboard layout (a) and preview of the interface after selecting the first button (b)

The next group we investigated can be characterized as gesture, stroke or swipe based keyboards. They share a common feature which is a form of specific movement required to character entry. In the Quikwriting [16] there are 9 zones, arranged in 3×3 grid (see Fig. 5a). It has been introduced for mobile platforms. The interaction starts from the resting zone (the central zone) and the user moves to one of the eight outer zones, optionally to the neighbouring outer zones, and finishes in the central zone [16]. Entry and return corresponds to the character in the middle of the given zone. When optional movement is applied to the neighbouring zones it denotes consecutive characters in the given direction from the starting zone. This approach allows its user to write entire words in a single continuous gesture [16]. Figure 5c–d presents successively entered letters: ‘K’, ‘O’ and ‘T’.

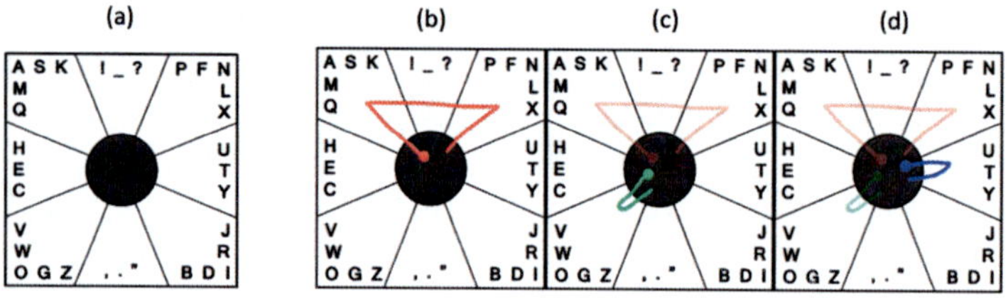

Fig. 5. The Quikwriting [16] keyboard (a) and examples of entering letters: ‘K’ (b), ‘O’ (c), ‘T’ (d)

The Quikwriting layout must have inspired the designers of the 8pen solution who had reduced the number of zones by 4, leaving the central zone and 4 directional zones (see Fig. 6a).

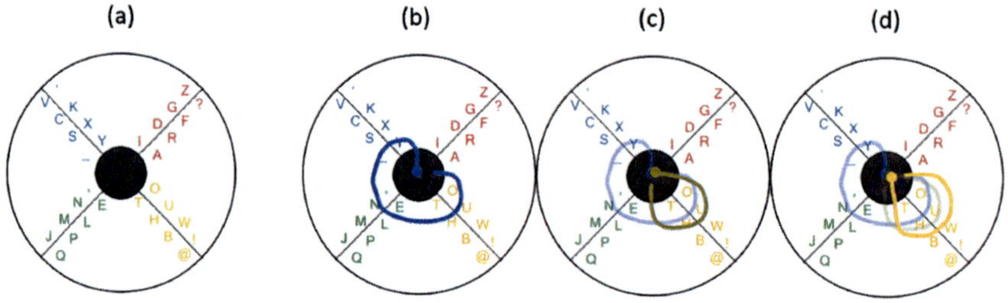

Fig. 6. The 8pen solution (a) and examples of entering letters: ‘K’ (b), ‘O’ (c), ‘T’ (d)

In the 8pen, the characters are grouped in fours on either side of the zone boundary. After leaving the central zone, user selects the zone with required character. The position of the letter on the edge (counting from the central zone) indicates how many sectors must be crossed in the clockwise or anticlockwise fashion (depends on which border the letter is situated). Examples are presented in Fig. 6c–d.

The Graffiti solution, originally invented for a personal digital assistant (PDA) by Palm Inc., employs a simplified character alphabet. It relies on the simplification of movement or elimination of additional strokes, like in the 'A' character. These simplifications facilitate the recognition task but requires from a user to learn the entry details. Figure 7a presents strokes required for specific characters and Fig. 7b shows part of our interface including the gesture hint sheet and the input window.

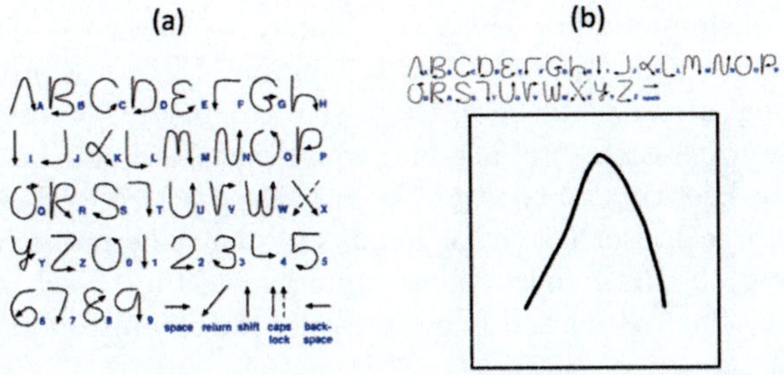

Fig. 7. The Graffiti solution (a) and examples of entering the 'A' character

The next group considered are keyboards that combine the pressing mechanism with the stroke technique. Figure 8a depicts the MessagEase approach. Characters highlighted in red are obtained by clicking, while the remaining ones are obtained by clicking and swiping in the appropriate direction to the one of the adjacent buttons. In the 5-Tiles, presented in Fig. 8b, the underlined characters on each button are obtained with a pressing, and the stroke gesture allows to select the remaining characters. Compared with the TapTap approach, two taps have been replaced with a single key press or a press with a swipe gesture.

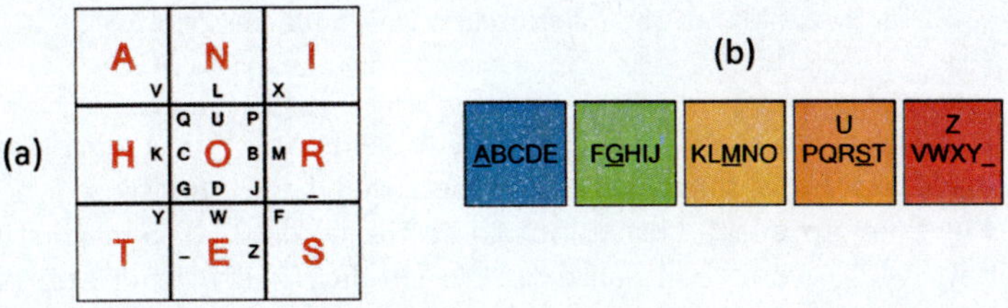

Fig. 8. MessagEase (a) and 5-Tiles (b) approaches

The above approaches have been supplemented with a one row keyboard with alphabetical arrangement of characters. It must be noted, that such keyboard requires movement in only one dimension.

4 Evaluation

To validate the gyroscope-based remote text entry we conducted user-based experiments. 10 volunteers participated in the study, all of them computer science students, participants of the Human-Computer Interaction course. It can be stated that all of them were good typist with fluent use of the QWERTY layout keyboard. Most of them have the opportunity to learn about other layouts but with no or little practice. At the beginning, each participant was asked to perform rehearsals, practicing cursor control on the screen. Then, user was presented with each method of text entry. Finally, we measured the time of text entry by each mode using the well known pangram 'the quick brown fox jumps over the lazy dog' and a sentence in the native language. The user had time to become familiar with each interface and could make any number of attempts. As the result the best time of typing of both sentences correctly was considered. The session with each user lasted for about an hour. The results are presented in Fig. 9 and Fig. 10. Both charts show typing speeds in the chars per minute (CPM) measure. The red dashed line represents the mean values.

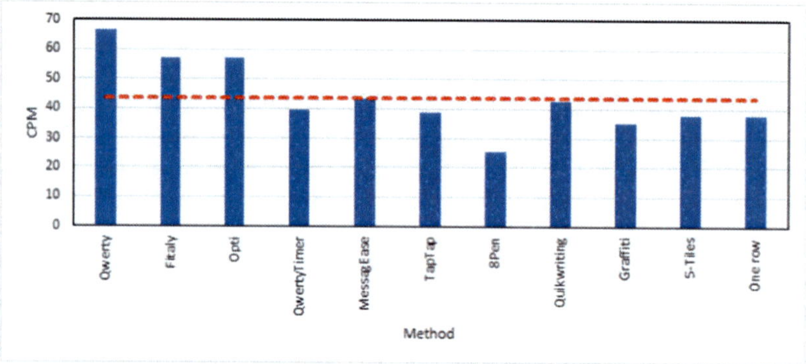

Fig. 9. Results of the experiments (methods perspective)

The high score for the QWERTY layout keyboard is not surprising. The QwertyTimer variant is burdened with the dwell time and scored below the average. The 8pen approach turned out to be the weakest. It is interesting since the result was 40% worse compared to the similar Quikwriting approach. All of the stroke or gesture based approaches performed below average with only the MessageEase close to the average. These results show that point and click technique seems to be best suited for the gyroscope-based text entry and 2-axes mapping procedure. As for the users, they show similar performance with the best participant faster by 22.3% from the slowest. The manner the experiments have been conducted, where users typed same sentences with unconstrained number of repetitions, was aimed at imitation a good familiarity with the interface without the need of a long practice.

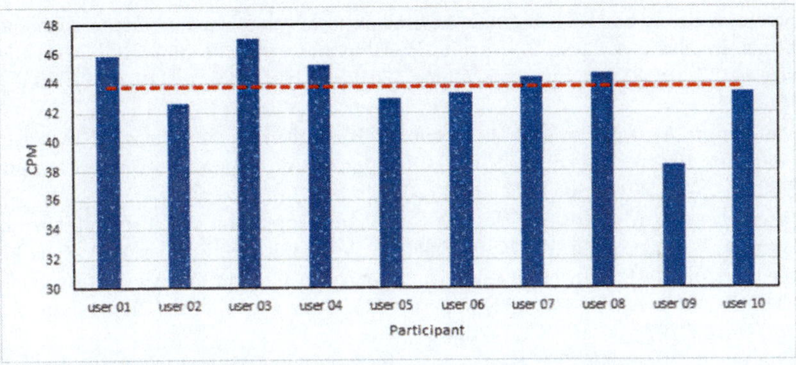

Fig. 10. Results of the experiments (users perspective)

5 Summary

In the paper, the problem of remote text entry using a gyroscope has been addressed. We focused on smartphones, since most modern devices are equipped with a built-in gyroscope. The article presents the interface where held in a hand smartphone is utilized as a virtual laser pointer. The rotary movements of the device are mapped to screen pointer motion. We examined various means of text entry and found that point and click approaches outperform those of stroke and gesture-based in the adopted setup. The solution may find application in casual interaction, large-screen environments or as an assistive tool dedicated to people with disabilities. In future, we plan to evaluate the gyroscope-based solution in head-operated interface.

Author contributions. A.N.: research concept, system architecture, methodology, supervision, literature review, data analysis, manuscript preparation. P.K.: software implementation, conducting experiments.

References

1. Małecki, K., Nowosielski, A., Kowalicki, M.: Gesture-based user interface for vehicle on-board system: a questionnaire and research approach. Appl. Sci. **10**(18), 6620 (2020)
2. Malik, M., Malik, M.K., Mehmood, K., Makhdoom, I.: Automatic speech recognition: a survey. Multimed. Tools Appl. (2020) https://doi.org/10.1007/s11042-020-10073-7
3. Ward, D.J., MacKay, D.J.C.: Fast hands-free writing by gaze direction. Nature **418**, 838 (2002). https://doi.org/10.1038/418838a
4. Ward, D.J., Blackwell, A.F., Mackay, D.J.C.: Dasher: a gesture-driven data entry interface for mobile computing human-computer interaction. Hum. Comput. Interact. **17**, 199–228 (2002)

5. Sreeja, S.R., et al.: BCI augmented text entry mechanism for people with special needs. In: Basu, A., Das, S., Horain, P., Bhattacharya, S. (eds.) IHCI 2016. LNCS, vol. 10127, pp. 81–93. Springer, Cham (2017). https://doi.org/10.1007/978-3-319-52503-7_7
6. Markussen, A., Jakobsen, M.R., Hornbæk, K.: Vulture: a mid-air word-gesture keyboard. In: Proceedings of the SIGCHI Conference on Human Factors in Computing Systems, CHI 2014, pp. 1073–1082 (2014)
7. Wierzchowski, M., Nowosielski, A.: Swipe text input for touchless interfaces. In: Burduk, R., Jackowski, K., Kurzyński, M., Woźniak, M., Żołnierek, A. (eds.) Proceedings of the 9th International Conference on Computer Recognition Systems CORES 2015. AISC, vol. 403, pp. 619–629. Springer, Cham (2016). https://doi.org/10.1007/978-3-319-26227-7_58
8. Dobosz, K., Popanda, D., Sawko, A.: Head-based text entry methods for motor-impaired people. In: Gruca, A., Czachórski, T., Deorowicz, S., Harężlak, K., Piotrowska, A. (eds.) ICMMI 2019. AISC, vol. 1061, pp. 3–11. Springer, Cham (2020). https://doi.org/10.1007/978-3-030-31964-9_1
9. Nowosielski, A., Forczmański, P.: Touchless typing with head movements captured in thermal spectrum. Pattern Anal. Appl. **22**, 841–855 (2019). https://doi.org/10.1007/s10044-018-0741-0
10. Lontis, E.R., Andreasen Struijk, L.N.: Mapping sensor activation time for typing tasks performed with a tongue controlled oral interface. In: 35th Annual International Conference of the IEEE EMBS, pp. 5911–5913 (2013). https://doi.org/10.1109/EMBC.2013.6610897
11. Han, H., Yoon, S.W.: Gyroscope-based continuous human hand gesture recognition for multi-modal wearable input device for human machine interaction. Sensors **19**(11), 2562 (2019). https://doi.org/10.3390/s19112562
12. Agrawal, S., Constandache, I., Gaonkar, S., Choudhury, R.R., Caves, K., DeRuyter, F.: Using mobile phones to write in air. In: Proceedings of the 9th International Conference on Mobile Systems, Applications, and Services (MobiSys 2011), New York, NY, USA, pp. 15–28. Association for Computing Machinery (2011). https://doi.org/10.1145/1999995.1999998
13. Deselaers, T., Keysers, D., Hosang, J., Rowley, H.A.: GyroPen: gyroscopes for pen-input with mobile phones. IEEE Trans. Hum. Mach. Syst. **45**(2), 263–271 (2015). https://doi.org/10.1109/THMS.2014.2365723
14. Amma, C., Georgi, M., Schultz, T.: Airwriting: hands-free mobile text input by spotting and continuous recognition of 3d-space handwriting with inertial sensors. In: 2012 16th International Symposium on Wearable Computers, Newcastle, UK, pp. 52–59 (2012). https://doi.org/10.1109/ISWC.2012.21
15. Borowska-Terka, A., Strumillo, P.: Person independent recognition of head gestures from parametrised and raw signals recorded from inertial measurement unit. Appl. Sci. **10**(12), 4213 (2020). https://doi.org/10.3390/app10124213
16. Perlin, K.: Quikwriting: continuous stylus-based text entry. In: Proceedings of the 11th Annual ACM Symposium on User Interface Software and Technology, pp. 215–216. ACM (1998)

Comparison of the Effects of Different Preprocessing Methods on Homogeneity Assessment of Digital Intraoral Radiographs of Root Canal Fillings

Jagoda Łebska[1], Barbara Obuchowicz[2], Rafał Obuchowicz[3], and Adam Piórkowski[1(✉)]

[1] Department of Biocybernetics and Biomedical Engineering, AGH University of Science and Technology, Mickiewicza 30 Avenue, 30-059 Kraków, Poland
pioro@agh.edu.pl

[2] Department of Conservative Dentistry with Endodontics, Jagiellonian University Collegium Medicum, Montelupich. 4, 31-155 Kraków, Poland

[3] Department of Diagnostic Imaging, Jagiellonian University Medical College, Kopernika 19, 31-501 Kraków, Poland

Abstract. Root canal treatment is a very common method in the treatment of pulpitis, however, filled root canals do not last forever, which means a tooth on which this has been performed will eventually have to be extracted. Therefore, the homogeneity of a filling is very important because the better it is, the longer the tooth will survive; however, there is no universal method of assessing the homogeneity of a filling. In this paper, the use of textural parameters for qualitative assessment of root canal fillings is proposed; these include the homogeneity of the gray levels in an image or long-run emphasis. The use of different preprocessing methods can be helpful to obtain even better results. Three different methods of preprocessing are tested: histogram equalisation, CLAHE and CLAHE + histogram equalisation.

Keywords: Digital intraoral radiography · Texture analysis · Homogeneity · Root canal treatment · CLAHE · Histogram equalisation

1 Introduction

Root canal treatment is used in cases of irreversible pulpitis. It consists of cleaning the pulp chamber and dental canals, thoroughly disinfecting them in order to remove all bacteria, and then filling the canals with sealing materials. To fulfill its function, the filling must fully adhere to the walls of the canal and be as homogeneous as possible in order to avoid the penetration of bacteria or other pathogenic agents into the tooth [5].

M. Choraś et al. (Eds.): CORES 2021, IP&C 2021, ACS 2021, LNNS 255, pp. 145–152, 2022.
https://doi.org/10.1007/978-3-030-81523-3_14

The success of this treatment largely depends on correct radiological evaluation because radiographs are performed at every stage from diagnosis to post-treatment examinations [10]. A good homogeneity assessment based on digital intraoral radiographs can help reduce the need for additional tests, which means the patient is exposed to less radiation. The use of textural parameters could allow qualitative assessment of such fillings, thus making it possible to determine the optimal methods for successful treatment [3].

Texture analysis has also been a subject of other studies in the field of dentistry [11]. In the study [9] textural entropy was presented as a potential feature for assessment of jaw bone healing process and study [2] shows the use of texture parameters in evaluation of dental implant stability. In the studies [1,4] texture analysis was used for illustration and characterization of healing process in bone regeneration. Using different preprocessing methods could help achieve better evaluation, thereby making the proposed homogeneity assessment method even more promising.

2 Materials and Methods

2.1 Used Data

The dataset used in this research consisted of 102 radiographs (DICOM format, RVG, pixel spacing: 0.019×0.019); these were divided into 3 sets of 34 for each of the tested preprocessing methods [12]. The three preprocessing methods were used: CLAHE [14], HEQ (histogram equalisation, [8]) and CLAHE+HEQ (Fig. 1).

2.2 Methods

Regions of Interest. The first step in the research was the selection of regions of interest. This was done with the help of MaZda software [17,18]. Every root canal was selected and labeled separately. The labels included the patient number, the number of the canal, and (for patients for which there were multiple photos) the number of the photo, e.g. P01K01 for the first canal of the first patient. In total, 94 canals were studied.

Selected Textural Parameters. Two textural parameters were selected for the homogeneity assessment. The first was the homogeneity of the gray levels in an image; this parameter is determined from a gray level co-occurrence matrix (GLCM) [6,7,13], which is given by the formula (1):

$$Homogenity = \sum_{i=0}^{N-1} \sum_{j=0}^{N-1} \frac{p(i,j)}{1+(i-j)^2} \tag{1}$$

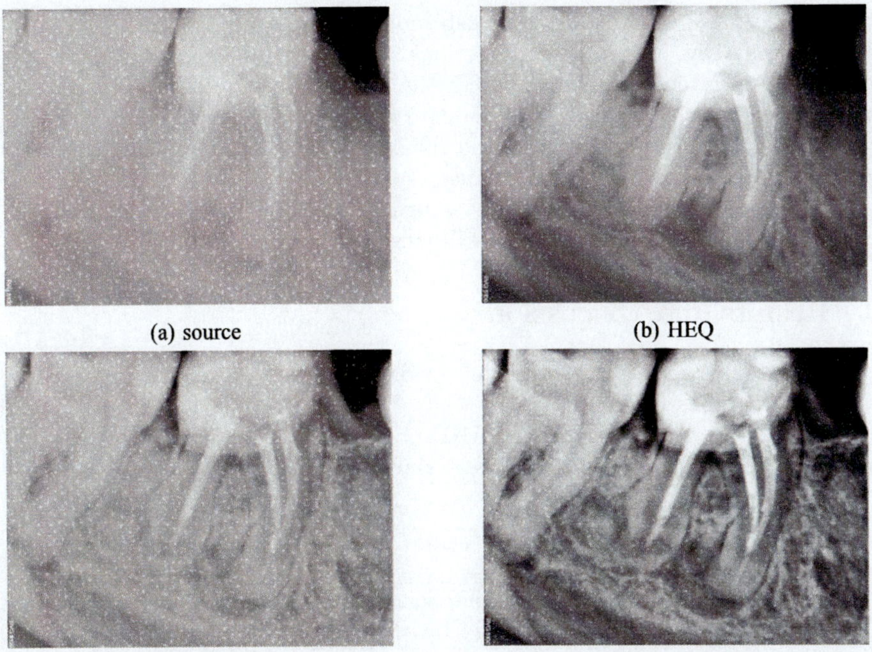

(a) source (b) HEQ

(c) CLAHE (d) CLAHE+HEQ

Fig. 1. The example of radiograph after HEQ (b), CLAHE (c) and CLAHE+HEQ (d) preprocessing.

where:

- N - size of the co-occurrence matrix;
- $p(i, j)$ - probability of the occurrence of pixels with brightness i and j at a given distance.

The second selected parameter was long-run emphasis, which is determined from a gray level run-length matrix (GLRLM) [15,16]. This gives information about the distribution of long homogeneous runs in an image and is given by the formula (2):

$$LongREmph = \frac{1}{A}\sum_{g=1}^{N}\sum_{\forall l} l^2 P(g,l) \tag{2}$$

where:

- A - number of homogeneous runs in the region of interest;
- $P(g, l)$ - number of runs of pixels with a value of g and length of l.

Both of these parameters were determined for the vertical and horizontal directions because it is in these directions that fillings are applied and discontinuities most often occur. For GLCM, the smallest distances between pairs of points were ignored in order to eliminate any inhomogeneities that could be caused by noise. After calculating the values of the parameters, the arithmetic mean for each of them was determined; subsequently, the values of the long-run emphasis were normalized to the range 0–1 by dividing each value by the maximum. Finally, the average of both parameters was determined [19].

3 Results and Discussion

All the obtained results are presented below in the tables (Tables 1, 2) and in the form of graphs (Figs. 3, 4).

The table above shows that CLAHE preprocessing achieves both the highest mean homogeneity and the largest standard deviation between canals. In

Table 1. Comparison of the homogeneities for all three preprocessing methods

	Homogeneity for HEQ	Homogeneity for CLAHE	Homogeneity for CLAHE+HEQ
Mean	0.397338	0.541488	0.34611409
Std	0.092749	0.104137	0.092347917

Table 2. Distribution of the values within one patient.

Patient	Mean	Std	Patient	Mean	Std
P01	0.423657339	0.142129028	P15	0.354690476	0.03669385
P02	0.378402722	0.095235145	P16	0.392874544	0.101762156
P03	0.256669789	0.005079705	P17	0.390368596	0.001213577
P04	0.440358876	0.010841445	P18	0.317643782	0.030120911
P05T01	0.360668799	0.038005763	P19	0.26595454	0.055048624
P05T02	0.317893772	0.023344115	P20	0.23798857	0.024907093
P06	0.268171945	0.014978925	P21	0.384419681	0.028742122
P07	0.293821839	0.010501876	P22	0.461360466	0.135068779
P08T01	0.432751729	0.106176899	P23	0.348281477	0.021692547
P08T02	0.330091076	0.032279236	P24	0.307672311	0.041898414
P09	0.43340344	0.040245039	P25	0.31465653	0.027887493
P10	0.276678039	0.05627211	P26	0.304153516	0.036642477
P11	0.328559195	0.053158939	P27T01	0.411594754	0.142924479
P12	0.288640079	0.027038891	P28	0.322179834	0.105423177
P13	0.323668316	0.000953234	P27T02	0.306012559	0.019285679
P14	0.388121688	0.1242744	P29	0.351830018	0.029520344

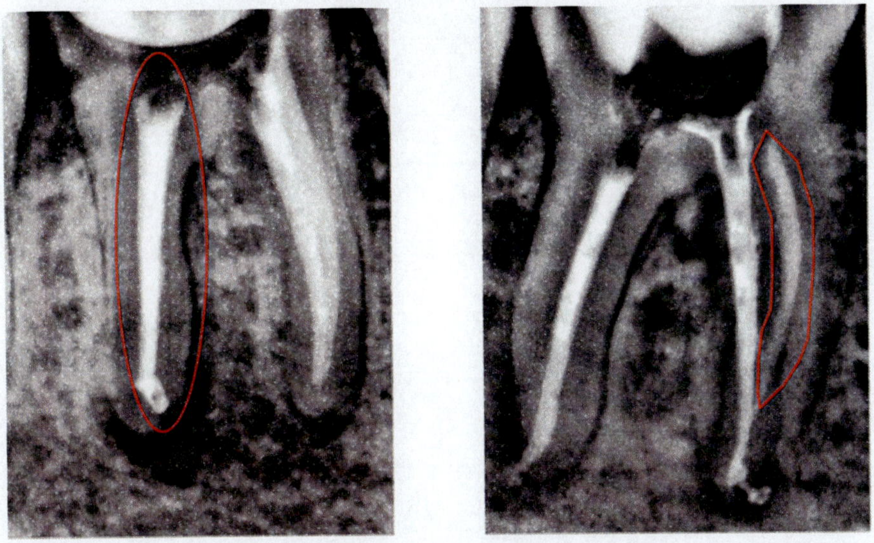

Fig. 2. Canals with the highest (a) and lowest (b) homogeneity of the filling.

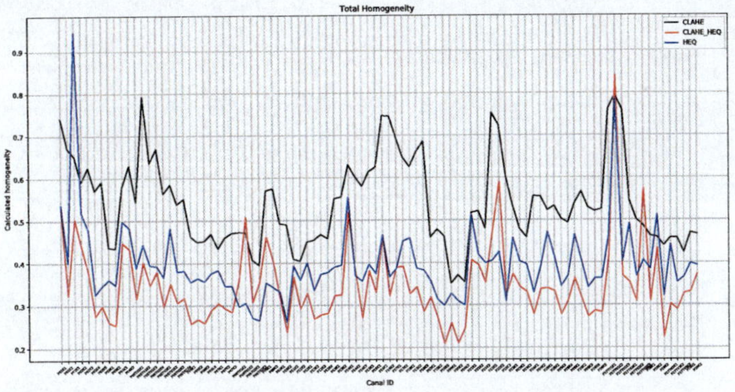

Fig. 3. Total homogeneity for all three methods of preprocessing.

contrast, CLAHE+HEQ preprocessing achieves the lowest mean homogeneity and the smallest deviation (Fig. 2). It can therefore be concluded that the most reliable method of testing homogeneity is CLAHE+HEQ preprocessing before analysis of photos. Better contrast results in more accurate testing and better detection of inhomogeneities.

In order to check how the canal homogeneity values are distributed within one patient, a table showing their mean values and standard deviations was constructed (Table 2).

This study shows that using different preprocessing methods can affect the results of homogeneity assessment. The best and most reliable option is

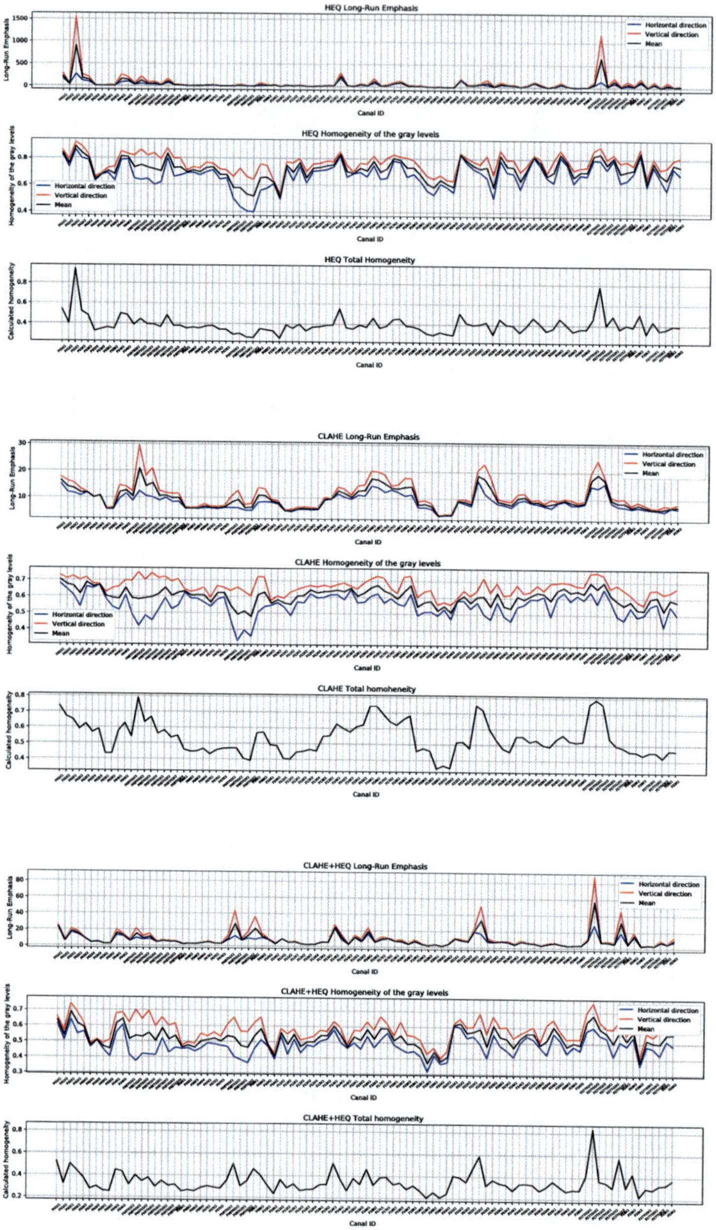

Fig. 4. Results for the radiographs after HEQ (a), CLAHE (b) and CLAHE+HEQ (c) preprocessing.

CLAHE+HEQ preprocessing because it is the most accurate method (Table 1). The use of textural parameters as a method of assessing homogeneity may have great applications in the future, but this would require further refinement to achieve better results.

Acknowledgement. This work was financed by the AGH – University of Science and Technology, Faculty of EAIIB, KBIB no 16.16.120.773.

References

1. Borowska, M., Bębas, E., Szarmach, J., Oczeretko, E.: Multifractal characterization of healing process after bone loss. Biomed. Signal Process. Control **52**, 179–186 (2019)
2. Borowska, M., Szarmach, J.: Evaluation of dental implant stability using radiovisiographic characterization and texture analysis. In: Pietka, E., Badura, P., Kawa, J., Wieclawek, W. (eds.) ITIB 2019. AISC, vol. 1011, pp. 304–313. Springer, Cham (2019). https://doi.org/10.1007/978-3-030-23762-2_27
3. Claus, E.B., Calvocoressi, L., Bondy, M.L., Schildkraut, J.M., Wiemels, J.L., Wrensch, M.: Dental X-rays and risk of meningioma. Cancer **118**(18), 4530–4537 (2012)
4. Girejko, G., Borowska, M., Szarmach, J.: Statistical analysis of radiographic textures illustrating healing process after the guided bone regeneration surgery. In: Pietka, E., Badura, P., Kawa, J., Wieclawek, W. (eds.) ITIB 2018. AISC, vol. 762, pp. 217–226. Springer, Cham (2019). https://doi.org/10.1007/978-3-319-91211-0_19
5. Goldberg, M.: Root canal treatment (RCT): from traditional endodontic therapies to innovating pulp regeneration. J. Dent. Oral Disord. Ther. **4**(2), 1–6 (2016)
6. Hall-Beyer, M.: GLCM texture: a tutorial. Natl. Council Geogr. Inf. Anal. Remote Sens. Core Curric. **3**, 75 (2000)
7. Haralick, R.M., Shanmugam, K., Dinstein, I.H.: Textural features for image classification. IEEE Trans. Syst. Man Cybern. **6**, 610–621 (1973)
8. Ketcham, D.J.: Real-time image enhancement techniques. Image Process. **74**, 120–125 (1976)
9. Kołaciński, M., Kozakiewicz, M., Materka, A.: Textural entropy as a potential feature for quantitative assessment of jaw bone healing process. Arch. Med. Sci. **11**(1), 78–84 (2015)
10. Kosińska, K., Sobecka, M.: Aspekty radiologiczne w leczeniu endodontycznym. Nowa Stomatologia **3**, 128–134 (2013)
11. Myciński, P., Dobroś, K., Kaczmarzyk, T., Zarzecka, J.: Assessment of conformity between clinical (radiological) and histopathological diagnoses of chronic inflammatory periapical lesions treated with apicoectomy. Folia Med. Cracov., 103–111 (2020)
12. Obuchowicz, R., Nurzynska, K., Obuchowicz, B., Urbanik, A., Piórkowski, A.: Use of texture feature maps for the refinement of information derived from digital intraoral radiographs of lytic and sclerotic lesions. Appl. Sci. **9**(15), 2968 (2019)
13. Pantic, I., Pantic, S., Basta-Jovanovic, G.: Gray level co-occurrence matrix texture analysis of germinal center light zone lymphocyte nuclei: physiology viewpoint with focus on apoptosis. Microsc. Microanal. **18**(3), 470 (2012)
14. Pizer, S.M., et al.: Adaptive histogram equalization and its variations. Comput. Vis. Graph. Image Process. **39**(3), 355–368 (1987)

15. Snitkowska, E.: Analiza tekstur w obrazach cyfrowych i jej zastosowanie do obrazów angiograficznych. Ph.D. thesis, The Institute of Control and Computation Engineering (2005)
16. Strzelecki, M., Materka, A.: Tekstura obrazów biomedycznych: Metody analizy komputerowej. Wydawnictwo Naukowe PWN (2017)
17. Szczypiński, P.M., Klepaczko, A., Kociołek, M.: QMaZda - software tools for image analysis and pattern recognition. In: 2017 Signal Processing: Algorithms, Architectures, Arrangements, and Applications (SPA), pp. 217–221. IEEE (2017)
18. Szczypiński, P.M., Strzelecki, M., Materka, A., Klepaczko, A.: MaZda - a software package for image texture analysis. Comput. Methods Progr. Biomed. **94**, 66–76 (2009)
19. Whitworth, J.: Methods of filling root canals: principles and practices. Endod. Top. **12**(1), 2–24 (2005)

Impact of Software Bug Report Preprocessing and Vectorization on Bug Assignment Accuracy

Lukasz Chmielowski[1,2(✉)] and Michal Kucharzak[2]

[1] Nokia Solutions and Networks sp. z o.o., ul. Rodziny Hiszpanskich 8, 02-685 Warszawa, Poland
lukasz.chmielowski@nokia.com

[2] Wroclaw University of Science and Technology, ul Wybrzeze Stanislawa Wyspianskiego 27, 50-370 Wroclaw, Poland

Abstract. Bug handling processes aimed at efficient defect resolution provisioning is an important part of a software development lifecycle and usually has a very formal process definition in modern and professional large software development organizations. Improvements of such process may include automated bug assignment, which is a task of selecting a correct development team for further investigations of a bug report. As bug reports contain lots of natural language descriptions, the bug assignment becomes a non-trivial task, especially in testing of large-scale projects or complex systems. This research focuses on natural language preprocessing and vectorization impact on accuracy of bug report assignment based on real data captured in large software development projects. Experimentation results cover stemming and lemmatization techniques applied for bug description preprocessing and *term frequency – inverse document frequency* (TF-IDF) parametrization as vectorization method.

Keywords: Bug assignment · Bug triaging · Bug report · Software bug · Text analysis

1 Introduction

Software testing is an inherent part of a software development lifecycle. It becomes more and more resource consuming with the growth of the system or product complexity. During software testing processes faults or bugs might occur and their proper handling is a crucial task in order to satisfy time-to-market requirements as well as to improve software quality. Bug report assignment to appropriate developers or development teams for its further handling is known as a "bug triage" [17,18], which may also include prioritizing of the bugs [1,19]. In software development practice bug tracking systems have been widely used to manage bugs and bug reports [11]. Such systems are used by users to report problems or bugs in an organized and process-defined manner. Bug report repositories may be used for data mining.

M. Choraś et al. (Eds.): CORES 2021, IP&C 2021, ACS 2021, LNNS 255, pp. 153–162, 2022.
https://doi.org/10.1007/978-3-030-81523-3_15

Discussed problem focuses on assigning a software bug to a correct development team automatically, based on details included in a software bug report. Such a report usually comprises natural language-based and free form text title and description as well as extra data or predefined parameters from a limited set of options, as software release or product type. Natural language title most likely provides the brief summary of the observed bug, whereas a written description brings a more detailed specification of the issue. The description shall give the developers and other parties specified information about any adverse event that occurred, e.g., test steps needed to reproduce the fault, comparison of actual vs expected results with reasoning, references to historical results or broken requirements etc.

In large software development projects, a quality and accuracy of a bug report description strongly depends on the system complexity. The same end-to-end effect or bug may be spotted by different tests in different contexts and the bug description often varies on test engineer's subjective perception, experience, correctness of observations, typos, grammar or linguistic style. Therefore, the accurate assignment of a software bug report based on natural language descriptions becomes a non-trivial task and requires studies on natural language processing towards deployment of ML-assisted automated tools. The paper focuses on impact of natural language preprocessing and vectorization methods of software bug report on bug assignment accuracy.

The main research contributions of this work comprise numerical experiments based on real data captured in a large-scale software development organization. Bug report repository contains descriptions of numerous faults reported in English by testing teams located in many different countires, what also influences the natural language style used in the free form texts titles and descriptions of the bugs.

2 Related Works

Typical text in fault reports usually comprises a free form text including also parts of internal log prints, source code, abbreviations or specific technical words. Natural language processing has been addressed in multiple studies devoted to software bug-related research. Software bug specific named entity recognition with the use of linear chain Conditional Random Fields was proposed in [20]. Work [12] described machine learning tool using xgboost classifier that automatically assigns a product and component for each new Mozilla Firefox bug based on free form texts. The tool uses title and comment of fault report processed with natural language pipeline. Authors of [10] present an approach with usage of stopwords removal, stemming and Naive Bayes classifier to predict reassignments of bug reports on data from various Mozilla and Eclipse products. Bug report assignment based on Bayesian learning has been also applied in [4]. Analysis of possible usage of *term frequency – inverse document frequency* (TF-IDF) and Naive Bayes classifier for predicting severity level of bugs was conducted in [6]. The authors proposed to create dedicated dictionary with critical terms, which should help in deciding on the choice of the severity level. Moreover, there are also plethora of research on applying machine learning to determine whether a bug is security related or not [2,5]. Another bug security classification paper by Microsoft [14] compared three techniques: Naive Bayes, logistic regression and boosted decision trees. Dual-output deep neural network for automated bug triaging on team and developer level has been introduced by University of Toronto and Intel experts in [3]. Automatic bug assignment and automatic fault localization was also a subject of PhD dissertation in Ericsson [8]. Support Vector Machine applied to finding duplication of bug reports is presented in [16]. In [9] *fastText* is used in order to solve a problem of qualifying tickets as bugs, feature requests, improvement requests, documentation requests, refactoring requests or others.

This paper focuses on comparison of preprocessing (stemming and lemmatization techniques) and vectorization (TF-IDF) impact on bug assignment accuracy with the use of logistic regression.

3 Natural Language Processing Pipeline

3.1 Typical Natural Language Processing Pipeline

Typical natural language pipeline consists of a couple of phases including preprocessing, vectorization and finally labelling, classification or clustering. The

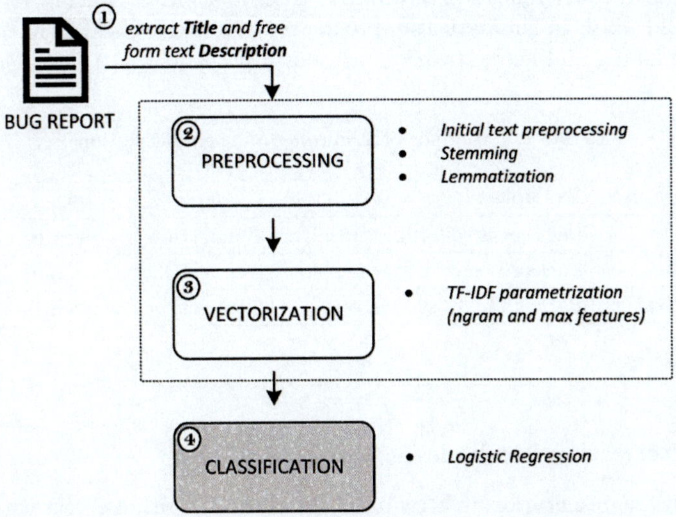

Fig. 1. Diagram of a bug assignment flow including natural language pipeline. Preprocessing and vectorization are subject of the numerical experiments

purpose of preprocessing is to clean a raw text by, e.g., removing special characters or removing inflectional suffixes. In next steps, vectorization creates representation of text data in a form of vector of numeric data. Eventually, numeric data sets are used by classifiers.

Figure 1 illustrates a generic diagram of an automatic bug assignment flow including typical natural language pipeline further discussed in the paper. An input text containing title and description fields ① is extracted directly from a bug report.

3.2 Preprocessing Methods

Preprocessing phase ② (see Fig. 1) prepares the natural language-based title and description for further processing by the following initial operations:

1. change the letters to lowercase,
2. remove special characters,
3. remove stop words.

Commonly used standard methods like changing letter to lowercase were applied (step 1). In step 2, special characters (e.g., $, #, etc.) are removed to leave only alphanumerical characters and underscore symbol. Next in step 3, stop words, like `the`, `or` are removed.

Depending on the variant of the experiment, **stemming** or **lemmatization** were additionally applied in preprocessing phase. The purpose of the stemming processes is to reduce inflection in words to their root forms. Stem is not required to be a valid word. In lemmatization, root word is the canonical form [7]. Examples of stemming and lemmatization are shown in Table 1.

Table 1. Examples of stemming and lemmatization

Operation used	Example 1	Example 2
None	Package is missing at receiver	Connection has been established
Stemming	Packag is miss at receiv	Connect ha been establish
Lemmatization	Package be miss receiver	Connection have be establish

3.3 Vectorization Methods

As numeric representation is a required input to algorithms, vectorization ③ is used to transform natural language text to its vecotr-based numerical form. Typical numerical representations are bags of words (multiset of words), binary frequency (set of words), term frequency (TF) or term frequency – inverse document frequency (TF-IDF) [15], where a term may contain one or more words. Moreover, natural language texts may be represented with the use of n-gram sequences, i.e., consecutive n neighbor items (e.g., words) are treated as a single term. For the sake of simplicity, a single term is defining a single feature. TF representation is shown in Eq. (1), where t is a term and d is a document (e.g., a bug report). TF-IDF representation is presented in formula (2), where IDF component is defined by Eq. (3), where P represents a corpus, i.e., a set of all documents d.

Additionally, a number of terms may be constrained by an extra vectorization parameter called `max features`, which limits feature space.

Finally, after preprocessing and vectorization, classification ④ of the bug report is completed with the use of Logistic Regression.

$$\mathrm{TF}_{t,d} = \frac{\text{no. of occurrence of } t \in d}{\text{total length of } d} \tag{1}$$

$$\text{TF-IDF}_{t,d} = \mathrm{TF}_{t,d} \times \mathrm{IDF}_t \tag{2}$$

$$\mathrm{IDF}_t = log_{10} \frac{\text{no. of documents } \in \ P}{\text{no. of documents of } P \text{ containing term } t} \tag{3}$$

4 Numerical Experiments

4.1 Research Questions

The goal of the numerical experiments is to evaluate the preprocessing techniques and TF-IDF vectorization parametrization in the context of the following research questions:

- What is an impact of stemming and lemmatization on bug assignment accuracy?
- What is an impact of different settings of n-gram and max features of TF-IDF on bug assignment accuracy?
- What is an impact of additional information from bug report related to product type on bug assignment accuracy?

4.2 Description and Experimental Protocol

The data used for the research come from five biggest company departments. Due to trade secrets no details about quantity of samples could be provided, but information about distribution of data and split into train and test set is shown in Table 2. Training set contains data from a 12 months range, whereas test set comprises newer bug reports collected during consecutive 1 month. Titles and descriptions of the bug reports contain a lot of technical words, abbreviations and software-specific configuration parameters names or values. All of the performed experiments used the same train and test set and bug assignment accuracy is verified further with the use of Logistic Regression. Initial text preprocessing is described in Sect. 3.2 and is a common step in all the experiments. Following measures are calculated: accuracy, precision, recall, F1. The reasons for selecting the metrics:

- accuracy is a fundamental global metric in classification problems,
- precision is a fundamental metric in information retrievals systems,
- recall that determines sensitivity,
- F1 is an additional balanced metric between precision and recall.

Table 2. Distribution of data between company departments

Department name	Percent of reports from training set	Percent of reports from test set
Department A	27.3%	19.7%
Department B	25.7%	32.5%
Department C	16.4%	17.5%
Department D	15.8%	13.5%
Department E	14.9%	16.9%

Table 3. Comparision of results when changing way of preprocessing

Preprocessing option	Acc	Precision	Recall	F1
Lemma	0.728	0.742	0.728	0.728
Lemma (+product)	0.735	0.749	0.735	0.736
Stem	0.722	0.733	0.722	0.723
Stem (+product)	0.728	0.741	0.728	0.730
None	0.720	0.732	0.720	0.722
None (+product)	0.725	0.737	0.720	0.722

4.3 Results and Lessons Learned

Table 3 contains the experimentation results showing accuracy, weighted precision, weighted recall, weighted F1 measures for different options of preprocessing. Lemmatization and stemming techniques are compared against the basic preprocessing without those options. Additionally, besides free form title and natural language description, product type identification was considered as an extra feature described in the bug report. The results show that the bug assignment accuracy increases in case product type is in use, by about 1.0%, 0.8% and 0.7% for lemmatization, stemming and without those methods, respectively. Regardless the use of additional dimensions, i.e., product type from the bug report, lemmatization guarantees the best results, exceeding the stemming by about 1% and basic preprocessing by 1.4%.

Impact of vectorization by the means of TF-IDF parametrization on bug assignment accuracy is shown in Table 4 and in Fig. 2. The best results are

Table 4. Comparision of results when changing way of vectorizing using TF-IDF

Vectorizing option	Acc	Precision	Recall	F1
Max features: 0.5k, n-gram range: 1–1	0.641	0.656	0.641	0.64
Max features: 0.5k, n-gram range: 1–2	0.605	0.62	0.605	0.604
Max features: 0.5k, n-gram range: 1–3	0.562	0.579	0.562	0.56
Max features: 1k, n-gram range: 1–1	0.681	0.703	0.681	0.685
Max features: 1k, n-gram range: 1–2	0.652	0.668	0.652	0.653
Max features: 1k, n-gram range: 1–3	0.648	0.663	0.648	0.648
Max features: 2k, n-gram range: 1–1	0.699	0.716	0.699	0.701
Max features: 2k, n-gram range: 1–2	0.691	0.709	0.691	0.693
Max features: 2k, n-gram range: 1–3	0.688	0.705	0.688	0.689
Max features: 4k, n-gram range: 1–1	0.715	0.73	0.715	0.717
Max features: 4k, n-gram range: 1–2	0.714	0.729	0.714	0.715
Max features: 4k, n-gram range: 1–3	0.697	0.712	0.697	0.699
Max features: 8k, n-gram range: 1–1	0.733	0.747	0.733	0.735
Max features: 8k, n-gram range: 1–2	0.718	0.732	0.718	0.719
Max features: 8k, n-gram range: 1–3	0.714	0.731	0.714	0.716
Max features: 16k, n-gram range: 1–1	0.732	0.745	0.732	0.734
Max features: 16k, n-gram range: 1–2	0.733	0.746	0.733	0.735
Max features: 16k, n-gram range: 1–3	0.726	0.74	0.726	0.727
Max features: 32k, n-gram range: 1–1	0.732	0.746	0.732	0.734
Max features: 32k, n-gram range: 1–2	0.737	0.751	0.737	0.739
Max features: 32k, n-gram range: 1–3	**0.738**	**0.753**	**0.738**	**0.739**
Max features: 64k, n-gram range: 1–1	0.729	0.743	0.729	0.73
Max features: 64k, n-gram range: 1–2	0.738	0.753	0.738	0.739
Max features: 64k, n-gram range: 1–3	0.737	0.751	0.737	0.738

Table 5. Detailed results for experiment with usage of TF-IDF max features: 32k, n-gram range: 1–3

Department	Precision	Recall	F1
Department A	0.560	0.711	0.627
Department B	0.787	0.864	0.824
Department C	0.759	0.683	0.719
Department D	0.829	0.629	0.716
Department E	0.846	0.671	0.749

obtained from calculations with parameters set to max features: 32k, n-gram range: 1–3; and 64k with n-gram range 1–2. Generally, incrementing the maximum number of features yields improvements in accuracy; after reaching 32k features, the results stabilises and no additional gain is observed. Adding bigrams and trigrams for values of max features up to 16k usually affects the accuracy, whereas for bigger number of features (32k or 64k), increasing n-grams supports result improvements.

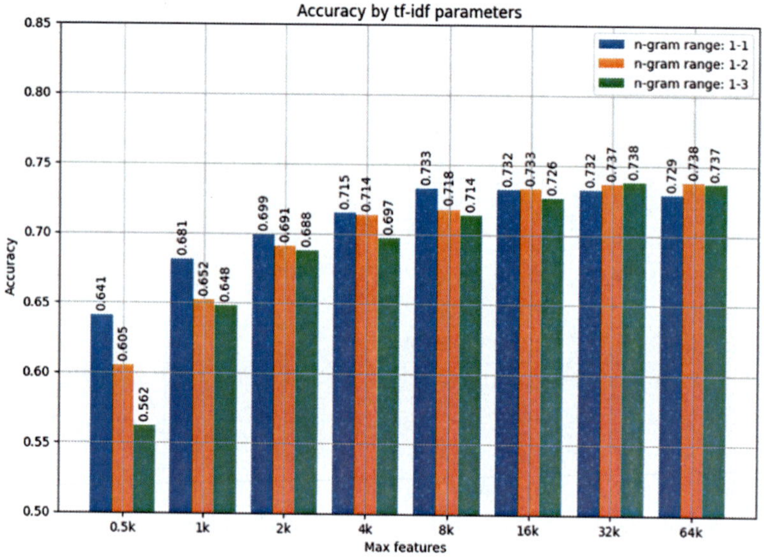

Fig. 2. Accuracy by TF-IDF parameters

Detailed view of results of a single experiment with lemmatization and information about product type and TF-IDF with max features: 32k, n-gram range: 1–3 are presented in Table 5. Comparing the results between classes (departments or development groups), it is noticeable that for Department A, the metrics are significantly lower than for the other departments. This might be related to changes in distribution between train and test set shown in Table 2; and to the fact that keywords (terms) related to the department are not as much specific as to others.

5 Summary

The paper discussed the results of machine learning simulations of predictions of assigning bug reports to department which should be responsible for further handling the reported problem. Experiments were carried out for different settings of preprocessing and vectorization. The following metrics were used: accuracy, weighted precision, weighted recall, weighted F1 measures. Although no significant difference is observed when switching between stemming and lemmatization, a slight improvement is noticeable in while the latter is applied along with the extra feature in the form of product type identification. Nevertheless, the use of preprocessing methods still increases the accuracy of the bug report assignment in comparison to only initial text preprocessing. Usage of bigrams and trigrams in TF-IDF vectorization phase yields accuracy improvement comparing to unigrams; also increasing the max features parameter value improves the results, however the results stabilises at 32k and applying bigger range of features does not guarantee further growth in accuracy.

Future works will comprise a comparison of additional vectorization techniques including *fastText*, *Skip-Gram*, *CBOW* [13] and their impact on bug report accuracy, as well as imact of different classification methods based on preprocessed and vectorized inputs.

Acknowledgement. This work has been carried out in cooperation between NOKIA and Wroclaw University of Science and Technology in context of a Ph. D. grant under the fourth edition of the "Implementation Doctorate Programme".

References

1. Alazzam, I., Aleroud, A., Al Latifah, Z., Karabatis, G.: Automatic bug triage in software systems using graph neighborhood relations for feature augmentation. IEEE Trans. Comput. Soc. Syst. **7**(5), 1288–1303 (2020). https://doi.org/10.1109/TCSS.2020.3017501
2. Behl, D., Handa, S., Arora, A.: A bug mining tool to identify and analyze security bugs using Naive Bayes and TF-IDF. In: 2014 International Conference on Reliability Optimization and Information Technology (ICROIT), pp. 294–299 (2014). https://doi.org/10.1109/ICROIT.2014.6798341
3. Choquette-Choo, C.A., Sheldon, D., Proppe, J., Alphonso-Gibbs, J., Gupta, H.: A multi-label, dual-output deep neural network for automated bug triaging. In: 2019 18th IEEE International Conference on Machine Learning and Applications (ICMLA), pp. 937–944 (2019). https://doi.org/10.1109/ICMLA.2019.00161
4. Cubranic, D., Murphy, G.: Automatic bug triage using text categorization. In: SEKE (2004)
5. Goseva-Popstojanova, K., Tyo, J.: Identification of security related bug reports via text mining using supervised and unsupervised classification. In: 2018 IEEE International Conference on Software Quality, Reliability and Security (QRS), pp. 344–355 (2018). https://doi.org/10.1109/QRS.2018.00047
6. Gujral, S., Sharma, G., Sharma, S., Diksha: Classifying bug severity using dictionary based approach. In: 2015 International Conference on Futuristic Trends on Computational Analysis and Knowledge Management (ABLAZE), pp. 599–602 (2015). https://doi.org/10.1109/ABLAZE.2015.7154933
7. Jabeen, H.: Stemming and Lemmatization in Python (2018). https://www.datacamp.com/community/tutorials/stemming-lemmatization-python
8. Jonsson, L.: Machine learning-based bug handling in large-scale software development. Ph.D. thesis, Linköping Studies in Science and Technology (2018)

9. Kallis, R., Di Sorbo, A., Canfora, G., Panichella, S.: Ticket tagger: machine learning driven issue classification. In: 2019 IEEE International Conference on Software Maintenance and Evolution (ICSME), pp. 406–409 (2019). https://doi.org/10.1109/ICSME.2019.00070
10. Lamkanfi, A., Demeyer, S.: Predicting reassignments of bug reports - an exploratory investigation. In: 2013 17th European Conference on Software Maintenance and Reengineering, pp. 327–330 (2013). https://doi.org/10.1109/CSMR.2013.42
11. Liu, K., Beng Kuan Tan, H., Zhang, H.: Has this bug been reported? In: 2013 20th Working Conference on Reverse Engineering (WCRE), pp. 82–91 (2013). https://doi.org/10.1109/WCRE.2013.6671283
12. M. Castelluccio, S.L.: Teaching machines to triage Firefox bugs. https://hacks.mozilla.org/2019/04/teaching-machines-to-triage-firefox-bugs/
13. Mikolov, T., Chen, K., Corrado, G., Dean, J.: Efficient estimation of word representations in vector space (2013)
14. Pereira, M., Kumar, A., Cristiansen, S.: Identifying security bug reports based solely on report titles and noisy data. In: 2019 IEEE International Conference on Smart Computing (SMARTCOMP), pp. 39–44 (2019). https://doi.org/10.1109/SMARTCOMP.2019.00026
15. Roul, R.K., Sahoo, J.K., Arora, K.: Modified TF-IDF term weighting strategies for text categorization. In: 2017 14th IEEE India Council International Conference (INDICON), pp. 1–6 (2017). https://doi.org/10.1109/INDICON.2017.8487593
16. Tsuruda, A., Manabe, Y., Aritsugi, M.: Can we detect bug report duplication with unfinished bug reports? In: 2015 Asia-Pacific Software Engineering Conference (APSEC), pp. 151–158 (2015). https://doi.org/10.1109/APSEC.2015.33
17. Zaidi, S.F.A., Lee, C.G.: Learning graph representation of bug reports to triage bugs using graph convolution network. In: 2021 International Conference on Information Networking (ICOIN), pp. 504–507 (2021). https://doi.org/10.1109/ICOIN50884.2021.9333902
18. Zaidi, S.F.A., Lee, C.G.: One-class classification based bug triage system to assign a newly added developer. In: 2021 International Conference on Information Networking (ICOIN), pp. 738–741 (2021). https://doi.org/10.1109/ICOIN50884.2021.9334002
19. Zhang, W.: Efficient bug triage for industrial environments. In: 2020 IEEE International Conference on Software Maintenance and Evolution (ICSME), pp. 727–735 (2020). https://doi.org/10.1109/ICSME46990.2020.00082
20. Zhou, C., Li, B., Sun, X., Guo, H.: Recognizing software bug-specific named entity in software bug repository. In: 2018 IEEE/ACM 26th International Conference on Program Comprehension (ICPC), pp. 108–10,811 (2018)

On Machine Learning for Autism Prediction from Functional Connectivity

Moises Silva[1(✉)] and Manuel Graña[2]

[1] Universidad Mayor de San Andrés, La Paz, Bolivia
[2] Computational Intelligence Group UPV/EHU, San Sebastian, Spain

Abstract. Autism Spectrum Disorder (ASD) has claimed big attention in order to find biomarkers in brain connectivity alterations detected in resting state functional magnetic resonance imaging (rs-fMRI) data by the induction of classification models from functional connectivities. In this paper we provide a comprehensive exploration of the impact of feature extraction/selection methods, classification model, connectivity measure and atlas parcelation on the predictive performance. We find, that principal component analysis (PCA) and factor analysis (FA) methods provide a significant improvement.

Keywords: Autism spectrum · Brain functional connectivity

1 Introduction

A wide variety of machine learning or pattern recognition studies to the prediction of autism spectrum disorders (ASD) have been reported on the basis of a plethora of information sources, including structural, diffusion and functional magnetic resonance imaging (MRI), as well as electroencephalography (EEG), additional demographic and clinical data [17], and behavioral measurements captured by computer vision or other body measurement approaches [8].

Brain connectivity analysis based on resting state functional MRI (rs-fMRI) can be done on the basis of brain parcellations into a set of regions of interest (ROIs) which can be defined either by anatomical guidelines or by data driven unsupervised segmentation [2]. rs-fMRI connectivity analysis has been accepted as source of information for the discovery of biomarkers of psychiatric disorders [24] such as schizophrenia and ASD. Predictive analysis approach to the search of brain connectivity based biomarkers is illustrated in Fig. 1. Connectivity matrices are extracted from rs-fMRI data as follows: a parcellation of the brain is defined; the time series corresponding to the voxels in each region of the parcellation are aggregated into one representative time series often by averaging; the connectivity matrix is built computing the similarity among the representatives of each pair of regions in the parcellation. Hence the connectivity matrix is always a symmetric matrix which can be interpreted as the adjacency matrix of a graph representing the relations among brain regions. The connectivity matrix is then

M. Choraś et al. (Eds.): CORES 2021, IP&C 2021, ACS 2021, LNNS 255, pp. 163–172, 2022.
https://doi.org/10.1007/978-3-030-81523-3_16

used as the raw data for machine learning processes which may involve feature extraction/selection and the training/testing of classification models often in a cross-validation scheme. Finally, the cross-validation classification performance results may be used to identify biomarkers in the connectivity matrix. We deal with the prediction of ASD vs. TD using the functional connectivity matrix computed from rs-fMRI data as the sole source of information for classification. We have carried out extensive cross-validation experiments on different choices of connectivity measure, brain atlas, feature selection/extraction algorithms and classification model building.

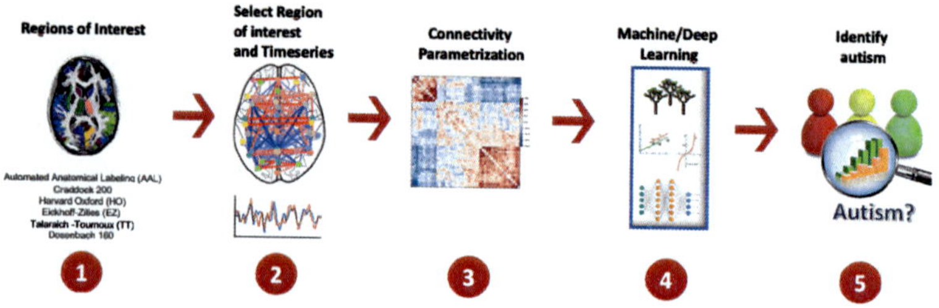

Fig. 1. Functional connectome predictive analysis pipeline steps after rs-fMRI data preprocessing (not shown): 1) given a parcellation of the brain, 2) obtain the representative time series of each region by averaging the time series of voxels within the region, 3) build the connectivity matrix by computing a similarity measure between each pair of representative time series, 4) carry out cross-validation experiments, using Machine Learning algorithms for feature extraction and classifier training; 5) report test results on the prediction of the ASD vs TD.

2 Materials and Methods

The software and the data needed to replicate the experiments reported in this paper have been published[1].

2.1 The Dataset

The dataset analyzed in the study is the Autism Brain Imaging Data Exchange (ABIDE)[2] providing a publicly available dataset of rsfMRI acquisitions of subjects diagnosed with ASD and ND, data of 1112 subjects. This dataset collects data from 10 sites as detailed in Table 1, with 1112 subjects. We have excluded cases with diagnosis as Asperger or PDD-NOS according to the DSM_IV_TR. We have selected and processed 884 subjects (ASD n = 408, TDn = 476). The selection of the subjects varies among studies for various reasons.

[1] https://github.com/mmscnet/Impact-feature-extraction-in-Autism.
[2] http://fcon_1000.projects.nitrc.org/indi/abide/.

2.2 Brain Parcellations

We have used the pre-processed acquisitions available as part of the Pre-processed Connectome Project (http://preprocessed-connectomes-project.org/abide/) [4]. The raw rs-fMRI data has been processed using the Configurable Pipeline for the Analysis of Connectomes (C-PAC) (http://fcp-indi.github.io/) in order to obtain the corrected and spatially normalized rs-fMRI volumes. On the one hand, we consider in this paper the anatomically guided parcellations such as the Tailarach and Tournoux (TT), Eickhoff-Zilles [11], Harvard Oxford (HO) [9], and theAutomated Anatomical Labeling (AAL) [18,22] defined fromn the brain segmentation of selected control populations. On the other hand, we consider parcellations are produced from the segmentation of the rs-fMRI time series of the brain volume using clustering techniques, such as the Dosenbach [10] and Craddok [6] parecellations.

2.3 Connectivity Matrices

Recalling Fig. 1, the first step of our computational pipeline is the estimation of the connectivity matrices from the time series representatives of the brain parcellations. We have considered five similarity metrics to build them available from the nilearn python package (https://nilearn.github.io/modules/generated/nilearn.connectome.ConnectivityMeasure.html). At the root of these computations is the robust estimation of the covariance matrix of the time series. We use the Ledoit-Wolf shrinkage estimator [15] following methodological recommendations in [1,7]. We consider the following connectivity measures: The covariance matrix computed using the Ledoit-Wolf shrinkage estimator [15], the Pearson Correlation Coefficient (PCC) among each pair of ROI time series, which is computed as the normalization of the covariance matrix [25], the precision computed as the inverse of the covariance matrix, the partial correlation obtained regressing out all other connections for each pair of regions [19], the tangent space representation of the matrices obtained by whitening them [23].

2.4 Classifier Model Building Methods

We have used several of the classifier building methods that are available from the open and free source Python library scikit-learn v0.22 (https://scikit-learn.org/). Specifically we have used the following: Random Forest (RF), K-Nearest Neighbors (KNN), Gaussian Naive Bayes (GNB), Support Vector Classifier (SVC) sometimes with variable selection procedure based on an ANOVA analysis [5], and we consider both sparse (ℓ_1) and non-sparse (ℓ_2) regularization terms, Logistic regression (LR) with both sparse (ℓ_1) and non-sparse (ℓ_2) regularizations, Least absolute shrinkage and selection operator (LASSO) [20], Ridge Classifier (RC), Bayesian Ridge Classifier (BRC) [21], Multi-Layer Perceptron (MLP) [13] with both the adam and the L-BFGS solvers with 5 and 10 hidden layers.

2.5 Feature Extraction

We have examined the impact of several dimensional reduction procedures which are either feature extraction or feature selection techniques. Feature extraction usually involve some transformation of the feature space where the meaning of the original variables is lost. Feature selection preserves some of the original variables discarding others. As feature extraction techniques we have applied the following ones available in the scikit-learn Python package: Probabilistic Principal Component Analysis (PCA), Isometric Mapping (Isomap), Local Linear Embedding (LNE), Multi-Dimensional Scaling (MDS), and Factor Analysis (FA).

2.6 Feature Selection

As a feature selection process we propose an approach that uses the PCC [16] between each connection value across subjects and the class label variable valued $\{-1, 1\}$, we compute the empirical distribution of the absolute values of these connection-label correlations. Finally, we select the connections falling in the selected upper percentile, discarding low correlated connections. We results results with percentiles 90%, 95%, 96%, 97%, 98%, and 99%.

2.7 Classification Performance Report

We carry out repetitions of the 10-fold cross-validation, where feature extraction/selection parameters are always computed only on the training dataset and applied in the test dataset to avoid double dipping issues [3,14]. We report the area under the receiver operating characteristic (ROC) (AUC) [12] as a more general and robust measure of the classifier performance. The accuracy is determined by the actual decision threshold applied to classify the test samples, while the ROC plots the balance of false positives (FP) versus TP across the entire range of decision threshold values. Similar to [7] we report best results of the median, 5% and 95% percentile values of the cross-validation repetitions results instead of the average value as a better description of their distribution.

3 Results and Discussion

Figure 2 summarizes the impact of feature selection/extraction methods versus the classifiers. It can be easily appreciated that the effect of feature extraction is greater than that of the classifier of choice, specifically the factor analysis, PCA and correlation based selection are quite comparable feature computation methods. Figure 3 summarizes the impact of feature computation versus the brain parcelation applied to extract brain regions. Results show the greater effect of feature computation methods, with the same qualitative results as above. Figure 4 summarizes the impact of classifiers versus the chosen connectivity measure, the results for the tangent and correlation measures show the greatest variances. Other than that theclassifiers do not have a clear ranking, except for the lower performance of k-NN and random forest.

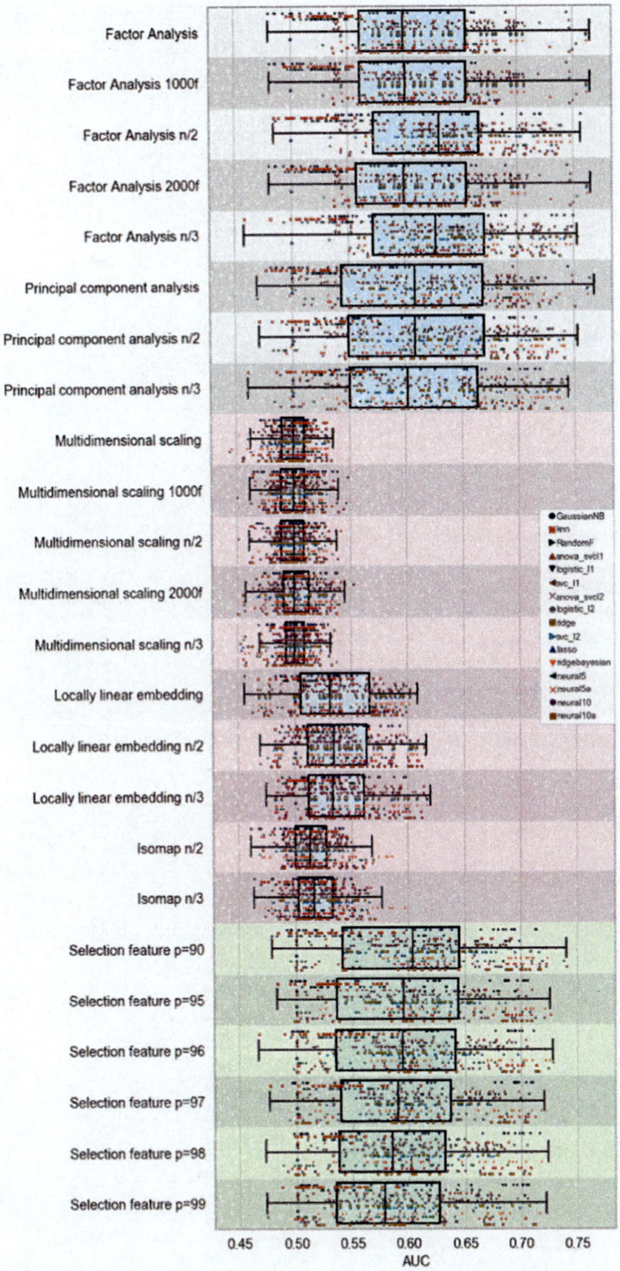

Fig. 2. Impact of feature selection versus classifiers.

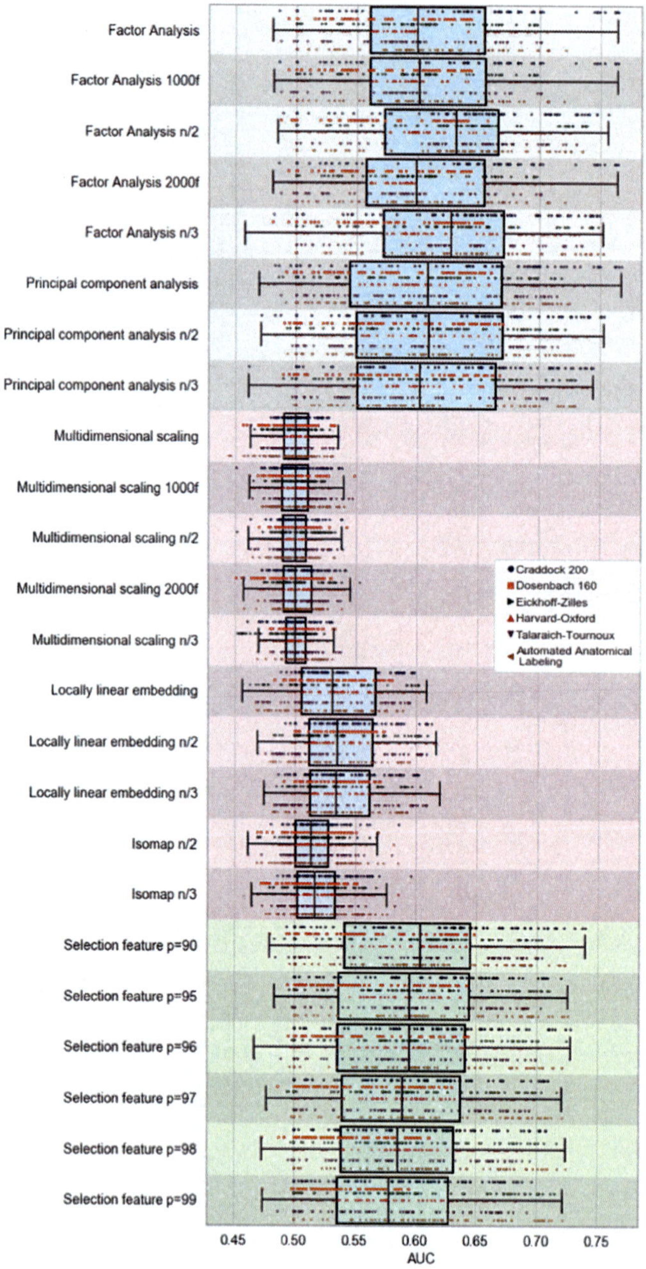

Fig. 3. Impact of features versus the atlas parcelation of the brain

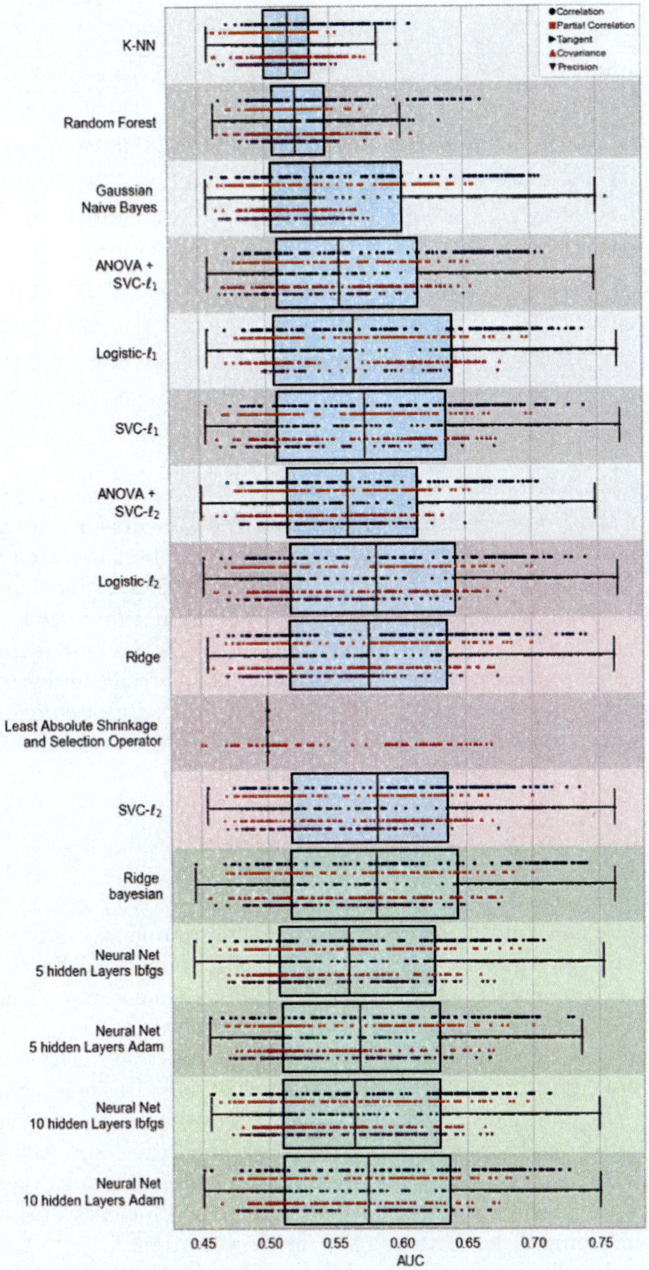

Fig. 4. Impact of classifiers versus the connectivity measure

Table 1. Best median AUC scores found in cross-validation repetitions, with corresponding settings (parcellation, feature extraction, classifier, and connectivity measure) that achieved it.

Settings				AUC percentiles		
Parcel.	Feat. extr.	Classifier	Conn. meas.	5%	Median	95%
cc200	fa	logistic ℓ_1	tangent	0.733	0.765	0.803
cc200	pca	logistic ℓ_2	tangent	0.739	0.765	0.801
cc200	fa	svc ℓ_1	tangent	0.739	0.764	0.803
cc200	fa1000	logistic ℓ_1	tangent	0.733	0.765	0.803
cc200	pca	svc ℓ_1	tangent	0.735	0.767	0.805
cc200	fa2000	logistic ℓ_1	tangent	0.733	0.765	0.803
cc200	fa2000	svc ℓ_1	tangent	0.739	0.764	0.803
Best reported to date [7]				0.66	0.711	0.756

Table 1 gives best median AUC scores found, together with the 5% and 95% percentiles of the cross-validation results in our experiments for comparison with the most comprehensive results to date [7], where the best reported results are 0.66, 0.711, and 0.756 for the 5%, 50%, 95% percentiles of median AUC distribution across all repetitions of the cross-validation experiments. The best connectivity measure appears to be the tangent, while the best feature extractions are the factor analysis and PCA. Best brain parcelation corresponds to cc200. Surprisingly, the logistic regression gives best results comparable with SVC. However, it must be noted that classifiers are quite unstable, giving very diverse results for different feature extraction.

4 Conclusions

The predictive approach to the analysis of brain connectivity from rs-fMRI data is gaining importance on the recent studies. In this approach, brain connectivity biomarkers are confirmed by the predictive performance in the classification between target populations. Up to this date there is no comprehensive study of the effect of feature extraction/selection approaches, hence we have carried out this study on the ABIDE I dataset, finding that some feature extraction procedures provide a boost on the performance of the classifiers across several connectivity matrix building approaches, namely principal component analysis (PCA) and factor analysis (FA). Also, the selection of the connections based on the Pearson's correlation (PCC) among connection value across subjects and the subject label provides enhanced results with the added benefit of preserving the anatomical meaning of the features used for classification.

Future work will be addressed to the extension of the computational experiments to the full extent of the ABIDE II dataset. Other connectomics datasets collecting subjects and controls from connectivity analysis regarding other diseases will also be considered.

References

1. Abraham, A., et al.: Deriving reproducible biomarkers from multi-site resting-state data: an autism-based example. NeuroImage **147**, 736–745 (2017). https://doi.org/10.1016/j.neuroimage.2016.10.045. http://www.sciencedirect.com/science/article/pii/S1053811916305924
2. Arslan, S., Ktena, S.I., Makropoulos, A., Robinson, E.C., Rueckert, D., Parisot, S.: Human brain mapping: a systematic comparison of parcellation methods for the human cerebral cortex. NeuroImage **170**, 5–30 (2018). https://doi.org/10.1016/j.neuroimage.2017.04.014. http://www.sciencedirect.com/science/article/pii/S1053811917303026. Segmenting the Brain
3. Button, K.S.: Double-dipping revisited. Nat. Neurosci. **22**(5), 688–690 (2019). https://doi.org/10.1038/s41593-019-0398-z
4. Cameron, C., et al.: The neuro bureau preprocessing initiative: open sharing of preprocessed neuroimaging data and derivatives. Front. Neuroinformat. **7** (2013). https://doi.org/10.3389/conf.fninf.2013.09.00041
5. Cox, D.R.: Principles of Statistical Inference. Cambridge University Press, New York (2006)
6. Craddock, R.C., James, G., Holtzheimer III, P.E., Hu, X.P., Mayberg, H.S.: A whole brain fMRI atlas generated via spatially constrained spectral clustering. Hum. Brain Mapping **33**(8), 1914–1928 (2012). https://doi.org/10.1002/hbm.21333. https://onlinelibrary.wiley.com/doi/abs/10.1002/hbm.21333
7. Dadi, K., et al.: Benchmarking functional connectome-based predictive models for resting-state fMRI. Neuroimage **192**, 115–134 (2019). https://doi.org/10.1016/j.neuroimage.2019.02.062
8. d'Anjou, A., Graña, M., Moutet, F.: Brief survey about the search for biomarkers and computer aided diagnosis of autism spectrum disorder. Technical report, Zenodo (2018). https://doi.org/10.5281/zenodo.1408202
9. Desikan, R.S., et al.: An automated labeling system for subdividing the human cerebral cortex on MRI scans into gyral based regions of interest. NeuroImage **31**(3), 968–980 (2006). https://doi.org/10.1016/j.neuroimage.2006.01.021. http://www.sciencedirect.com/science/article/pii/S1053811906000437
10. Dosenbach, N.U.F., et al.: Prediction of individual brain maturity using fMRI. Science **329**(5997), 1358–1361 (2010). https://doi.org/10.1126/science.1194144. https://science.sciencemag.org/content/329/5997/1358
11. Eickhoff, S.B., et al.: Assignment of functional activations to probabilistic cytoarchitectonic areas revisited. NeuroImage **36**(3), 511–521 (2007). https://doi.org/10.1016/j.neuroimage.2007.03.060. http://www.sciencedirect.com/science/article/pii/S1053811907002303
12. Fawcett, T.: ROC graphs with instance-varying costs. Pattern Recogn. Lett. **27**(8), 882–891 (2006). https://doi.org/10.1016/j.patrec.2005.10.012. http://www.sciencedirect.com/science/article/pii/S0167865505003053. ROC Analysis in Pattern Recognition
13. Hinton, G.: Connectionist learning procedures. Artif. Intell. **40**, 185–234 (1989)
14. Kriegeskorte, N., Simmons, W.K., Bellgowan, P.S.F., Baker, C.I.: Circular analysis in systems neuroscience: the dangers of double dipping. Nat. Neurosci. **12**(5), 535–540 (2009). https://doi.org/10.1038/nn.2303
15. Ledoit, O., Wolf, M.: A well-conditioned estimator for large-dimensional covariance matrices. J. Multivariate Anal. **88**(2), 365–411 (2004). https://doi.org/10.1016/S0047-259X(03)00096-4. http://www.sciencedirect.com/science/article/pii/S0047259X03000964

16. Pearson, K.: Notes on regression and inheritance in the case of two parents. Proc. R. Soc. Lond. **58**, 240–242 (1895)
17. Puerto, E., Aguilar, J., López, C., Chávez, D.: Using multilayer fuzzy cognitive maps to diagnose autism spectrum disorder. Appl. Soft Comput. **75**, 58–71 (2019). https://doi.org/10.1016/j.asoc.2018.10.034. http://www.sciencedirect.com/science/article/pii/S156849461830591X
18. Rolls, E.T., Huang, C.C., Lin, C.P., Feng, J., Joliot, M.: Automated anatomical labelling atlas 3. NeuroImage 116189 (2019). https://doi.org/10.1016/j.neuroimage.2019.116189. http://www.sciencedirect.com/science/article/pii/S1053811919307803
19. Smith, S.M., et al.: Network modelling methods for fMRI. NeuroImage **54**(2), 875–891 (2011). https://doi.org/10.1016/j.neuroimage.2010.08.063. http://www.sciencedirect.com/science/article/pii/S1053811910011602
20. Tibshirani, R.: Regression shrinkage and selection via the lasso. J. R. Stat. Soc. Ser. B (Methodol.) **58**(1), 267–88 (1996)
21. Tipping, M.: Sparse Bayesian learning and the relevance vector machine. J. Mach. Learn. Res. **1**, 211–244 (2001)
22. Tzourio-Mazoyer, N., et al.: Automated anatomical labeling of activations in SPM using a macroscopic anatomical parcellation of the MNI MRI single-subject brain. NeuroImage **15**(1), 273–289 (2002). https://doi.org/10.1006/nimg.2001.0978. http://www.sciencedirect.com/science/article/pii/S1053811901909784
23. Varoquaux, G., Baronnet, F., Kleinschmidt, A., Fillard, P., Thirion, B.: Detection of brain functional-connectivity difference in post-stroke patients using group-level covariance modeling. In: MICCAI (2010)
24. Woodward, N.D., Cascio, C.J.: Resting-state functional connectivity in psychiatric disorders. JAMA Psychiatry **72**(8), 743–744 (2015). https://doi.org/10.1001/jamapsychiatry.2015.0484
25. Zalesky, A., Fornito, A., Bullmore, E.: On the use of correlation as a measure of network connectivity. NeuroImage **60**(4), 2096–2106 (2012). https://doi.org/10.1016/j.neuroimage.2012.02.001. http://www.sciencedirect.com/science/article/pii/S1053811912001784

The Impact of Data Preprocessing on the Accuracy of CNN-Based Heart Segmentation

Julia Lasek(✉)

Medical Simulation Technologies sp. z o.o,
Miechowska 5b m 4, 30–055 Krakow, Poland
j.lasek@mstech.eu
http://www.mstech.eu

Abstract. Whole heart segmentation significantly improves the diagnostic value of computed tomography images. Since manual segmentation is very time consuming, efforts have been made to automate this process, especially using deep learning methods such as convolutional neural networks (CNN). This paper considers how preprocessing of computed tomography with the Statistical Dominance Algorithm would affect the results of segmentation obtained with CNNs. Segmentation results were compared to original CTs and processed images. Compared to unprocessed data, improvements in segmentation accuracy were obtained after image processing. Using three-fold cross-validation, the average Dice similarity coefficient that was achieved was 0.811 for unprocessed images and 0.863 for processed images.

Keywords: Heart segmentation · Convolution neural network · Computed tomography · Medical segmentation

1 Introduction

Computed tomography (CT) is one of the most important diagnostic tools for cardiovascular disease. The diagnostic value of CT images is significantly affected by segmentation of the whole heart, which means dividing an image into different regions to distinguish distinct ones. Due to the fact that manual segmentation is very time consuming (even 6–10 h for each image [22]), efforts have been made to automate this process in recent years. Especially good results are observed in works based on whole heart segmentation using deep learning methods. However, whole heart segmentation is still a challenging issue due to the high variability of subjects, such as differences in cardiac anatomy, pathological and physiological changes, and the transition of heart shape during systole and diastole [20,21]. Another problem associated with CT images is the variability in image appearance and quality; this might depend on the type of scanner, acquisition parameters, motion artifacts, differently contrasted heart chambers, or artifacts

M. Choraś et al. (Eds.): CORES 2021, IP&C 2021, ACS 2021, LNNS 255, pp. 173–180, 2022.
https://doi.org/10.1007/978-3-030-81523-3_17

caused by the presence of artificial elements in the heart. Segmentation of the heart plays a significant role in clinical diagnostic procedures, data analysis and medical simulations [5,14]. In recent years, highly accurate automatic segmentation has been achieved. In this work, an attempt has been made to examine how processing input data using the Statistical Dominance Algorithm will affect the final segmentation of the four cardiac chambers.

2 Related Works

The promising capabilities of deep learning methods have made them a significant option for image segmentation, especially for medical image segmentation. Consequently, in recent years an increase in the number of publications that examine medical image segmentation using deep learning methods has been observed [4,9,11]. Ye et al. proposed a deeply supervised 3D U-Net which applies multi-depth fusion to the original network [18]. Wang et al. developed a framework that combines this segmentation with 2D U-Nets and shape context estimation [16]. A new network architecture based on a conditional generative adversarial network, called R2Unet-GAN, for whole heart auto segmentation was presented by Le et al. [10]. Yuan et al. proposed a novel fully convolutional architecture based on the UNet++ by adding spatial and channel attention mechanisms [19]. The subject area of the attention mechanism was also introduced by Wahg et al. where three modified attention mechanism were compared for the heart segmentation task [17]. Another work based on U-Net was presented by Li et al. where the original 4-layer network was deepened to eight layers, thus achieving high accuracy of left ventricle segmentation [7]. A comparison and review of multiple cardiac segmentation techniques can be found in [2,8].

3 Materials and Methods

The proposed method of improving four-chamber segmentation accuracy is based on preprocessing CTs with the SDA algorithm. Details of the algorithm and the neural network on which it was tested are described below.

3.1 Data

The dataset includes 23 Coronary Computed Tomography Angiography (CCTA) scans (9,934 images in total); each subject includes from 159 to 764 CTs. Sixteen subjects were chosen for the training set; 3 for the validation set; and 4 for the test set. The images were obtained with different scanner types and were stored in anonymized DICOM format. The images have 512×512 pixel resolution and 12-bit depth (example of such image is presented in Fig. 1a). Manual segmentation was performed by qualified computed tomography data analysts using the 3D Slicer software [3,6] (see Fig. 1b for example). The initial segmentation consisted of 13 separate elements (including the left ventricle, right ventricle, left atrium, right atrium, myocardium of LV, aorta, pulmonary artery, lungs, coronary sinus, superior vena cava, inferior vena cava, coronary artery, pulmonary vein) and was processed into a binary mask of the four heart chambers (Fig. 1c).

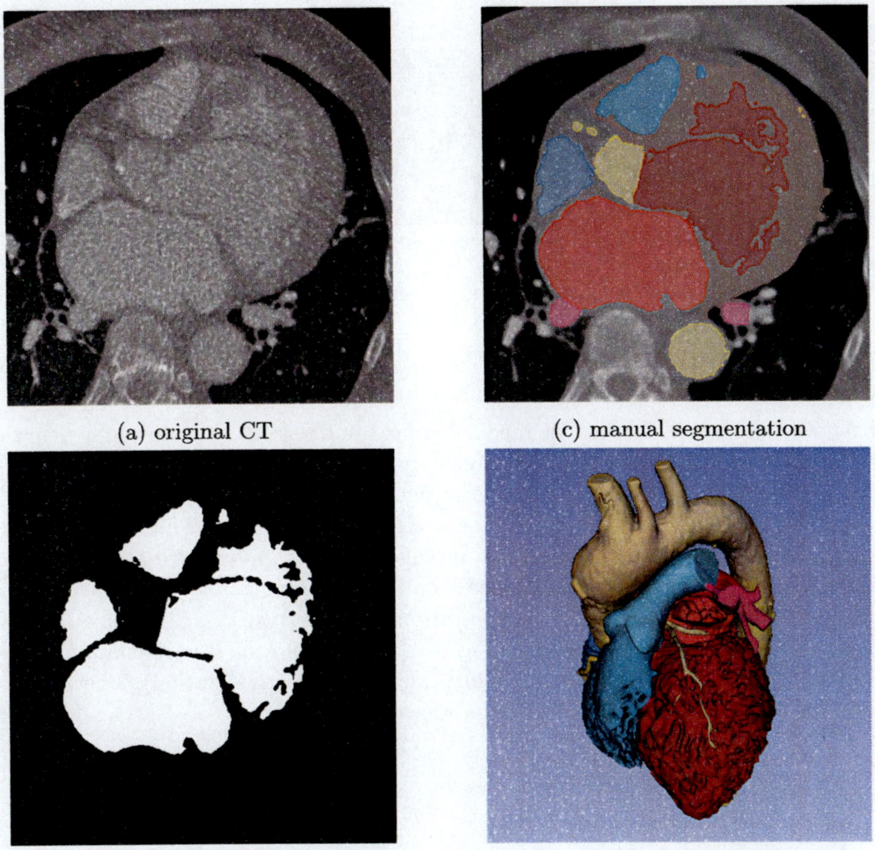

(a) original CT

(c) manual segmentation

(b) binary mask of 4 chambers

(d) 3D visualisation

Fig. 1. An example of data used in this work

3.2 SDA

The Statistical Dominance Algorithm (SDA) is used to preprocess images for object segmentation or edge detection [13]. It has performed very well in several medical image segmentation cases [1,12]. After applying the algorithm, the image is transformed from the brightness domain to the statistical information domain. In a region of a given radius, SDA determines the number of neighbours with greater or equal brightness compared to the central pixel. In this work, the radius and threshold were the same for each set in order to be able to use the same model parameters later with new subjects: $r=25$ and $threshold = 100$. The value of r was selected according to the study [12] (Fig. 6d) and *threshold* value was fixed based on Hounsfield units for the structures appearing on cardiac CTs. The images obtained using the SDA algorithm were subtracted from the original CTs, as shown in Fig. 2.

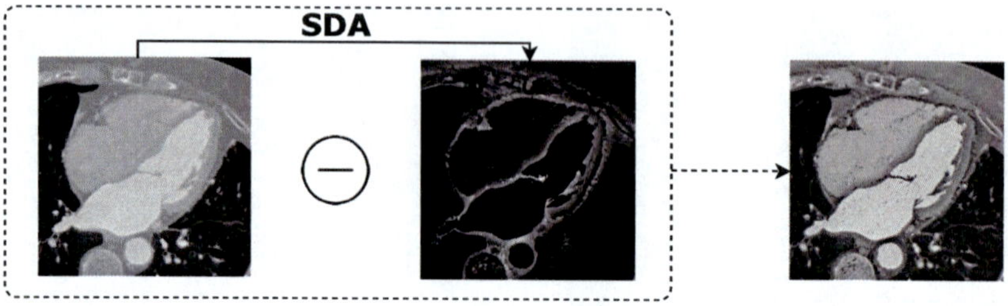

Fig. 2. Scheme of preprocessing of input data with SDA algorithm.

3.3 U-Net

The prevailing architecture in medical image analysis is convolutional neural networks (CNN). A more sophisticated type of CNN is U-Net, which was originally invented for biomedical image segmentation by Ronneberger et al. [15] and has been successfully applied for such purposes [9,16]. Therefore, U-Net was implemented in this paper as a one of the most appropriate architectures with which to test the impact of medical image preprocessing on segmentation results. A detailed schematic representation of the network architecture implemented in this work is depicted in Fig. 3. Each convolutional part contains two convolutional layers (Convolution, ReLU and Batch Normalization). There were separate models for processed and unprocessed data.

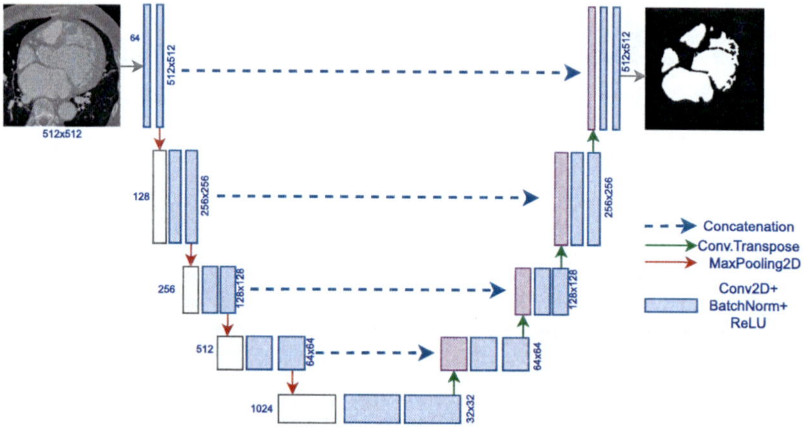

Fig. 3. Schematic drawing of U-Net architecture. Boxes represent cross-sections of feature maps and arrows represent operations, specified by the legend.

Adaptive Moment Estimation (ADAM) is used as the optimizer (learning rate = 10^{-5}) in all training process, and the number of epochs is fixed at 24 for both U-Nets. The batch size used was 1, which is justified due to the memory constraint. The values of the aforementioned parameters were selected after experimentation. The network was implemented by Tensorflow and Keras. All of the experiments were performed with Intel CPU and NVIDIA GeForce RTX 3060.

3.4 Evaluation

Segmentation evaluation is based on the number of pixels correctly and/or incorrectly assigned to a given image region. To measure the accuracy of segmentation, the Dice similarity coefficient (DSC) was used. It measures the spatial overlap between the predicted and manual results, which can be explained as

$$DSC = \frac{2\,|X \cap Y|}{|X| + |Y|} \tag{1}$$

where X means predicted results and Y means manual segmentation. The DSC was separately calculated for each test subject. The three-fold cross-validation was used.

4 Results

The Dice coefficient for each subject after three-fold cross-validation is compared in Table 1. Visual results of the segmentation are shown in Fig. 4.

Table 1. Comparing the Dice coefficient of the original data and images processed with SDA

	Dice original	Dice SDA
Subject 1	0,860	0,860
Subject 2	0,816	0,847
Subject 3	0,816	0,867
Subject 4	0,753	0,876
Mean	0,811	0,863

From the results obtained, it can be concluded that the segmentation is more accurate after applying SDA. The Dice coefficient after 24 epochs was 0.811 for the original data and 0.863 for the data processed with SDA. Based on Table 1, it can be concluded that the Dice coefficient was higher than or equal to in all of the subjects. In the images that are compared in Fig. 4 it can be seen that the SDA images were better segmented, particularly at the right and left ventricle.

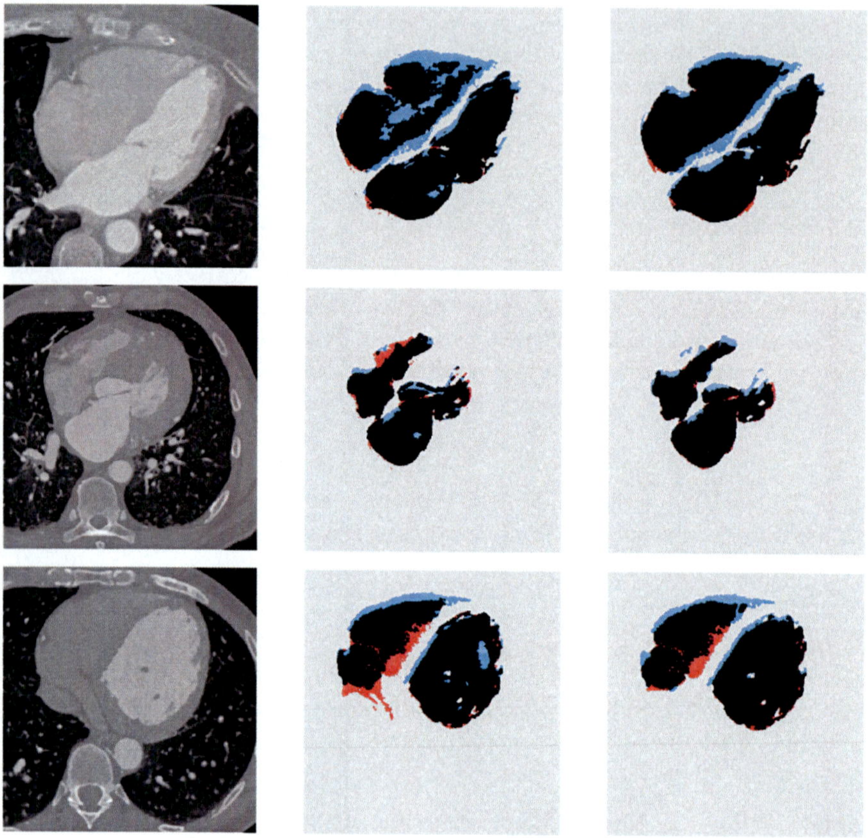

(a) original CT (b) segmentation on unprocessed data (c) segmentation on data processed with SDA

Fig. 4. Visual comparison of segmentation. Left column shows original pictures; middle column shows results without SDA; right column shows segmentation after SDA. Black colour of segmentation stands for TP (true positive), blue–FN (false negative), red–FP (false positive), grey–true negative.

5 Conclusions

This paper proposes SDA as a preprocessing step before image segmentation using convolution neural networks. An average segmentation result of 0.863 Dice was obtained for the processed data. It has been shown that image preprocessing using SDA can increase the accuracy of automated heart chamber segmentation with U-Net. Considering the wide possibilities of SDA and neural networks, such a solution could also influence the quality of segmentation of other structures. Future work will elaborate further on the diagnostic value of medical image segmentation with different types of deep learning methods and on various structures.

Acknowledgement. This work was supported by the European Union under the Regional Operational Programme for Malopolska Region 2014–2020, project no: RPMP.01.02.01-12-0027/19.

References

1. Bielecka, M., Obuchowicz, R., Korkosz, M.: The shape language in application to the diagnosis of cervical vertebrae pathology. PLOS ONE **13**, e0204,546 (2018). https://doi.org/10.1371/journal.pone.0204546
2. Chen, C., et al.: Deep learning for cardiac image segmentation: a review. Front. Cardiovasc. Med. **7** (2020). https://doi.org/10.3389/fcvm.2020.00025
3. Fedorov, A., et al.: 3D slicer as an image computing platform for the quantitative imaging network. Magn. Reson. Imaging **30**, 1323–41 (2012). https://doi.org/10.1016/j.mri.2012.05.001
4. Hesamian, M.H., Jia, W., He, X., Kennedy, P.: Deep learning techniques for medical image segmentation: achievements and challenges. J. Digit. Imaging **32** (2019). https://doi.org/10.1007/s10278-019-00227-x
5. Kempny, A., Piórkowski, A.: CT2TEE - a novel, internet-based simulator of transoesophageal echocardiography in congenital heart disease. Kardiol. Pol. **68**(3), 374–379 (2010)
6. Kikinis, R., Pieper, S.D., Vosburgh, K.G.: 3D Slicer: a platform for subject-specific image analysis, visualization, and clinical support. pp. 277–289. Springer, New York (2014). https://doi.org/10.1007/978-1-4614-7657-3_19
7. Li, C., et al.: An 8-layer residual u-net with deep supervision for segmentation of the left ventricle in cardiac CT angiography. Comput. Methods Programs Biomed. **200**, 105,876 (2020). https://doi.org/10.1016/j.cmpb.2020.105876
8. Litjens, G., et al.: State-of-the-art deep learning in cardiovascular image analysis. JACC: Cardiovasc. Imaging **12**, 1549–1565 (2019). https://doi.org/10.1016/j.jcmg.2019.06.009
9. Liu, T., Tian, Y., Zhao, S., Huang, X., Wang, Q.: Automatic whole heart segmentation using a two-stage u-net framework and an adaptive threshold window. IEEE Access **7**, 1 (2019). https://doi.org/10.1109/ACCESS.2019.2923318
10. Lou, Z., Huo, W., Le, K., Tian, X.: Whole heart auto segmentation of cardiac CT images using u-net based GAN, pp. 192–196 (2020). https://doi.org/10.1109/CISP-BMEI51763.2020.9263532
11. Nurzynska, K.: Deep learning as a tool for automatic segmentation of corneal endothelium images. Symmetry **10**(3) (2018). https://doi.org/10.3390/sym10030060
12. Nurzynska, K., Mikhalkin, A., Piorkowski, A.: CAS: cell annotation software - research on neuronal tissue has never been so transparent. Neuroinformatics **15**, 365–382 (2017)
13. Piorkowski, A.: A statistical dominance algorithm for edge detection and segmentation of medical images. In: Information Technologies in Medicine. Advances in Intelligent Systems and Computing, vol. 471, pp. 3–14. Springer (2016)
14. Piórkowski, A., Kempny, A.: The transesophageal echocardiography simulator based on computed tomography images. IEEE Trans. Biomed. Eng. **60**(2), 292–299 (2013)
15. Ronneberger, O., Fischer, P., Brox, T.: U-net: convolutional networks for biomedical image segmentation. Lecture Notes in Computer Science, vol. 9351 (2015)

16. Wang, C., Smedby, O.: Automatic whole heart segmentation using deep learning and shape context, pp. 242–249. Springer (2018). https://doi.org/10.1007/978-3-319-75541-0_26
17. Wang, W., Ye, C., Zhang, S., Xu, Y., Wang, K.: Improving whole-heart CT image segmentation by attention mechanism. IEEE Access **8**, 14579–14587 (2020). https://doi.org/10.1109/ACCESS.2019.2961410
18. Ye, C., Wang, W., Zhang, S., Wang, K.: Multi-depth fusion network for whole-heart CT image segmentation. IEEE Access **PP**, 1 (2019). https://doi.org/10.1109/ACCESS.2019.2899635
19. Yuan, X., Zhu, Y., Wang, Y.: Attention based encoder-decoder network for cardiac semantic segmentation. In: 2020 Chinese Automation Congress (CAC), pp. 4578–4582 (2020). https://doi.org/10.1109/CAC51589.2020.9326844
20. Zhuang, X.: Challenges and methodologies of fully automatic whole heart segmentation: a review. J. Healthcare Eng. **4**, 371–408 (2013). https://doi.org/10.1260/2040-2295.4.3.371
21. Zhuang, X., et al.: Evaluation of algorithms for multi-modality whole heart segmentation: an open-access grand challenge. Med. Image Anal. **58**, 101,537 (2019). https://doi.org/10.1016/j.media.2019.101537
22. Zhuang, X., Shen, J.: Multi-scale patch and multi-modality atlases for whole heart segmentation of MRI. Med. Image Anal. **31** (2016). https://doi.org/10.1016/j.media.2016.02.006

Comparison of Selected Acoustic Signal Parameterization Methods in the Problem of Machine Recognition of Classical Music Styles

Maciej Walczyński(✉) and Jacek Fica

Wrocław University of Science and Technology, Wrocław, Poland
maciej.walczynski@pwr.edu.pl

Abstract. The purpose of this study was to review and implementation of different forms of data extraction and their classification of classical music styles in a machine learning process. The most popular parameters describing audio signals were considered: spectrogram, melo-scale spectrogram and Mel-Frequency Cepstral Coefficients (MFCC). The influence of the machine learning algorithm on the obtained effectiveness of classical music style classification. In this paper, we analyze CNN (Convolutional Neural Network) and Recurrent Neural Network (RNN). All proposed algorithms were implemented in Python with various libraries to support the work with neural networks and audio signals such as TensorFlow, Keras, Librosa.

Keywords: Machine learning · CNN · RNN · MFCC · Computional musicology

1 Introduction

Music theory allows us to divide music into genres. Nowadays we distinguish categories such as pop, rock, and classical music [1,2]. Each of these genres is characterized by its own unique musical properties. Among them characteristics such as tonality, rhythm, harmony and structure of a musical piece. Music classification is widely used today, especially in popular streaming services, where it attempts to match songs to listener preferences. Just as the above-mentioned musical characteristics are identified by human computer programs classify through the use of machine learning algorithms with machine learning algorithms with satisfactory results [3]. The purpose of such activities is to retrieve Music Information Retrieval (MIR) [4,5] for the purpose of not only classifying musical genres, but also in many different fields related to audio signals such as acoustic fingerprint [6] and instrument tagging [7]. For obvious reasons due to popularity, some music genres are less well studied, while others are better studied. This is particularly evident this is particularly evident in popular music and classical music. Classical music is most often divided into eras [8], so the division is based on the date

M. Choraś et al. (Eds.): CORES 2021, IP&C 2021, ACS 2021, LNNS 255, pp. 181–190, 2022.
https://doi.org/10.1007/978-3-030-81523-3_18

of composition of the work. Theorists distinguish 5 main genres: Renaissance, Baroque, Classicism, Romanticism and Modernism. Each of these times differs in such properties as instrumentation, harmony, structure of a musical piece, rhythm or dynamics. Each of these musical characteristics changes over time. For example, the number and types of instruments used in Renaissance music (lyre, viola, recorder, lute, trumpets) are completely different from those used in the Baroque (harpsichord, violin, organ, cello). Music information retrieval (MIR) consists of two basic steps. The first is the selection of an appropriate database and the extraction of the desired features based on the musical works. The second step is the selection of an appropriate algorithm machine learning processing algorithm that best fits the acquired data. the extracted data. In the literature, one can search in vain for considerations of exactly the same problem. There are widely found works discussing methods of parameterization and classification of musical works by genre. One can find works on computational musicology and classification of works of piano literature.

2 Database Development

2.1 Main Assumptions

Selecting the right database is the first key problem that to be tackled in the machine learning process. For this reason, it was necessary to define main assumptions about its content. It was decided that the classification would be carried out into 5 main eras, defined in classical music: renaissance, baroque Classicism, Romanticism, and Modernism. As mentioned in the introduction, one of the most important properties that distinguish individual styles is the instrumentation. This means, that the instruments in the recordings had to be in keeping with the era to which the work belonged. For this reason, performances consisting only of one instruments, such as the piano, had to be omitted. Another important issue was the quality of the recordings. The sound files should be encoded in lossless and uncompressed digital recording format to avoid data loss. This assumption resulted in rejection of the following formats MP3 and OGG Vorbis due to the lossy audio codecs used. Preferred formats were WAV and AIFF. Another key consideration was the sampling rate and bit resolution, which could not be worse than 44.1 kHz and 16 bits.

Selection of Musical Pieces. In order to properly differentiate the works into eras, it was necessary to assign eras to the corresponding composers. It is difficult to determine the exact time frame of a given period in classical music. For this reason, in order to avoid misclassification of a work to a given epoch, we had to omit composers whose works fall at the turn of two musical periods. For example, we omitted such famous musicians such as Ludwig van Beethoven or Franz Schubert were omitted [9].

In order to create the database, the MusicNet collection and the albums from the Naxos label were used to create the database. The MusicNet database contains 330 recordings of classical music pieces. Each recording contained information about the composer. Thanks to that, it was possible to filtered into eras, according to the first figure. After this procedure, only 97 recordings remained. For this reason, the collection of recordings of the music label Naxos was used. This allowed the database to grow to 460 tracks.

3 Data Extraction

For automatic classification, the most popular features that can be extracted from audio files. These include:

- spectrogram,
- mel-scale spectrogram,
- mel-cepstral coefficients.

For each song, the above features were extracted based on the first two minutes of the recording. If the music file was shorter, samples were added at the end with zero values, so that all features were of the same length.

3.1 Spectrogram

The first feature that is often encountered when analyzing music files is the spectrogram. It is used both for sound file classification purposes, speech processing and in phonetics, for example, to analyze the laryngeal tone. It is a graph showing how the frequency spectrum of a signal changes as a function of time. During this process, signal bookmarking is very often performed. This is the division of the signal into overlapping frames. This process is intended to increase the temporal resolution. This is due to the fact that the smaller the length of the window, the better to distinguish frequency changes between events in the signal.

Music signals are among the strongly non-stationary data because of the continuous changes in frequency characteristics with duration. For this reason, a single Fourier transform is not performed on the entire file, but the signal is divided into short segments. This is called windowing. In order to avoid spectrum leakage due to discontinuities, different types of time windows are used at the ends of the time window. Among the most popular are the hamming window, the hanning window, and the blackmann window. All signals were processed using parameterization methods implemented in the Python library librosa.

3.2 Mel Scale Spectrogram

One of the basic scales used in acoustics is the mel scale. It is related to pitch and the fact that the human ear does not perceive sounds linearly. An increase in frequency does not respond linearly to the human auditory experience. This means that it is easier to hear the difference 300 Hz 600 Hz than 7000 Hz 7300 Hz. It is this non-linearity of human hearing that the Melodic Scale aims to expose. It is determined by subjective changes in pitch. In order to avoid scale dependence on loudness, a level of 40dB SPL is used in the definition. The process of obtaining mel-scale spectrograms can be divided into (Fig. 1):

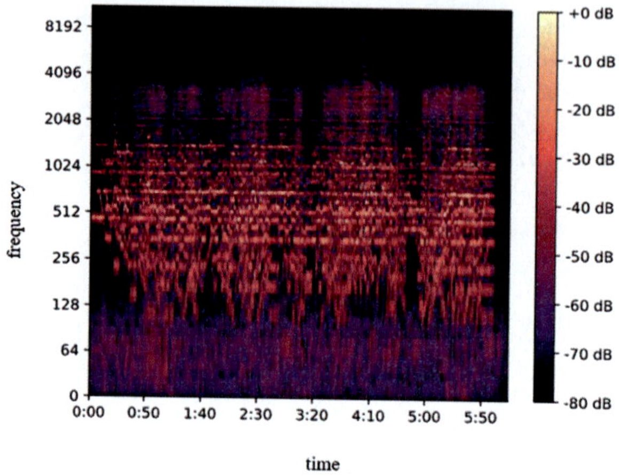

Fig. 1. Example spectrogram of one of the analyzed works

- dividing the signal into frames,
- calculation of the Fourier transform,
- squaring the short-time Fourier transform,
- conversion of the frequency scale to mel scale.

Below is a sample graph created using the above procedures:

3.3 Mel-Cepstral Parameters

Mel-Frequency Cepstrum Coefficients (MFCCs) are one of the most common forms of data extraction in the problem of automatic classification of audio signals. They are obtained by a cepstrum from a mel-scale signal. The process of obtaining the above coefficients consists of 5 steps:

- dividing the signal into frames,
- calculation of the Fourier transform,
- squaring the short-time Fourier transform,
- conversion of the frequency scale to mel scale,
- cosine transform.

The final step in obtaining the mel spectral coefficients to the inverse Fourier transform. This is done in order to transition from the signal spectrum to the cepstrum. To do this the obtained coefficients are subjected to logarithm and then multiplied by Discrete Cosine Transform (DCT) (Fig. 2).

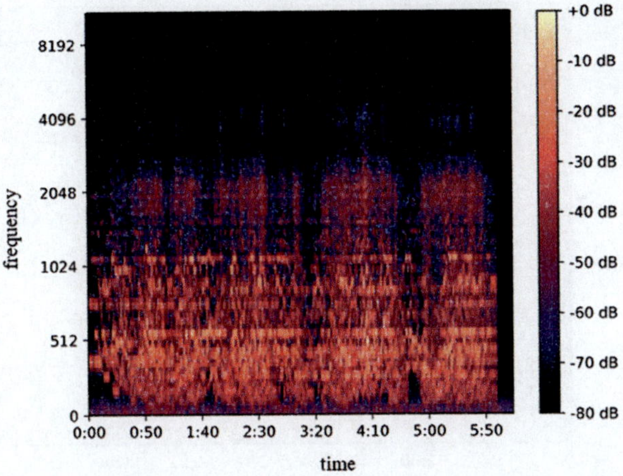

Fig. 2. Example mel scale spectrogram of one of the analyzed works

3.4 Chromagram

A chromagram is a graph showing the distribution of individual pitches in a piece of music as a function of time. It is directly related to the twelve-note musical scale (uniformly tempered). The graph shows the distribution of individual notes belonging to this scale. It does not distinguish which octave a given sound belongs to. The only thing that matters is the name of the sound. According to the accepted norm the pitch of the A1 sound 440 Hz.

4 Computational Experiments

4.1 Data Preprocessing

The first efficiency trials of neural networks showed low efficiencies and high overtraining. This means that the weights of individual neurons were adapted to classify learning data instead of validation data. For this reason, there were large differences between these efficiencies. For this reason, the existing database was processed and divided into files of two minutes in length. This allowed the input data to grow to 583 files. The division of the files into The following pie chart shows the breakdown of the files into individual epochs (Fig. 3).

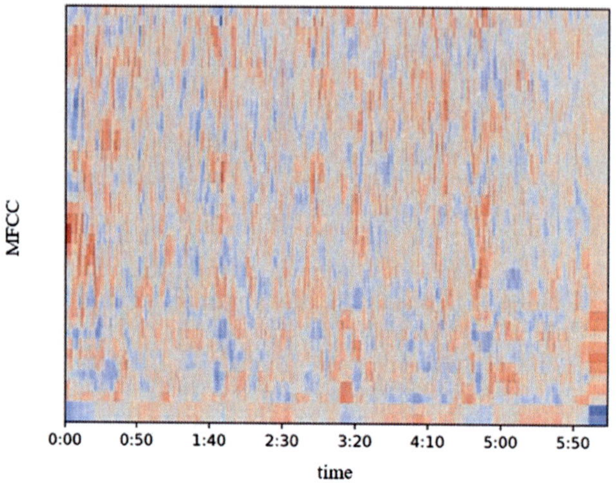

Fig. 3. Example graph showing MFCC parameters for one of the music pieces analyzed

Next, spectrograms, mel-scale spectrograms, mel-cepstral coefficients, and chromagrams were generated for each file according to the assumptions given in previous Chapter. The next step was to load them as matrices along with changing the colors to grayscale colors. This reduced the dimension of the data from a three-dimensional matrix to a two-dimensional matrix. Images without additional information such as axes or axes were also loaded. The file database was divided into training and validation files in a ratio of 80:20. This selection was based on randomization, meaning that the number of validation files totaled 20% for the entire database rather than for each classified epoch. The number of music files belonging to the Baroque era was also reduced so as not to overload the database in one direction, which could result in unsatisfactory machine learning results. The neural network was taught using the fit function included in the keras package. The function parameters were assumed to be - batch size = 128 - epochs = 20 The batch size parameter determines the size of the neural network batch file, that is, each run of the training cycle will be based on 128 samples of the training part. One song is taken as a sample, so 76 spectrograms of size 128 by 64. After one batch file passes through the neural network, the previously mentioned back propagation period takes place, which adjusts the values of the activation function of the perceptrons before processing the next piece of data. The epochs parameter specifies the number of deep learning epochs. In other words, it determines the number of repetitions of the entire learning cycle. Due to the complexity of the algorithms involved in the deep learning process, the values As the algorithms that comprise the deep learning process are complex, the values of the of parameter values and finally deciding on the one which provides the best results.

4.2 Neural Network Training and Test Results

Based on the extracted data and CNN neural network architectures, performance tests were performed on the machine learning algorithms to classify classical music styles. The following table shows the results for the training set and the validation set.

Table 1. Performance values for the training and validation database for each machine learning algorithm and each form of data extraction

Algorithm	RNN		CNN	
Data Set	Training	Validation	Training	Validation
Spectrogram	58%	52%	66%	57%
Mel spectrogram	37%	31%	90%	57%
MFCC	89%	41%	80%	52%
Chromagram	23%	26%	80%	31%

From the above Table 1, it can be concluded that the convolutional network (CNN) achieves better classification efficiency results. The best results were obtained from the spectrograms for both algorithms, 52% for RNN and 47% for CNN respectively. The convolutional network also achieved good results based on mel spectrograms (57%) and mel-cepstral coefficients (52%). However, for both of these forms there was quite a bit of overtraining, i.e., quite a large difference in performance between the training and validation bases. The CNN algorithm performed the worst with chromagrams. It is most likely caused by the loss of data about the instruments on which a given piece of music is played, that is, about their characteristic frequency properties. The RNN achieved the best results as already mentioned for input data in the form of spectrograms. Another result was 41% for mfcc, where it is worth noting that there was also overtraining of the network. Based on mel spectrogams and chromagrams, the algorithm achieved efficiencies of 31% and 26%, respectively. Numerical results do not show everything, so a better form of presenting the results of learning neural networks are error matrices (Confusion matrix). Thanks to them you can read which categories were classified better and which worse for a given algorithm and a given form of data extraction. The vertical axis shows the correct values of the test file data and for them we read their classifications by the program on the horizontal axis. The color scale helps to make the number of classified files more visible.

Classification error matrices for all algorithms and extraction forms are given in the following figures (Fig. 4 and Fig. 5).

It can be read from them that spectrograms were the best form of data extraction and classification. The error matrices confirm the obtained efficiency seen in the first table. It is worth noting that the smallest classification error occurred for works from the Baroque and Renaissance eras. This is probably

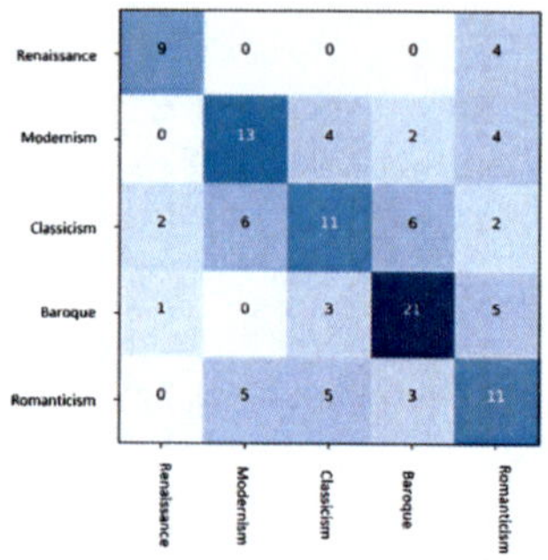

(a) RNN and input data in the form of spectrograms

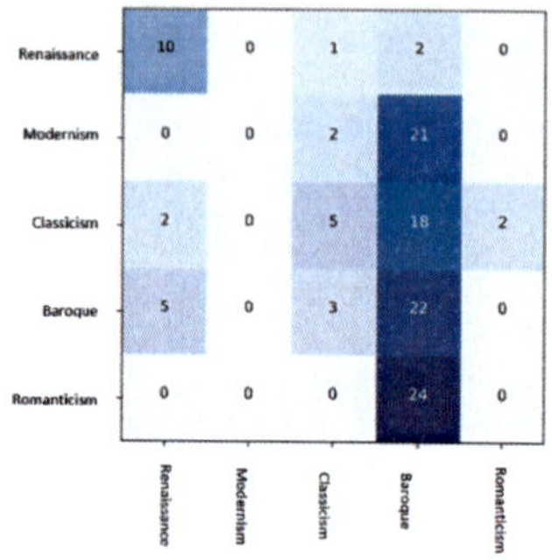

(b) RNN and input data in the form of mel spectrograms

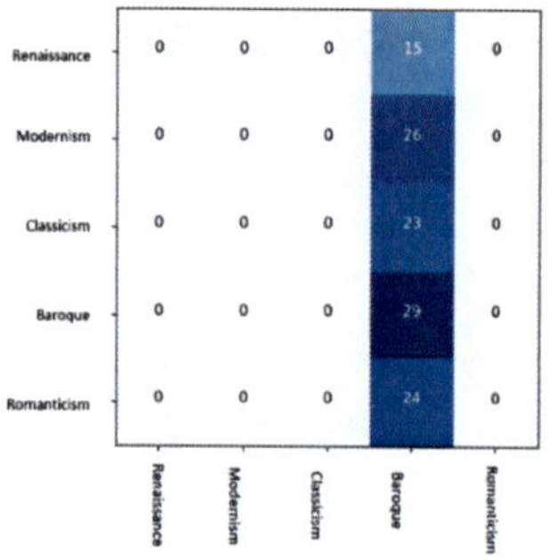

(c) RNN and input data in the form of chromagrams

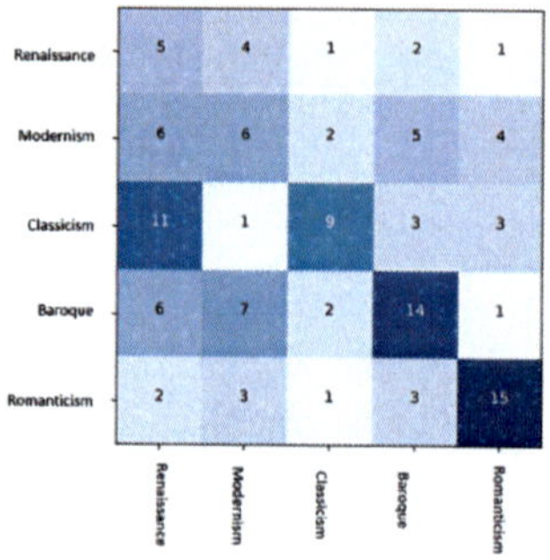

(d) RNN and input data in the form of MFCC

Fig. 4. Error matrices for RNNs and different parameterization methods

influenced by the set of instruments in the given pieces. For example, Renaissance music makes extensive use of the lute and flute, which are not as common in any other era. Both RNN and CNN had a problem with the proper classification of chromagrams as shown by both the performance and classification error matrices. It was unable to correctly classify the input data, so for the recursive algorithm it assigned them to a single epoch, while for the CNN they were mostly classified to modernism and baroque.

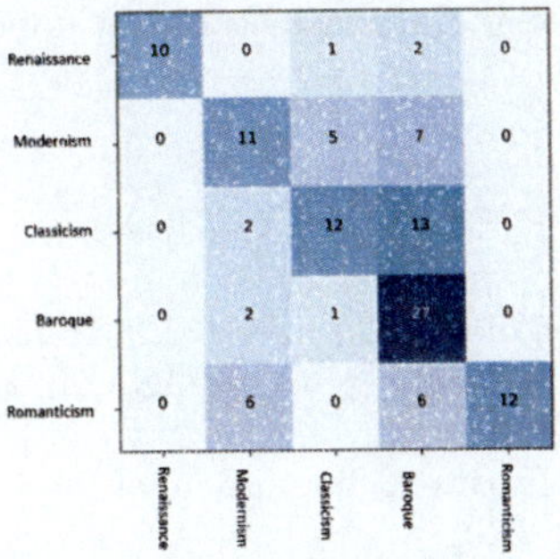

(a) CNN and input data in the form of spectrograms

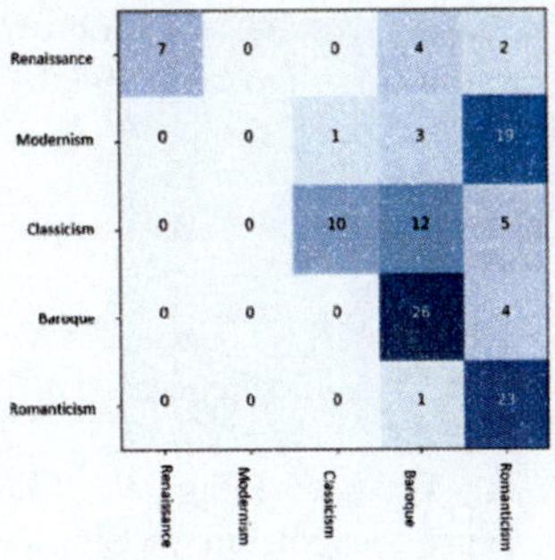

(b) CNN and input data in the form of mel spectrograms

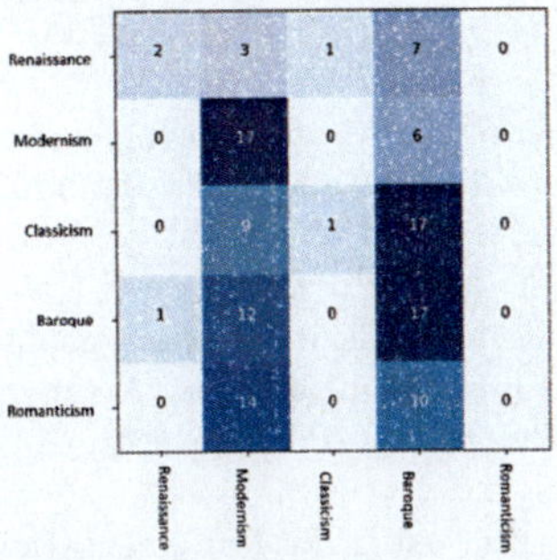

(c) CNN and input data in the form of chromagrams

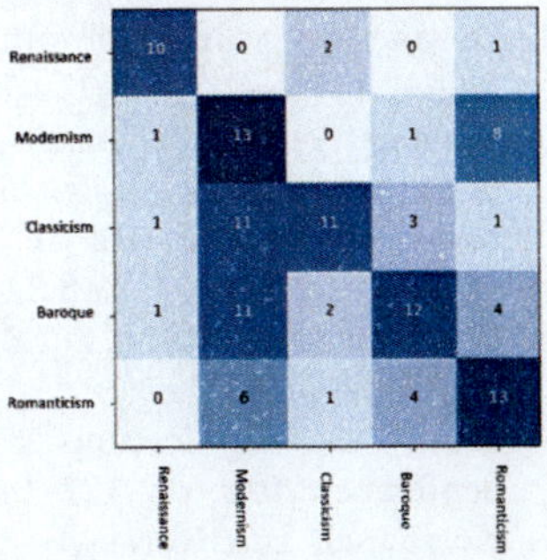

(d) CNN and input data in the form of MFCC

Fig. 5. Confusion matrices for CNNs and different parameterization methods

5 Summary

There are not many works devoted to the selection of parameters in the process of automatic classification of epochs, which include musical works belonging to the classical music canon.The authors focus on the problem of automatic classification of music genres. However, this is a much more general issue. In this approach, the authors Kostek B. and Rosner A. [10] obtain results of 72% for a collection containing 8000 music tracks belonging to 13 styles. In one of the earlier work [11] authors achieved better results for Baroque music, but the music material had symbolic form and was stored in mxml format. For that reason different methods of parametrization were used in this case. Other authors also report achieving excellent efficacy of up to 94.4% [12]. However, the issue concerns composer recognition based on symbolic notation in midi format using n-grams and modified cortical algorithms. In this paper, we have presented an automated identification methods for the recognition of classical music epochs using spectrogram, mel spectrogram, MFCC and chromagram along with RNN and CNN algorithms. Our proposed system uses music database stored in non-symbolic form (in contrast to the compared works) [13,14].

References

1. Tzanetakis, G., Cook, P.: Musical genre classification of audio signals. IEEE Trans. Speech Audio Process. **10**(5), 293–302 (2002). https://doi.org/10.1109/TSA.2002.800560
2. Ng, W.W.Y., Zeng, W., Wang, T.: Multi-level local feature coding fusion for music genre recognition. IEEE Access **8**, 152713–152727 (2020)
3. Micchi, G.: A neural network for composer classification. In: International Society for Music Information Retrieval Conference (ISMIR 2018), Paris, France (2018). ffhal-01879276f
4. Downie, J.S.: Music information retrieval. Ann. Rev. Info. Sci. Tech. **37**, 295–340 (2003). https://doi.org/10.1002/aris.1440370108
5. Vishnupriya, S., Meenakshi, K.: Automatic music genre classification using convolution neural network. In: 2018 International Conference on Computer Communication and Informatics (ICCCI), Coimbatore, India, pp. 1–4 (2018). https://doi.org/10.1109/ICCCI.2018.8441340
6. Walczyński, M., Ryba, D.: Effectiveness of the acoustic fingerprint in various acoustical environments. In: Signal Processing, Algorithms, Architectures, Arrangements, and Applications, SPA 2019: Conference Processing, Poznań, 18th–20th September 2019, pp. 137–141. IEEE, Danvers (2020)
7. Koszewski, D., Kostek, B.: Musical instrument tagging using data augmentation and effective noisy data processing. J. Audio Eng. Soc. **68**, 57–65 (2020). https://doi.org/10.17743/jaes.2019.0050
8. van't Spijker, B.C.: Classifying Classical Piano Music Into Time Period Using Machine Learning (2020)
9. Taenzer, M., Abeßer, J., Mimilakis, S., Weiß, Ch., Müller, M., Lukashevich, H.: Investigating CNN-Based Instrument Family Recognition for Western Classical Music Recordings (2019). https://doi.org/10.5281/zenodo.3258829
10. Rosner, A., Kostek, B.: Automatic music genre classification based on musical instrument track separation. J. Intell. Inf. Syst. **50**, 1–22 (2018). https://doi.org/10.1007/s10844-017-0464-5
11. Walczyński, M., Grzybała, P.: Selected methods of parametrization in problem of automatic classification classical music from the Renaissance era against the classical works from other eras. In: Signal Processing, Algorithms, Architectures, Arrangements, and Applications: Conference Proceedings, SPA 2020, Poznań, 23rd–25th September 2020, pp. 151–156. IEEE, Danvers (2020)
12. Hajj, N., Filo, M., Awad, M.: Automated composer recognition for multi-voice piano compositions using rhythmic features, n-grams and modified cortical algorithms. Complex Intell. Syst. **4**, 55–65 (2018). https://doi.org/10.1007/s40747-017-0052-x
13. Wołkowicz, J., Kulka, Z., Kešelj, V.: N-gram-based approach to composer recognition. Arch. Acoust. **33**(1), 43–55 (2008)
14. Pollastri, E., Simoncelli, G.: Classification of melodies by composer with hidden Markov models. In: Proceedings of first International Conference on Web Delivering of Music, pp. 88–95. IEEE (2001)

The Influence of Bit-Depth Reduction on Correlation of Texture Features with a Patient's Age

Patrycja Mazur(✉)

Department of Biocybernetics and Biomedical Engineering, AGH University of Science and Technology, Mickiewicza 30 Av., 30–059 Krakow, Poland

Abstract. As we age, irreversible changes occur in the human spine. This process is affected not only by aging but also by diseases, lifestyle and genetics. The relationship between a patient's age and bone condition can be valuable diagnostic information. This is made possible by texture analysis, which provides information about the structure and properties of imaged tissue; however, it depends on many factors, such as image format or image noise. The aim of this work was to test whether the analyzed textures are also sensitive to image quantization. To achieve this, 183 cervical spine radiographs were analyzed to extract features that correlate with patients' age. The research made it possible to distinguish features that are or are not sensitive to bit-depth reduction. Moreover, significantly lower correlation values of texture features with age were observed for X-rays of female patients than for those of males.

Keywords: Texture analysis · Cervical spine · X-ray · Radiographs

1 Introduction

Changes in the spine are an integral part of the aging process of the body, which significantly affects quality of life. The course and dynamics of these changes depends not only on genetic conditions but also on lifestyle, hormonal changes or diseases of the skeletal system [1,6,10].

At a young age, a vertebra's cortical bone is thin and dense. It covers cancellous bone, which consists of irregularly shaped trabeculae that are oriented horizontally and vertically for better load distribution. However, with increasing age there is selective loss of trabeculae, mainly transversely oriented [18]. As a result, the vertebrae become wider, shorter and less flexible. This can lead to diseases such as osteopenia or osteoporosis [4,12].

Evaluation of spine X-rays can be done by means of textural analysis, which is possible with all digitally stored medical images. Texture is a source of information about the structure and properties of the imaged tissue. It is a kind of pattern that consists of elements of a certain brightness, shape, color, etc. [17].

M. Choraś et al. (Eds.): CORES 2021, IP&C 2021, ACS 2021, LNNS 255, pp. 191–198, 2022.
https://doi.org/10.1007/978-3-030-81523-3_19

The analysis of biomedical images is affected by many factors, such as image format or noise [5,9]. Bit-depth reduction can also have a large impact on texture analysis. Reducing the number of bits eliminates some texture details, thus generalizing the information stored in the image.

2 Related Work

Assessment of the progression of spinal lesions in the context of a patient's age allows better diagnosis and selection of appropriate treatment, therefore this topic is important for medical specialists [7,8]. One published study presented a scale that makes it possible to determine the biological age of a patient's cervical spine [19]. The accuracy of this scale was about 72.6%. However, the study did not include important factors in assessing the condition of the spine, such as osteoporosis.

The research carried out so far has revealed that some algorithms that are used to compute textures may be sensitive to quantization [2]. Limitation of gray level reduces noise, which improves classification of images [13]. Therefore, as the gray levels decrease, the classification ability increases. Due to the fact that there are many texture features, not all of them have been verified. In addition, it has not been observed whether there are any texture features that are insensitive to bit-depth reduction.

3 Materials and Methods

The collected 183 radiographs of the cervical spine were acquired from AgfaCR25 or AgfaCR35 equipment [9]. The pixel spacing value for all images was 0.1×0.1 mm. Due to hormonal factors, which have a large impact on bones, the analysis was carried out separately for X-rays of female (100 images) and male patients (83 images). In both cases, the group of patients was diverse in terms of age (Fig. 1). It is worth emphasizing that the radiographs were performed for medical indications, therefore the various types of the patients' diseases may have influenced the results.

Input images in DICOM format were scaled to 8-bit images with linear mapping of minimum and maximum values to a scale of 0–255. Ultimately, it was the images in this format that were subjected to textural analysis using qMaZda software [14,15]. The first step of the analysis was definition of the region of interest (ROI), which was the cervical vertebral bodies (Fig. 2); however, due to the presence of artifacts or noise, not all of them were always selected. For each ROI, 10412 texture features were calculated: 2076 features for each quantization level (8, 7, 6, 5 and 4 bits/pixel) and 32 features computed on the basis of LBP algorithm [3].

In order to evaluate the relationship between a patient's age and the texture values obtained in qMaZda software, statistical analysis was performed by calculating Pearson's r correlation coefficient.This made it possible to distinguish only those features that significantly correlated with age. Moreover, all texture features were calculated for different quantization levels. This tested whether the bit-depth reduction of images has an impact on this type of research.

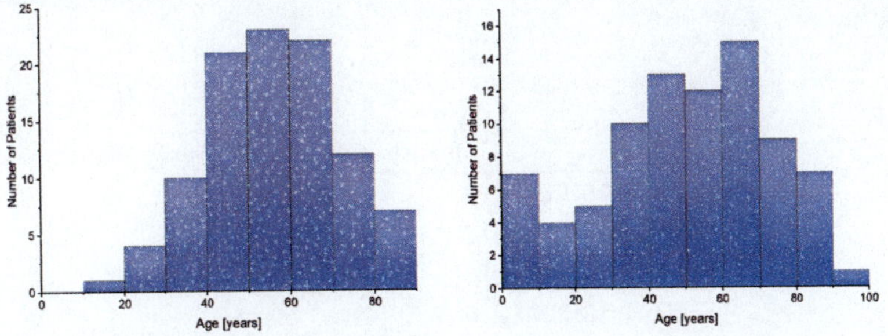

(a) (b)

Fig. 1. Histograms of patients age: (a) females, (b) males

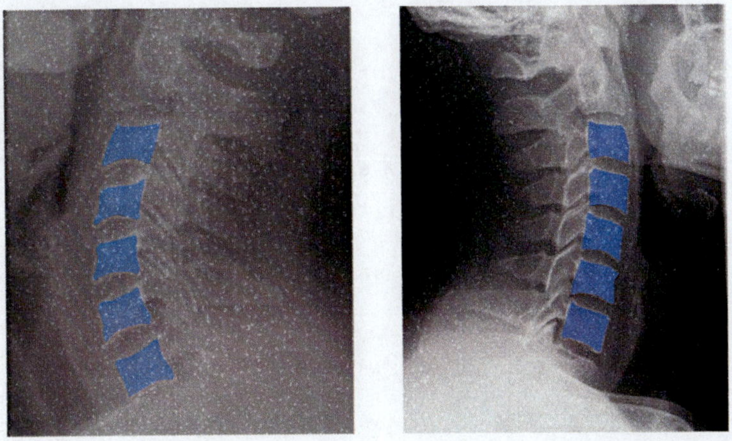

(a) (b)

Fig. 2. X-rays with selected ROI (a) 40-year-old woman, (b) 27-year-old man

3.1 X-Rays of Women's Cervical Spine

The statistical analysis identified the texture features that significantly correlate with a patient's age. For women, these were the traits that were calculated using the Gabor transform [16] (Fig. 3). The Pearson correlation coefficients of these features are shown in Table 1.

Statistically significant results were obtained only for the direction defined as ‘H’ (horizontal); the remaining directions (Z, V, N) did not reveal any relationships. It is worth noting that quantization had no effect on the results in this case. Gray-level reduction caused neither an increase nor decrease in correlation.

Table 1. Pearson correlation coefficient values (women)

Features	Quantization	Pearson’s r
YM8Gab16H8Mag	8	−0.45
YM7Gab16H8Mag	7	−0.45
YM6Gab16H8Mag	6	−0.45
YM5Gab16H8Mag	5	−0.45
YM4Gab16H8Mag	4	−0.45

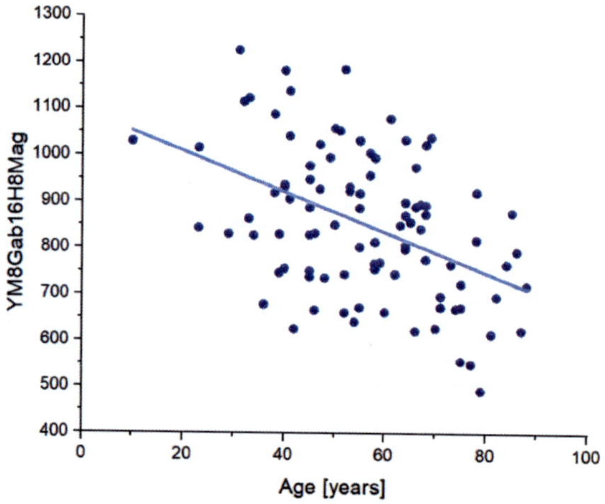

Fig. 3. Change of the YM8Gab16H8Mag value, depending on the patient’s age (women)

3.2 X-Rays of Men’s Cervical Spine

In the case of the male cervical spine X-rays, statistical analysis yielded quite different results than in the case of the X-rays of females. Most importantly, quantization-sensitive features correlated with age and those for which Pearson’s r did not change significantly with bit depth reduction were identified (Table 2).

Texture features that were not significantly affected by bit-depth reduction were computed on the basis of the gray-level co-occurrence matrix (GLCM) using the *SumOfSqs*, *SumAverg* and *SumVarnc* functions [16]. It is also important that the features correlated notably with age, regardless of the direction of the algorithm: horizontal (H), vertical (V), 135° (N) and 45° (Z). Pearson's r values for one of the GLCM features as a function of different quantization values are shown in Table 3.

A completely different situation occurred in the case of features calculated on the basis of the GLCM algorithm using the *SumEntrp* function and the gray-level run-length matrix (GRLM) using the *MGLevNonUn* function [16] (Fig. 4). For these texture features, bit-depth reduction had a very big impact on the results (Table 3).

Table 2. Texture features that correlate with patient's age (men)

Features insensitive to bit-depth reduction	Features sensitive to bit-depth reduction
YS(x)[a]Glcm**H**(y)[b]SumOfSqs	YN(x)Glcm**H**(y)SumEntrp
YS(x)Glcm**H**(y)SumAverg	YN(x)Glcm**V**(y)SumEntrp
YS(x)Glcm**H**(y)SumVarnc	YN(x)Glcm**N**(y)SumEntrp
YS(x)Glcm**V**(y)SumOfSqs	YN(x)Glcm**Z**(y)SumEntrp
YS(x)Glcm**V**(y)SumAverg	YN(x)Grlm**H**MGLevNonUn
YS(x)Glcm**V**(y)SumVarnc	YN(x)Grlm**V**MGLevNonUn
YS(x)Glcm**N**(y)SumOfSqs	YN(x)Grlm**N**MGLevNonUn
YS(x)Glcm**N**(y)SumAverg	YN(x)Grlm**Z**MGLevNonUn
YS(x)Glcm**N**(y)SumVarnc	
YS(x)Glcm**Z**(y)SumOfSqs	
YS(x)Glcm**Z**(y)SumAverg	
YS(x)Glcm**Z**(y)SumVarnc	

[a]x = {8,7,6,5,4} - quantization
[b]y = {1,2,3,4,5}

Table 3. Pearson's r value change for the YS(x)GlcmH4SumOfSqs and YN(x)GlcmH1SumEntrp feature, depending on bit-depth reduction (men)

Quantization (bits/pixel)	Pearson's r (YS(x)GlcmH4SumOfSqs)	Pearson's r (YN(x)GlcmH1SumEntrp)
8	0.68	−0.09
7	0.69	−0.06
6	0.67	0.53
5	0.67	0.60
4	0.66	0.60

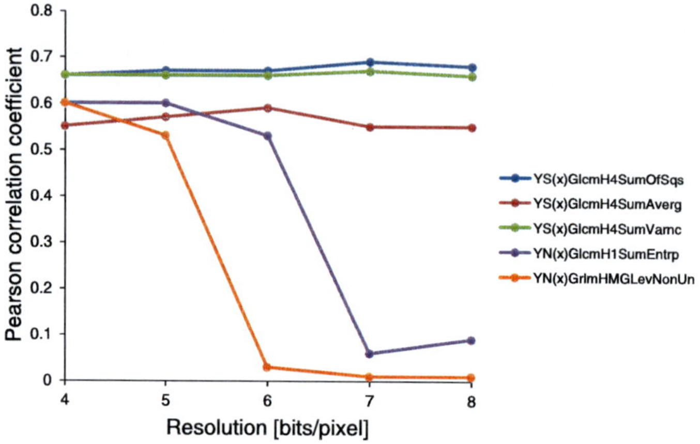

Fig. 4. Comparison of absolute values of Pearson's r depending on bit-depth reduction for selected texture features (men)

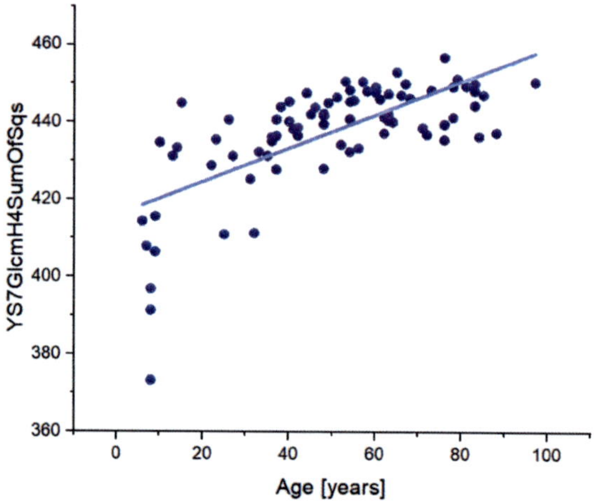

Fig. 5. Change of the YS7GlcmH4SumOfSqs value depending on the patient's age (men)

The highest correlation with age was obtained for the YS7GlcmH4SumOfSqs feature (Fig. 5). It is also worth noting that the Pearson correlation coefficients for the analysis of male patient X-rays are considerably higher than for the analysis of female patient X-rays.

4 Conclusions

Textural analysis of cervical spine radiographs made many observations possible. Features that significantly correlated with the age of patients were extracted.

Due to hormonal factors, analysis was performed separately for men and women. Moreover, the influence of bit-depth reduction on the values of some features was determined.

In the case of female patient X-rays, the features that correlated best with age were those calculated using the Gabor transform. Correlation with age was negative and amounted to -0.45. It was observed that quantization did not affect this feature, therefore the value of Pearson's r remained the same despite the bit-depth reduction. For females, no features sensitive to gray-level reduction were isolated. This does not mean that there were no such features, but their correlation values were too low to be worth taking into account.

Analysis of the male patient X-rays revealed completely different dependencies. In this case, it was possible to distinguish features that are both sensitive and insensitive to bit-depth reduction. The Pearson correlation coefficient remained similar for GLCM features calculated using the SumOfSqs, SumAverg and SumVarnc functions. The differences between the highest and lowest correlations were about 0.03. It is worth noting that for these features the direction of the algorithm also did not significantly affect the results.

The features calculated also from the GLCM matrix but using the SumEntrp function showed completely different properties. In this case, bit-depth reduction had a very large impact on the results. The difference between the highest and lowest Pearson' r values was as much as 0.54. It was noted that higher correlation values were observed for quantization of 6, 5 and 4 bits/pixel. In contrast, for 8 and 7 bits/pixel, the r values indicated no correlation with the age of patients. The features calculated on the basis of the GRLM algorithm showed the same relationships. As in the case of the features that were insensitive to gray-level reduction, the direction of the algorithm did not affect the results.

An interesting fact is that completely different texture features that correlated significantly with age were extracted for women and men, hence the conclusion that analyses of this kind should be carried out by gender. Moreover, the obtained Pearson correlation coefficients had much higher values for the X-rays of male patients.

The texture analysis of X-ray images allowed many observations. However, it is worth carrying out further studies with more data and using texture feature maps [11]. An interesting approach would be to examine the impact of individual diseases on the aging process of the spine. This would make it possible to create a pattern for the early detection of abnormalities.

References

1. Bielecka, M., Obuchowicz, R., Korkosz, M.: The shape language in application to the diagnosis of cervical vertebrae pathology. PloS One **13**(10), e0204,546 (2018)
2. Clausi, D.A.: An analysis of co-occurrence texture statistics as a function of grey level quantization. Can. J. Remote. Sens. **28**(1), 45–62 (2002)
3. Haralick, R.M., Shanmugam, K., Dinstein, I.H.: Textural features for image classification. IEEE Trans. Syst. Man Cybern. **6**, 610–621 (1973)

4. He, L.: Radiomics based on lumbar spine magnetic resonance imaging to detect osteoporosis. Acad. Radiol. **28**, e165–e171 (2020)
5. Kociołek, M., Strzelecki, M., Obuchowicz, R.: Does image normalization and intensity resolution impact texture classification? Comput. Med. Imaging Graph. **81**, 101,716 (2020)
6. Korkosz, M., Gluszko, P., Marcinek, P.: Comment on bone mineral density and bone turnover markers in a group of male ankylosing spondylitis patients. JCR: J. Clin. Rheumatol. **8**(6), 359–360 (2002)
7. Leslie, W., Aubry-Rozier, B., Lix, L., Morin, S., Majumdar, S., Hans, D.: Spine bone texture assessed by trabecular bone score (TBS) predicts osteoporotic fractures in men: the Manitoba bone density program. Bone **67**, 10–14 (2014)
8. Leslie, W.D., et al.: Lumbar spine texture enhances 10-year fracture probability assessment. Osteoporos. Int. **25**(9), 2271–2277 (2014)
9. Mazur, P., Obuchowicz, R., Piórkowski, A.: The influence of age on morphometric and textural vertebrae features in lateral cervical spine radiographs. In: Information Technology in Biomedicine, pp. 71–80. Springer (2021)
10. Nurzynska, K., et al.: Automatical syndesmophyte contour extraction from lateral c spine radiographs. In: Polish Conference on Biocybernetics and Biomedical Engineering, pp. 164–173. Springer (2017)
11. Obuchowicz, R., Nurzynska, K., Obuchowicz, B., Urbanik, A., Piórkowski, A.: Caries detection enhancement using texture feature maps of intraoral radiographs. Oral Radiol. **36**(3), 275–287 (2020)
12. Palanivel, D.A., Natarajan, S., Gopalakrishnan, S., Jennane, R.: Multifractal-based lacunarity analysis of trabecular bone in radiography. Comput. Biol. Med. **116**, 103,559 (2020)
13. Strzelecki, M., Kociołek, M., Materka, A.: On the influence of image features wordlength reduction on texture classification. In: International Conference on Information Technologies in Biomedicine, pp. 15–26. Springer (2018)
14. Szczypiński, P.M., Klepaczko, A., Kociołek, M.: QMaZda–software tools for image analysis and pattern recognition. In: 2017 Signal Processing: Algorithms, Architectures, Arrangements, and Applications (SPA), pp. 217–221. IEEE (2017)
15. Szczypiński, P.M., Strzelecki, M., Materka, A., Klepaczko, A.: MaZda–a software package for image texture analysis. Comput. Methods Programs Biomed. **94**(1), 66–76 (2009)
16. Szczypiński, P.: qMaZda manual. http://eletel.p.lodz.pl/pms/Programy/qmazda.pdF
17. Tamura, H., Mori, S., Yamawaki, T.: Textural features corresponding to visual perception. IEEE Trans. Syst. Man Cybern. **8**(6), 460–473 (1978)
18. Twomey, L., Taylor, J.: The lumbar spine: structure, function, age changes and physiotherapy. Aust. J. Physiother. **40**, 19–30 (1994)
19. Wierzbicki, V., Pesce, A., Marrocco, L., Piccione, E., Colonnese, C., Caruso, R.: How old is your cervical spine? Cervical spine biological age: a new evaluation scale. Eur. Spine J. **24**(12), 2763–2770 (2015)

Advances in ML for Security and Networks

Deep Learning vs. Traditional Approaches to Malware Traffic Classification – A Comparative Study

Jacek Krupski, Damian Rybicki, Waldemar Graniszewski(✉), and Marcin Iwanowski

Warsaw University of Technology, Institute of Control and Industrial Electronics, ul. Koszykowa 75, 00-662 Warszawa, Poland
{jacek.krupski,damian.rybicki,waldemar.graniszewski, marcin.iwanowski}@ee.pw.edu.pl

Abstract. In the paper, decision trees and Bayes classifiers are applied to detect malware traffic in computer networks and compared with the state-of-the-art convolutional neural network reference model. The experiments have been conducted using the known USTC-TFC2016 data set. The results obtained are close to 100% accuracy and are better than the original results produced by the reference method.

Keywords: Malware traffic detection · Decision tree · Bayes classifier · Deep learning · Cybersecurity

1 Introduction

Owing to the wide opening of various computer systems to the Internet, the risk of misuse of the connectivity has recently grown dramatically. It resulted in the increasing importance of cybersecurity issues. While in the early 1980s, attack sophistication was relatively low and demanded mainly social engineering skills and high intruder knowledge, the first incidents were primarily connected with password guessing and cracking. Then later, with a timeline, the tools used became highly developed, and the attacker's skills were not so essential [5]. Due to an exponential increase of computer attacks in the last decades, Intrusion Detection Systems play an essential role in supporting Security Operation Centers or similar Response Teams in organizations. Those systems use different methods such as statistical and machine learning techniques. Most of the efforts are put in this area to develop methods and algorithms to classify the computer network traffic to detect and separate the malware activities and, consequently, protect the whole system. Various approaches are used in malware traffic detection methods. Some of them are following the most recent advances in the machine learning field, which are deep neural networks. In the current research, we have been inspired by one of such methods – study revealed in [10,11].

M. Choraś et al. (Eds.): CORES 2021, IP&C 2021, ACS 2021, LNNS 255, pp. 201–209, 2022.
https://doi.org/10.1007/978-3-030-81523-3_20

This research[1] aims to compare their results with traditional methods, i.e., Bayes and decision trees classifiers, to verify if we can achieve comparable results using the same data set as [11]. Despite the rapidly increasing popularity of deep neural networks, the traditional classifiers are still in common use. The principal factor that determines which approach is suitable for the particular purpose is the input's data complexity. Deep neural techniques exceed the classic methods in complex class separation boundaries in multidimensional feature spaces. They are, however, much more computationally demanding as far as learning and prediction are concerned. Classic approaches are much faster and – sometimes can produce better results in terms of accuracy, especially in the case of not-too-complex data. The current study's goal is to compare the neural classifier [11] applied to malware traffic classification with two classic approaches [4]: Naïve Bayes and decision tree classifiers.

The rest of the paper is organized as follows. Section 2 shows the background and related works focusing on malware traffic classifications and using the same data set. It also describes details of the reference method and reference data set [11]. Section 3 presents alternative classical approaches, i.e., Bayes and decision trees algorithms which we implement in the current paper. Classification results and comparison with a deep learning approach are presented in Sect. 4. Finally, in Sect. 5 we conclude from our studies and propose future works.

2 Related Works and Reference Method

The current research has been motivated by one of the first applications of deep neural networks to malware detection, which is [11]. The authors create and make use of their data set USTC-TFC2016. The above data set gained popularity and was applied to test both deep-learning [6,7,9,12], and classic machine learning [1–3] methods. The decision tree and Naïve Bayes approach were also described in [1]. Additionally, the work does the tests on the same data set. Nevertheless, these are misleading indicators as our research is distinct. First and foremost, we process raw traffic data rather than the selected features of the traffic.

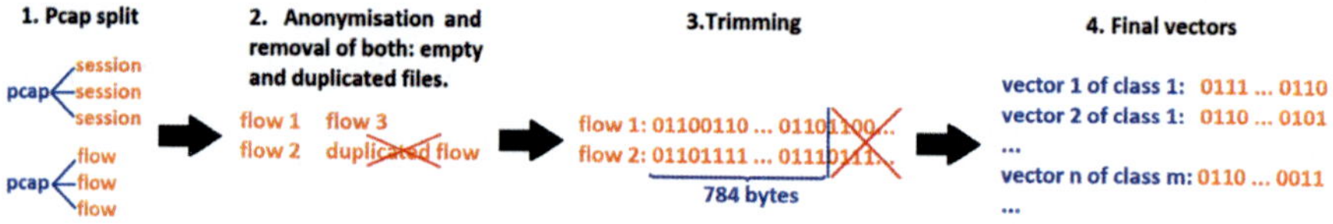

Fig. 1. Data processing steps, inherited from the reference paper.

[1] The research was funded by POB Cybersecurity and data analysis of Warsaw University of Technology within the Excellence Initiative: Research University (IDUB) program.

Another work in the malware detection topic highlights that [11] achieves better results for traffic containing all network layers rather than using the payload data only [13]. Additionally, they point out that network traffic files may consist of information that can bias the classification results. The doubt raised in the article assures us that there is a need to review the classification method proposed and tested on the USTC-TFC2016.

The method used in this paper as a reference has been introduced in [11]. It combines the computer network traffic data processing with the deep neural network, applied initially to classify images. To adapt the 1-D computer network data to the neural network's input, previously used to classify images, the data samples were converted to the 2-D format. The traffic data was represented by the 2-D arrays of size 28×28 bytes. The method, in four steps, transforms captured traffic files (pcap files) into gray-scale images (see Fig. 1). In our paper, we follow the technique until the unified vectors in the third step are formed. We omitted step number four - conversion to images. We also conduct our tests on the USTC-TFC2016 data set.

The first step of the algorithm starts from pcap file splitting with the SplitCap tool. Pcap files contain all details of the captured traffic, including payloads. Important traffic indicators are included in protocols headers, i.a. MAC addresses, IP addresses, and port numbers of computers of both: the sender and the receiver. In the method, there are four split options: 1) flow with information from headers of all protocol data units (PDUs), of the 7-layer ISO Open Systems Interconnection Reference Model (OSI); 2) flow with only L7 OSI PDU header; 3) session with all OSI PDU (L2-L7); 4) session with only L7 OSI PDU. The flow stands for the unidirectional traffic capture, consisting of the 5-tuple, for the sender and the receiver: MAC addresses, IP addresses, port numbers, and flow length. Flow is less detailed than pcap, whereas the session information is a bidirectional traffic exchange split from the original pcap file. Then, in the second step, the authors randomize not only MAC addresses but also IP addresses. Later, empty and duplicate files are removed.

In the third step, all packets are trimmed to a unified size of 784 bytes. This specific size is related to the LeNeT-5 Convolutional Neural Network (CNN), which is known to be used along with the MNIST data set – a collection of thousands of handwritten digits of size 28×28 pixels (with a total number of pixels equal to 784). While trimming, all packets shorter than 784 bytes are padded with zeros. The vectors, of a uniform size of 784 bytes, with a traffic class assigned to each of them, are the elements of the reference data sets. They make the input for both the original method and our experiments described in the following sections. All steps common for our and the reference paper methods are shown in Fig. 1. In the original approach, the 784-byte vector is next transformed into a digital image by wrapping to a matrix of 28 by 28 bytes. The image is a gray level one with possible 256 gray levels.

In the discussed paper, there are two scenarios in which the model was tested. In scenario A, there is a binary classifier that decides whether the sample is benign or malware. Then, two classifications take place: the first one that assigns the malware to the particular category, and the second one in which benign files are matched. Unlike scenario A, scenario B works like a typical antivirus program using one classification, which should assign samples to one out of twenty different classes. These classes are either benign or malware ones.

3 Application of Classic Classifiers

In the current study, we compare the original, deep-neural approach [11], described in the previous section, with the two popular, classic classifiers: decision tree and Naïve Bayes. We follow the original workflow in its preparatory steps, obtaining the vector consisting of 784 bytes taken from the original traffic sample. Since, however, the input to the classifier is 1-dimensional, the original shape of the data was preserved (the 1-D to 2-D transformation has not been done). Next, we replace the original neural network-based model with the two classifiers, and finally, we process the results obtained in a way described in [11].

3.1 Naïve Bayes

The Naïve Bayes classifier is a classic tool and one of the oldest methods for supervised learning. It is based on Bayes probabilistic theory and uses the probabilistic models of data. It aims at producing, in the prediction phase, a vector of probabilities that estimates the likelihood of belonging an input vector to particular classes. The naive model assumes that input variables, elements of the input vector are independent. This assumption is not fulfilled in all situations. In most of the cases, at least, some input variables depend on another. Despite this, the classifier works well for not-too-complicated data. The usual way of modeling the probabilities is based on Gaussian probability distributions. The discriminating function, in this case, is equivalent to the Gaussian function, with parameters – average value and variance – based on the training data in the process of learning.

The likelihood of the features is assumed to be Gaussian:

$$P(x_i \mid y) = \frac{1}{\sqrt{2\pi\sigma_y^2}} \exp\left(-\frac{(x_i - \mu_y)^2}{2\sigma_y^2}\right), \tag{1}$$

where x_i stands for the i-th vector of raw traffic data (consisting of 784 features), y is the output class of the vector, μ_y is the mean of the values in x associated with class y, and σ_y – the Bessel corrected variance of the values in x associated with class y. The parameters σ_y and μ_y are estimated using maximum likelihood.

3.2 Decision Tree

This study's second method is based on decision trees that belong to classic predictive modeling approaches. They are applied to both: classification and regression tasks, depending on the target variable. In our case, classifying the traffic data into one of several given classes belongs to the classification one. A decision tree consists of tests performed on the data attributes that consecutively split the input set into smaller subsets. Tests that are automatically selected by the tree induction algorithm are chosen using the selection criteria based on the class homogeneity within resulting subsets. The decision trees have been already applied to the network traffic classification tasks [1,8] and its collections – random forests [1–3,13].

In the case under study, a learning set of traffic samples consists of labeled samples of the traffic. Each sample is created by taking a list of consecutive bytes of the traffic flow, while its label is related to the traffic class. Tests located in the tree are investigating particular bytes of the sample by checking whether the value is higher/lower than a given threshold. Keeping in mind that single bytes are related to the features, the decision tree is the classifier and a feature selector. Finally, the resulting classification tree contains in its node tests on particular, automatically selected bytes that differentiate between various traffic categories.

Tests of significant features in each vector of tested traffic were performed, and a class was predicted in less than 14 comparisons. Each comparison was made on pixel brightness which takes values in the range of 0–255. In our case, entropy for the information got significantly better performance than Gini index. The other decision tree parameter, which was picked by tuning, was tree depth. The best performance was achieved with a depth value equal to 14.

4 Classification Results

4.1 Evaluation Scenario

There were four tests conducted to check the efficiency of the decision tree. To test the Naïve Bayes classifier, another four tests were made. Each quadruple of tests consisted of the same tests as in the reference paper [11], referred to four split options: 1) all layers (data units, PDUs) – *Flow All* 2) flow with only layer 7 OSI PDU header – *Flow L7* 3) session with all OSI PDU – *Session All* 4) session with only L7 OSI PDU – *Session L7*. Also, we followed, for each classifier, the same scenarios as in the reference: binary classifier, 10-class classifier for benign and malware traffic, and 20-class classifier.

The data set USTC-TFC2016 used in this study is composed of 10 pcaps of unwanted traffic: Cridex (a worm), Geodo (a trojan), Htbot (a backdoor), Miuref (a trojan), Neris (a botnet), Nsis-ay (a botnet), Shifu (a trojan), Tinba (a trojan), Virut (a botnet) and Zeus (a trojan). In the data set, there is also a group of 10 pcaps of benign traffic: BitTorrent, FTP, Facetime, Gmail, MySQL, Outlook, SMB, Skype, Weibo, and WorldOfWarcraft. To comply with the reference method, we did not alter the data set into a balanced one.

Evaluation metrics such as accuracy (A), precision (P), recall (R), and $F1-$ score are used for reporting results. These metrics were evaluated and described in the related work. Method comparison was based on each classifier's overall accuracy, and values of P, R, and $F1$ are presented as the results of every class of traffic.

$$P = \frac{TP}{TP + FP}, \quad R = \frac{TP}{TP + FN}, \tag{2}$$

$$A = \frac{TP + TN}{TP + FP + TN + FN}, \quad F1 = 2 \times \frac{P \times R}{P + R} \tag{3}$$

where TP – true positive is the number of positive samples correctly classified, TN – true negative is the number of negative samples correctly classified, FP – false positive is the number of negative samples incorrectly classified, and FN – false negative is the number of positive samples incorrectly classified.

Table 1. Comparison of accuracy of the three classifiers (%) - session with all OSI PDU

Metric	Average	Binary classifier	10-class classifier		20-class classifier	Classifier
			Normal	Malware		
Accuracy	99,41	100	99,94	98,52	99,17	CNN
Accuracy	100,00	100	100	100	99,98	Decision tree
Accuracy	98,95	100	100	99,99	96,87	Gaussian Naïve Bayes

Accuracy is one of the most intuitive metrics. It is simply a ratio of correctly predicted samples to the total observations. $F1$ score is the weighted average of Precision and Recall. It considers both FP and FN, thus is usually more useful than accuracy, especially if we have an uneven class distribution. Precision is the ratio of TP – observations to the total positively predicted samples ($TP + FP$). Recall is the ratio of TP – observations to the all samples in actual class ($TP + FN$).

4.2 Results Comparison

Following the evaluation scenario described in the previous section, the set of the test was performed. All 784 feature vectors were split into train/test subsets with the proportion 0.8/0.2. In order to get reliable results, a 20-fold split was performed. Furthermore, the final results are average results of 20 tests – one per draw. Table 1 shows the overall accuracy comparison of three classifiers for a different number of classes prediction performed in the related work [11]. Average accuracy in decision tree classifier is better than CNN network evaluation. A decision tree is better in all three cases, and Gaussian Naïve Bayes obtained slightly worse accuracy in the 20-classes scenario.

Table 2 shows the accuracy for different input data. Similarly, as in the related work, the best accuracy was gained for the session with PDUs from all layers. In our case, for normal sessions with only L7 PDUs, obtained results are slightly worse.

Table 2. Accuracy of four representations with two classes of traffic - the decision tree

	FlowAll	FlowL7	SessionAll	SessionL7
Benign traffic	100	62	100	92
Malware traffic	100	100	100	100

Table 3. Comparison of precision, recall and F1 value of 20-classes classifiers (%) - session with all OSI PDU

DATA	CNN			Decision tree			Gaussian Naïve Bayes		
	P	R	F1	P	R	F1	P	R	F1
Cridex	100,0	100,0	100,0	99,9	100,0	100,0	100,0	100,0	100,0
Geodo	100,0	99,9	99,9	100,0	100,0	100,0	100,0	100,0	100,0
Htbot	99,8	100,0	99,9	100,0	100,0	100,0	100,0	100,0	100,0
Miuref	100,0	100,0	100,0	100,0	100,0	100,0	99,8	100,0	99,9
Neris	96,3	92,9	94,6	100,0	100,0	100,0	76,7	99,4	86,6
Nsis	99,8	99,0	99,4	100,0	99,9	100,0	100,0	99,3	99,6
Shifu	99,9	99,9	99,9	100,0	100,0	100,0	100,0	100,0	100,0
Tinba	100,0	100,0	100,0	100,0	99,9	99,9	99,9	99,8	99,9
Virut	90,7	95,6	93,1	99,9	99,9	99,9	98,5	69,9	81,8
Zeus	100,0	100,0	100,0	100,0	100,0	100,0	100,0	100,0	100,0
BitTrt	100,0	100,0	100,0	100,0	100,0	100,0	100,0	100,0	100,0
Facetime	100,0	100,0	100,0	100,0	100,0	100,0	100,0	100,0	100,0
FTP	100,0	100,0	100,0	100,0	100,0	100,0	100,0	100,0	100,0
Gmail	98,4	99,3	98,8	100,0	100,0	100,0	100,0	100,0	100,0
MySQL	100,0	100,0	100,0	100,0	100,0	100,0	100,0	100,0	100,0
Outlook	99,2	98,1	98,7	100,0	100,0	100,0	100,0	100,0	100,0
Skype	99,8	100,0	99,9	99,9	100,0	100,0	100,0	100,0	100,0
SMB	100,0	100,0	100,0	100,0	100,0	100,0	100,0	100,0	100,0
Weibo	100,0	100,0	100,0	100,0	100,0	100,0	100,0	100,0	100,0
Wow	100,0	99,9	99,9	100,0	100,0	100,0	100,0	100,0	100,0

Tables 3 and 4 show comparison of the precision, recall and F1-score of each traffic class in three classification methods. Table 3 shows that eighteen classes achieved very high metrics value. The remaining two classes show a bit lower metrics value. Also, in the Gaussian Naïve Bayes classifier, these classes obtain slightly worse results than CNN. Table 4 shows that classical method can achieve higher or equal results in case of 10-classes comparison.

Table 4. Comparison of precision, recall and F1 value of 10-classes classifiers (%) - session with all OSI PDU

DATA	CNN			Decision tree			Gaussian Naïve Bayes		
	P	R	F1	P	R	F1	P	R	F1
Cridex	100,0	100,0	100,0	100,0	100,0	100,0	100,0	100,0	100,0
Geodo	100,0	100,0	100,0	100,0	100,0	100,0	100,0	100,0	100,0
Htbot	99,8	99,8	99,8	100,0	100,0	100,0	100,0	100,0	100,0
Miuref	100,0	100,0	100,0	100,0	100,0	100,0	100,0	100,0	100,0
Neris	97,1	91,1	94,0	100,0	100,0	100,0	100,0	100,0	100,0
Nsis	99,7	99,3	99,5	100,0	100,0	100,0	100,0	100,0	100,0
Shifu	99,9	99,8	99,8	100,0	100,0	100,0	100,0	100,0	100,0
Tinba	99,9	100,0	99,9	100,0	100,0	100,0	99,9	100,0	99,9
Virut	89,0	96,5	92,6	100,0	100,0	100,0	100,0	100,0	100,0
Zeus	100,0	100,0	100,0	100,0	100,0	100,0	100,0	100,0	100,0
BitTrt	100,0	100,0	100,0	100,0	100,0	100,0	100,0	100,0	100,0
Facetime	100,0	100,0	100,0	100,0	100,0	100,0	100,0	100,0	100,0
FTP	100,0	100,0	100,0	100,0	100,0	100,0	100,0	100,0	100,0
Gmail	99,9	99,7	99,8	100,0	100,0	100,0	100,0	100,0	100,0
MySQL	100,0	100,0	100,0	100,0	100,0	100,0	100,0	100,0	100,0
Outlook	99,6	99,9	99,7	100,0	100,0	100,0	100,0	100,0	100,0
Skype	100,0	100,0	100,0	100,0	100,0	100,0	100,0	100,0	100,0
SMB	100,0	100,0	100,0	100,0	100,0	100,0	100,0	100,0	100,0
Weibo	100,0	100,0	100,0	100,0	100,0	100,0	100,0	100,0	100,0
Wow	100,0	100,0	100,0	100,0	100,0	100,0	100,0	100,0	100,0

5 Conclusions

In the paper, decision trees and Bayes classifiers are applied to detect malware traffic in computer networks and compared with a state-of-the-art convolutional neural network reference model. We decided to take two simple, nonparametric classifiers, such as the two mentioned. Like the deep learning tool from the referenced method, classic classifiers can efficiently process unchanged raw traffic data. Testing these classifiers on the USTC-TFC2016 data set and for the same data vector length, i.e., 784 bytes, have obtained 100 % accuracy – better than the reference study [11]. The main conclusion that one can draw is that less sophisticated machine learning models are explicitly not worse than deep learning networks or even are more useful for some malware detection problems.

References

1. Adeke, J.M., Chen, J., Zhang, L., Mensah, R.N., Tong, K.: An efficient approach based on parameter optimization for network traffic classification using machine learning. In: 2020 7th International Conference on Dependable Systems and Their Applications (DSA), pp. 99–107. IEEE (2020)
2. Chen, L., Jiang, Y., Kuang, X., Xu, A.: Deep learning detection method of encrypted malicious traffic for power grid. In: 2020 IEEE International Conference on Energy Internet (ICEI), pp. 86–91. IEEE (2020)
3. Chen, Y., Li, Z., Shi, J., Gou, G., Liu, C., Xiong, G.: Not afraid of the unseen: a siamese network based scheme for unknown traffic discovery. In: 2020 IEEE Symposium on Computers and Communications (ISCC), pp. 1–7 (2020). https://doi.org/10.1109/ISCC50000.2020.9219734
4. Duda, R.O., Hart, P.E., Stork, D.G.: Pattern Classification. Wiley, Hoboken (2001)
5. Lazarevic, A., Kumar, V., Srivastava, J.: Intrusion Detection: A Survey, pp. 19–78. Springer, Boston (2005). https://doi.org/10.1007/0-387-24230-9_2
6. Lin, K., Xu, X., Gao, H.: TSCRNN: a novel classification scheme of encrypted traffic based on flow spatiotemporal features for efficient management of IIoT. Comput. Netw. **190**, 107974 (2021)
7. Niu, D., Zhang, J., Wang, L., Yan, K., Fu, T., Chen, X.: A network traffic anomaly detection method based on CNN and XGBoost. In: 2020 Chinese Automation Congress (CAC), pp. 5453–5457. IEEE (2020)
8. Tang, Z.Z., Zeng, X., Guo, Z., Song, M.: Malware traffic classification based on recurrence quantification analysis. IJ Netw. Secur. **22**(3), 449–459 (2020)
9. Wang, B., Su, Y., Zhang, M., Nie, J.: A deep hierarchical network for packet-level malicious traffic detection. IEEE Access **8**, 201,728–201,740 (2020)
10. Wang, Z.: The applications of deep learning on traffic identification. BlackHat USA **24**(11), 1–10 (2015)
11. Wang, W., Zhu, M., Zeng, X., Ye, X., Sheng, Y.: Malware traffic classification using convolutional neural network for representation learning. In: 2017 International Conference on Information Networking (ICOIN), pp. 712–717 (2017). https://doi.org/10.1109/ICOIN.2017.7899588
12. Zhang, L., Li, B., Liu, Y., Zhao, X., Wang, Y., Wu, J.: FPGA acceleration of CNNs-based malware traffic classification. Electronics **9**(10), 1631 (2020)
13. Zhao, S., Chen, S., Sun, Y., Cai, Z., Su, J., Su, C.: Identifying known and unknown mobile application traffic using a multilevel classifier. Secur. Commun. Netw. **2019**, 11 p. (2019). https://doi.org/10.1155/2019/9595081. https://www.hindawi.com/journals/scn/2019/9595081/

Using Machine Learning to Detect the Signs of Radicalization and Hate Speech on Twitter

Marcin Kuchczyński[1], Aleksandra Pawlicka[2(✉)], Marek Pawlicki[1], and Michał Choraś[1]

[1] UTP University of Science and Technology, Bydgoszcz, Poland
[2] ITTI Sp. z o.o., Poznań, Poland
apawlicka@itti.com.pl

Abstract. The paper deals with the issue of hate speech and radicalization. Oftentimes, they are spread by means of social media. Twitter lets one express their views in a relatively anonymous way; however, it seems to be a simple, yet effective tool for disseminating offensive or radical contents, too. The paper proposes an effective solution which applies machine learning for detecting signs of radicalization and hate speech in Twitter posts. The authors decided to use the Polish language, which due to the level of its complexity is known to pose a challenge for automated sentiment analysis. The authors also needed to create their own dataset of posts containing hate speech, as prior to the experiment, there existed no such datasets in the language. In the paper, the underlying technologies are first presented, then the course of experiment is described and the final conclusions are given thereafter.

Keywords: Machine learning · Sentiment analysis · Hate speech

1 Introduction

Although access to Internet has been deemed one of basic, human rights [2], cyberspace has also been perpetually used for malicious purposes, such as promoting terrorism, radicalizing people's behaviours and conducting hate speech campaigns. The fact that access to the Internet is more and more widespread (in the years 2000–2020 the number of people using the Internet raised by 1271% [13]), may allow one to assume that the amount of hate speech in it is also going to increase. To make matters worse, the Internet attacks and hateful campaigns ceased to be the domain of cyberspace only; increasingly often they trickle into the real world, by influencing the behaviours of the society or causing shifts in the financial domain, etc. Recently, the attacks have become even more politically-motivated and more radical; they aim at dividing people or even stimulate violent behaviours (the most recent examples: President Trump' false claims of the "stolen" election, which encouraged the storming of the United States Capitol

M. Choraś et al. (Eds.): CORES 2021, IP&C 2021, ACS 2021, LNNS 255, pp. 210–218, 2022.
https://doi.org/10.1007/978-3-030-81523-3_21

[14], mob killing in India caused by a WhatsApp rumour ([30], to name just a few). It is a burning issue of the modern world and finding solutions to it may help save thousands of lives.

The aim of the work presented in this paper was to create a tool – an application – based on an artificial neural network, which would be capable of detecting radicalization or hate speech in text, spread via Twitter. The model would enable detecting offensive words in real time and learn on the go. The user would be able to assess the classification outcome.

The paper is structured as follows: Firstly, the issue of hate speech and radicalization in social media is discussed. Then, the technologies selected for tackling the problem are presented, along with a brief state-of-the-art. Then, the creation of the dataset and the experimental setup have been presented, along with the results of the experiment, followed by the final conclusions.

2 Hate Speech, Machine Learning and Sentiment Analysis

2.1 Radicalization and Hate Speech

In this work, radicalization is understood as the whole process in which individuals manipulate and influence the views of their viewers/readers or followers, so that they become more and more extreme. It should be worrisome, as it may lead to real-life violence or terrorism. Communication plays a crucial role in the process. The rise of the Internet has made it far easier; in fact, the Net has become a very effective source of propaganda [4].

At present, the flood of online information is so severe and overwhelming that most citizens are unable to tell false news from the truth. Often, the extremists' outlets use artificial intelligence to spread propaganda, too [32]. The youth seems to be exceptionally prone to online radicalization [7].

When speaking about the social media which may take part in the radicalization attempts, one usually means:

- Facebook, which has often been used as the synonym for social media in general. Along with similar portals, it is a source of many divides in the society, and is responsible for the spread of many pseudoscientific groups (e.g., the "antivaccers", "flatearthers", climate change skeptics, etc. [5]); this kind of portals tend to keep their users in the so-called information bubble.
- Various chatrooms - according to FBI, extremists' groups have used them for recruitment, too [11],
- Video streaming platforms such as YouTube. They are said to appeal mostly to the youth. In 2018, it was revealed that YouTube contained a recommendation system that promoted various political views, from libertarianism to obvious nationalism [18],
- Online games; they have also been used for spreading propaganda, by means of the in-game chatrooms or forums. Terrorist organizations, such as ISIS or Hezbollah have been reported to use this medium for their purposes [28].

– Lastly, there are microblogs, such as Twitter. They provide a perfect space for expressing one's views and opinions whilst remaining relatively anonymous. Moreover, they are a simple, yet effective tool utilized for spreading propaganda, by means of false, fabricated accounts and bots.

Radicalization goes side by side with hate speech. According to the UN's definition, it is "any kind of communication in speech, writing or behaviour, that attacks or uses pejorative or discriminatory language with reference to a person or a group on the basis of who they are, in other words, based on their religion, ethnicity, nationality, race, colour, descent, gender or other identity factor [31]". Hate speech has become an inseparable element of people's everyday's lives; it is present in the political debates on television, discussions on the Internet forums, or even in newspapers and tabloids.

Creating an effective tool which would be able to spot attempts of radicalization or hate speech on the Internet in an automated way could prove useful, both for the admins of the site and for individual end-users. The following sections will present the tools the authors found to best suit this task.

2.2 Sentiment Analysis

Humans, based on their life experience, are usually able to decide whether a text is mean-spirited or not. At present, machines are capable of reaching similar conclusions by means of the so-called sentiment analysis. It consists in analyzing texts in order to decide whether the opinions or sentiment expressed in it are positive or negative. Sentiment analysis may thus be perceived as a kind of text classification, with sentiment being its classes. However, when compared to traditional classification (i.e., based on topic), it proves to be a much greater challenge. This is so, as instead of extracting just keywords, one needs a greater amount of implicit information [19]. Along with the development of technology, machine learning has become a widely-applied approach for sentiment analysis. The main problems of using this technique consist in the way of extracting complex features, and deciding which kind of values are of greater value. One of the proposed methods is concentrating on semantic features, as they enable disclosing the sematic relationships among words, which in turn may be helpful in performing the sentiment classification [10]. In order to make the words be "understandable" for algorithms and boost the analysis, word embeddings are applied.

2.3 Word Embeddings and Word2Vec

Word embedding is a method for feature learning, part of Natural Language Processing (NLP). In it, based on the context in which they appear, words are represented by real-valued vectors [16]. Representing word in such a way enables mapping every word in a vocabulary to a point in a vector space. According to [22]'s 'distributional hypothesis', words occurring in the same contexts are related or convey similar meaning. Based on this theory, it may be expected that if words are related in a semantic or syntactic way, their embeddings are nearer each other in the vector space, as opposed to unrelated words. By applying this technique, it is possible to capture the information on words from a great number of unlabelled corpora [10].

Word2Vec model [23] is one of the most well-known examples of word embeddings. It makes the words, which share similarities, be near each other in the embedding space [3]. It contains two learning algorithms: the Continuous Bag of Words (CBOW) and Skip-gram; the former predicts the word given its context, the latter predicts the context given a word [21]. The model is said to exceed other word embedding methods, leading to its being used in a variety of applications [15]. Finally, the proposed tool needed the most suitable neural network algorithm.

2.4 LSTM and Other Neural Networks Used for Sentiment Analysis

The Long-Short Term Memory (LSTM) algorithm is based on recurrent neural networks (RNNs). The network has proved useful in an array of applications – it is effective in processing data blocks (e.g., image)s as well as data sequences (such as speech or video). For this reason, LSTM has been used for handwriting analysis, speech recognition, processing texts or even detecting anomalies in network traffic. LSTM networks have been designed to solve the issue of vanishing gradient, which is typical to the traditional RNNs. It is also capable of learning much longer sequences [29].

LSTM is not the only neural network which is applied in sentiment analysis. Other methods which are used for this task are, e.g. Naïve Bayes or Random Forest. Naïve Bayes is the simplest of them, and is known to yield great results in some applications. In a comparative study [6], Naive Bayes and LSTM had very similar results; about 85–87%, which means the model is highly effective. Random Forest has proved successful at detecting hate speech, as shown in [25]; it has also been used to detect radicalization and extremism [24]. In addition, it needs fewer resources than LSTM. However, the LSTM network eventually became the tool of choice, taking into account its immense potential and the fact that Polish is a very complex language.

2.5 State of the Art

So far, there have been several attempts at detecting hate speech on Twitter using machine learning. The choice of the used technologies often depends on the characteristic features of the target language, and the available data. Some of the most remarkable trials were as follows. [9] applied the Linear SVM and Naive Bayes classifiers in order to find hateful language in Twitter posts. [12] compared the outcomes of Logistic Regression, Naïve Bayes, and Support Vector Machine (SVM). [26] compare several approaches to detecting hate speech in Spanish; they found that a combination of a LSTM+MLP neural network is the most effective. The authors of [1] have tested four models: CNN, gated recurrent units (GRU), CNN + GRU, and BERT, in order to find the one which would be most suitable for automatic hate speech detection in the Saudi "Twittersphere". Both BERT and the Neural Networks are used in the literature in a similar task, Fake News detection [8,17]. Finally, [27] have employed an LSTM. However, to our best knowledge, there have not been any experiments detecting hate speech in the Polish language using an LSTM.

3 Experimental Setup

3.1 The Data Used for Experiment

Machine analysis of the Polish language is said to be particularly difficult, as according to [20], Polish is the 3rd most difficult language to learn in the world, beside Chinese and Arabic. Apart from dozens of inflectional forms, a number of additional features make it even more complicated, e.g., its syntax (there are several possible ways of combining words, which makes it possible to interpret the resulting phrases in many ways), proverbs and phrasal verbs (which do not convey the literal meaning) and the fact that in Polish there are multiple homonyms (words which are spelled identically but have various meanings).

As there exist very few datasets of hate speech/offensive words in Polish, and they are usually not publicly available anyway, it has been decided to use Twitter for gathering the data suitable for the experiment. The tweets were downloaded using the Tweepy library. The data was gathered within two time frames. The first one was 6–14 August, 2020, the other – 20–27 October, 2020. Whilst during the former timeframe there were no radical situations recorded in Poland, the latter timeframe fell into the time of the so-called national women strike, related to the controversial abortion ban. At that time, the Internet and social media in Poland were flooded with politically-marked materials (both right- and left-wing), and the society witnessed a great increase in the amount of hate speech. Table 1 shows the phrases used for searching tweets.

Out of the gathered data, 3 files were created:

- bad_tweets.csv – containing 1128 tweets, 787 of them assessed by the authors as offensive and 341 – non offensive. They were searched on the 14th Aug., 2020.
- bad_tweets2.csv – containing 2186 tweets, searched on 27th Oct., 2020. 417 of them were labelled as offensive, 866 – non offensive, and 903 were left unlabelled.
- bad_tweets3.csv – a combination of the other two sets. Altogether, the final set contained about 60 thousand words.

Table 1. The phrases used for searching tweets.

ofiaro + losu	niemyty	mecz	Olimpia	kurza + twarz	zniszczcie	jazda + z + nimi
tępa + dzido	knurze	inteligencja	przegrali	obiecali	zabijcie	

3.2 The Tool's Architecture

The tool consisted of 4 layers; one input layer, one word embedding layer containing 100 neurons, one LSTM layer containing 128 neurons, and the output layer. The output represented two classes – either hate speech or non-hate speech. The input data consisted of the pandas.DataFrame object – a previously refined learning set. The word embedding layer was constructed from the Word2Vec model created by the National Corpus of Polish, Polish Academy of Sciences. The applied model contained 100-dimensional models and based on lemmata and the Skip-Gram neural network architecture. The model contained all the parts of speech; the learning algorithm was the Hierarchical Softmax. The LSTM's two default parameters were changed to experimentally–selected values, namely the dropout and recurrent_dropout were changed to 0.5. In order to check the network's efficiency, another model was created. This model had an additional inner LSTM layer containing 8 neurons.

The dataset is split into training and testing sets, tokenized with the Word2Vec; It is transformed into sequences. The embedding layer is created from the Word2Vec model.

The downloaded posts were structured as follows: tweet_author, text, date_of _publication. Then, a label of "offensive" was added. The initial text contained a great number of words and symbols which were insignificant as far as hate speech detection is concerned. They were tags, some special characters and links to external sites. Although emoticons might have had some value to the analyzed texts, they were also omitted in the model training process.

The next step consisted in feeding the existing Word2Vec model from the Polish Academy of Sciences' website to the tool.

The next stage consisted in creating a neural network model based on word embeddings and an LSTM layer. Initially, the algorithms were fed with 600 labelled tweets, then the number was raised to 1100. It has been observed that with the greater number of samples, the accuracy was far better for the testing set.

For the basic network, the influence of batch_size (i.e., the number of samples fed with each iteration in an epoch) was studied.

The data suggested that the greater the batch size, the faster the model learns. However, the results are sometimes significantly worse. Eventually, in the course the experiment, it turned out that for the model, the best batch_size = 32, and the number of epochs = 1000. Consequently, the two neural network models mentioned before have been trained using these parameters.

The model with one LSTM layer performed better than the one with two LSTM layers. In the training set, the difference between the metrics has been quite significant. In the testing set, the differences were less distinct.

The Fig. 1 shows the confusion matrix for the model with one LSTM layer, whilst Fig. 2 – with two LSTM layers. For all the 483 tested samples, the model with the single LSTM layer classified correctly 90.68% of the samples (438), whilst the one with two LSTM layers – 90.48% samples (437). Additionally, the learning time for the model with two LSTM layers was significantly longer.

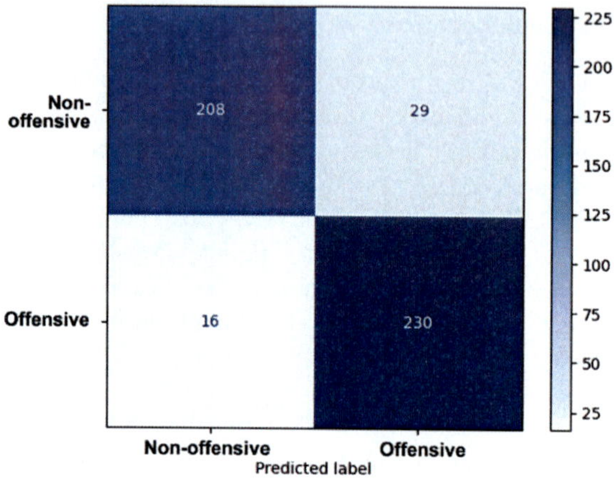

Fig. 1. The confusion matrix for the basic model.

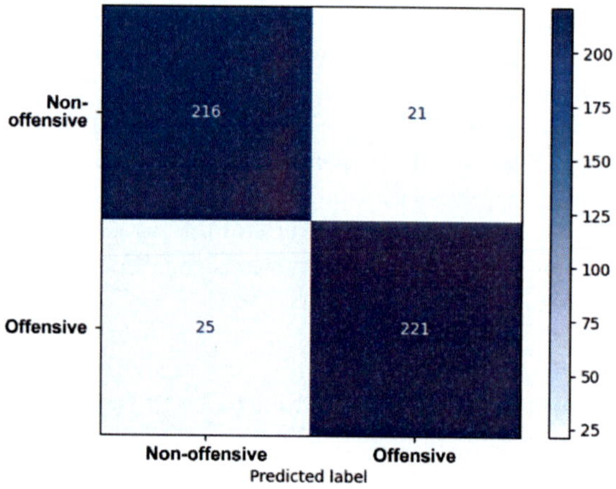

Fig. 2. The confusion matrix for the model with two LSTM layers.

4 Conclusions

This paper has presented the issue of hate speech and radicalization, which are ubiquitous on the Internet today, and which may not only offend people's feelings, but also influence opinions and behaviours, and even result in people being hurt or killed.

This work has a method of hate speech detection in the Polish language. The presented method relies on an artificial neural network, which allows classifying content as hate speech or radicalization. The words were vectorised using word2vec embedding created by the National Corpus of Polish, Polish Academy of Sciences. In order for it to serve its purpose, the text must be labelled correctly, in a reliable and objective manner. It is not an easy task, as one's life experience, knowledge, social background or even individual sense of humour may influence the results of the labelling process. Thus, labelling may prove to be the most crucial step of the process – even the best model may yield unexpected results if the training and testing sets are inconsistent and chaotic. Another conclusion concerns the choice of the data set. For the sake of this paper, the set of about two thousand posts was collected, labelled and used. The classes in the dataset were not in balance as far as the ratio of offensive to non-offensive posts was concerned. Using a more balanced dataset containing more labelled tweets could perhaps produce different results.

As it transpired, adding another LSTM layer did not give better results. On the contrary: it resulted in worsening the performance and making the learning time significantly longer.

In the future, more research should be carried out, using data sets which are larger and more balanced, to check whether it affects the results of the classification.

References

1. Alshalan, R., Al-Khalifa, H.: A deep learning approach for automatic hate speech detection in the Saudi Twittersphere. Appl. Sci. **10**(23), 8614 (2020). https://doi.org/10.3390/app10238614
2. Article 19: UN HRC maintains consensus on Internet resolution (2018). https://tinyurl.com/tp3p7pu3
3. Bengio, Y., Ducharme, R., Vincent, P., Jauvin, C.: A neural probabilistic language model. J. Mach. Learn. Res. **3**, 1137–1155 (2003)
4. Berger, J., Morgan, J.: The ISIS Twitter census defining and describing the population of ISIS supporters on Twitter. Technical report, The Brookings Project on U.S. Relations with the Islamic World, Washington (2015)
5. Bloomfield, E.F., Tillery, D.: The circulation of climate change denial online: rhetorical and networking strategies on Facebook. Environ. Commun. **13**(1), 23–34 (2019). https://doi.org/10.1080/17524032.2018.1527378
6. Bobriakov, I.: Sentiment analysis with naive bayes and LSTM. Data Science Central (2020). https://tinyurl.com/5mdzkf4h
7. Bradshaw, S., Howard, P.N.: The global disinformation order 2019 global inventory of organised social media manipulation. Technical report, Computational Propaganda Research Project (2019). https://tinyurl.com/mz9nf5j8
8. Choraś, M., et al.: Advanced machine learning techniques for fake news (online disinformation) detection: a systematic mapping study. Appl. Soft Comput. **101**, 107050 (2020)
9. De Souza, G.A., Da Costa-Abreu, M.: Automatic offensive language detection from Twitter data using machine learning and feature selection of metadata. In: 2020 IJCNN, pp. 1–6. IEEE (2020). https://doi.org/10.1109/IJCNN48605.2020.9207652
10. Fauzi, M.A.: Word2Vec model for sentiment analysis of product reviews in Indonesian language. In. J. Electr. Comput. Eng. (IJECE) **9**(1), 525 (2019). https://doi.org/10.11591/ijece.v9i1.pp525-530
11. Fbi: How Do Violent Extremists Make Contact? (2021). https://www.fbi.gov/cve508/teen-website/how
12. Gaydhani, A., Doma, V., Kendre, S., Bhagwat, L.: Detecting hate speech and offensive language on Twitter using machine learning: an N-gram and TFIDF based approach (2018)

13. Internet World Stats: Internet Usage Statistics; The Internet Big Picture; World Internet Users and 2021 Population Stats (2021). https://www.internetworldstats.com/stats.htm
14. Jacobo, J.: This is what Trump told supporters before many stormed Capitol Hill. ABC News (2021). https://tinyurl.com/w5aaar5c
15. Jang, B., Kim, I., Kim, J.W.: Word2vec convolutional neural networks for classification of news articles and tweets. PLOS One **14**(8), e0220,976 (2019). https://doi.org/10.1371/journal.pone.0220976
16. Khattak, F.K., Jeblee, S., Pou-Prom, C., Abdalla, M., Meaney, C., Rudzicz, F.: A survey of word embeddings for clinical text. J. Biomed. Inf. X **4**, 100,057 (2019). https://doi.org/10.1016/j.yjbinx.2019.100057
17. Kula, S., Choraś, M., Kozik, R.: Application of the BERT-based architecture in fake news detection. In: Conference on Complex, Intelligent, and Software Intensive Systems, pp. 239–249. Springer (2020)
18. Lewis, R.: Alternative influence; Broadcasting the reactionary right on YouTube. Data & Society (2018). https://tinyurl.com/4pys8w93
19. Liu, B.: Sentiment analysis and subjectivity. Handb. Nat. Lang. Process. **2**(2010), 627–666 (2010)
20. Lyons, D.: The 6 hardest languages For English speakers to learn. Babbel Magazine (2021). https://tinyurl.com/drb83774
21. Ma, L., Zhang, Y.: Using Word2Vec to process big text data. In: 2015 IEEE International Conference on Big Data (Big Data), pp. 2895–2897. IEEE (2015). https://doi.org/10.1109/BigData.2015.7364114
22. McDonald, S., Ramscar, M.: Testing the distributioanl hypothesis: the influence of context on judgements of semantic similarity. In: Proceedings of the Annual Meeting of the Cognitive Science Society, vol. 23 (2001)
23. Mikolov, T., Chen, K., Corrado, G., Dean, J.: Efficient estimation of word representations in vector space (2013). http://arxiv.org/abs/1301.3781
24. Mussiraliyeva, S., Bolatbek, M., Omarov, B., Medetbek, Z., Baispay, G., Ospanov, R.: On detecting online radicalization and extremism using natural language processing. In: 2020 21st International Arab Conference on Information Technology (ACIT), pp. 1–5. IEEE (2020). https://doi.org/10.1109/ACIT50332.2020.9300086
25. Nugroho, K., et al.: Improving random forest method to detect hatespeech and offensive word. In: 2019 ICOIACT, pp. 514–518. IEEE (2019). https://doi.org/10.1109/ICOIACT46704.2019.8938451
26. Pereira-Kohatsu, J.C., Quijano-Sánchez, L., Liberatore, F., Camacho-Collados, M.: Detecting and monitoring hate speech in Twitter. Sensors **19**(21), 4654 (2019). https://doi.org/10.3390/s19214654
27. Pitsilis, G.K., Ramampiaro, H., Langseth, H.: Effective hate-speech detection in Twitter data using recurrent neural networks. Appl. Intell. **48**(12), 4730–4742 (2018). https://doi.org/10.1007/s10489-018-1242-y
28. Ran: Extremists' Use of Video Gaming - Strategies and Narratives (2020)
29. Staudemeyer, R.C., Morris, E.R.: Understanding LSTM - a tutorial into long short-term memory recurrent neural networks (2019)
30. The Washington Post: How rumors on WhatsApp led to a mob killing in India — The Fact Checker. The Washington Post (2020)
31. United Nations Organization: United Nations Strategy and Plan of Action on Hate Speech (2020)
32. Westerlund, M.: The emergence of deepfake technology: a review. Technol. Innov. Manage. Rev. **9**(11), 39–52 (2019). https://doi.org/10.22215/timreview/1282

Clutter-Type Classification of the Telco BTS Units with the Use of Limited Input Information

Lukasz Nowak[1], Michal Panek[1,2](✉), and Ireneusz Jablonski[1,3]

[1] Nokia Solutions and Networks Sp. z o.o. ul. Rodziny Hiszpanskich 8, Regional Office Wroclaw, 02-685 Warszawa, Poland
michal.panek@nokia.com

[2] Department of Systems and Computer Networks, Faculty of Electronics, Wroclaw University of Science and Technology, Wybrzeze Wyspianskiego 27, 50-370 Wroclaw, Poland

[3] Chair of Electronic and Photonic Metrology, Faculty of Electronics, Wroclaw University of Science and Technology, ul. B. Prusa 53/55, 50-317 Wroclaw, Poland

Abstract. In the paper, the real-world problem of predictive modeling is demonstrated. The objective is to associate automatically a clutter type to a telco network element (NE) using a limited input information on its local configuration and performance. For experimental data recorded in operator networks preprocessing actions have been applied which revealed the feature importance in the context of separation of network state space into several baskets (in reference to the clutter type systematics used in telecommunication). Then, selected machine learning models have been designed and applied to solve the problem of binary and multiclassification. Obtained results for the case of real network data measured in two city areas (North America) exhibited that it is possible to differentiate reliably (accuracy close to 90%) between rural and urban characterization of network elements, and even identify three classes of clutters, i.e. rural, urban (including suburban), and dense urban, with reliability significant for the telecommunication applications. Reported results can be further improved and used then for automation of network analytics, its optimization and management. The main contribution concerns the problem algorithmization and data processing in application to automation in telecommunication.

Keywords: Cellular network · Clutter type · Predictive classification · Machine learning · Network automation and management

1 Introduction

Radio Access Network (RAN) forms a pillar of every cellular network. It consists of Base Transceiver Stations (BTS), often referred to as site. BTS is a physical entity, which performs two-way translation between digital information and radio frequency waveform for remote communication. For mobile devices it is an

M. Choraś et al. (Eds.): CORES 2021, IP&C 2021, ACS 2021, LNNS 255, pp. 219–229, 2022.
https://doi.org/10.1007/978-3-030-81523-3_22

equipment providing connection to cellular and external networks. Each BTS is equipped with antennas thanks to which a radio communication is performed. An area covered by distinguishable signal emitted from single BTS, from one or more of its' antennas, is called cell. However, in telecommunication jargon, cell is often referred to as a logical entity bounded to certain BTS, characterized by radio signal frequency (often denoted as carrier) and a set of related configuration parameters. Cellular network is formed by a mesh of cells providing uninterrupted coverage in broad area (e.g. country). Each BTS hosts from one to tens of the cells. End-user terminals are connected via radio interface to the cells to access the cellular network.

Each cell is aware of its surroundings by maintaining neighbor relations, which is a table holding network addresses of nearby cells. The size of cellular network differs significantly and spans from tens of BTSes for small networks to even more than 1 million BTSes for nationwide networks in the biggest countries [3].

According to [1], 55% of people worldwide are the residents of the urban areas in 2018. Modern cellular networks provide communication services to around 97% of world's population [4]. Thus, cellular networks are expected to operate in different geographical, social, economic, etc. conditions. Typically, this diversification is structured through the definition of the clutter type scheme, similarly like presented in [5], where e.g. 'Sea', 'Forest', 'Industrial areas', 'Parks' are listed among various clutter types. The other fact in the field of modern communication networks is the challenge associated to the introduction of zero-touch networks which requires realization of cognitive services for the complex network monitoring and management [6]. Thus, two dimensions cross in the point between internal and external complexity, respectively, due to the network and the surroundings nature, which condition analytical challenges. Distinguishing between the inherited properties of the network elements is strictly associated with their geographical location and parametrization. For example, due to different characteristics of both radio propagation conditions as well as population density and arrangement between urban and rural environments, it can be expected that cellular networks' parametrization and behavior may differ between aforementioned areas (as described in Sect. 2). For telco network analysis, e.g. the performance assessment or the network optimization [2], a knowledge on the type/class of terrain (due to systematics listed in [5]) a certain cell provides coverage for may be crucial. For a network operator, such an information is usually maintained manually or semi-automatically, as base transceiver station (BTS) is frequently not aware of its location or especially BTS doesn't know the features of its coverage area. Manual maintenance assumes a person or group of specialists responsible for assigning terrain type to each cell. Semi-automated mode may use, e.g. radio coverage planning tools for assigning clutter type to a cell. In practice, typically the third parties don't offer a map with details of location valid for network elements of given operator. What is more, even if this would be the case, one should consider the issue of security and raised the cost of service. Parametrization used by vendors and operators need to include the local characteristics of signal propagation, user densities, telecommunication service preferences, etc. which is measured by dedicated squads in a one-time scenario and can bring incomplete information on the dynamic changes of that conditions (significant for dynamic management of network resources like in 5G and beyond networks). The problem formulated above triggers the need to establish a methodology for measurement of local conditions significant for BTS operation, which in turn determines the strategy for BTS parametrization in order to maximize the efficiency of it work. In the paper, the method for depiction of local conditions valid for BTS operation is proposed and linked with the clutter type systematics, all with the use of limited amount of input data.

2 Problem Significance

For a mobile network operator, rural and urban environments differ substantially. First, rural environment is usually characterized by lower population density. For mobile operator it means that single cell of the mobile network can cover larger area than in a city, but still it serves a comparable number of users. Thus, network density in and out of the city will usually differ. Typically, in densely populated areas, operators tend to the cells profile which uses a significant portion or all of the available radio spectrum to provide service for great number of users. In low-populated areas, operators may limit the use of spectrum to lower radio frequencies (e.g. sub-1 GHz), as this range of frequencies provides better propagation conditions. In other words, an effective communication can be maintained on greater distances with the lower frequency. However, it is not a solid rule that low frequencies will be used mostly in rural areas. Other factors, such as spectrum available for certain operator, locally prohibited frequencies or different approach to building network may results in a mixture of frequencies present both in rural and urban areas. Second, traffic generated by end users in rural and urban areas may differ. One example could be a rural area in distant location, where no fixed network connections are possible. It may happen, that high-resolution video streaming services will be more required in rural area than in urban environment, where such a service may be delivered via fixed network more often. Traffic may also differ in terms of its patterns. In rural or sub-urban areas, traffic during a day may be lower due to movements of their residents to the working zones located typically within the city centers. At evenings, the traffic may be greater in residential areas than in business centers. Another example of a difference is a traffic steering policy, which for rural cells may be formulated differently than in urban ones. Here it may happen that due to low network density, the operator may decide to steer more users to lower frequencies, to avoid coverage problems. On the other hand, it may be more relevant to keep the users in high frequencies in the city as much as possible, so that uplink interference impact is limited (due to lower range of propagation on higher frequencies). For a mobile network operator engineers, access to both configuration data as well as data on physical location and coverage maps enables to deduce or directly read on the clutter type assigned to each cell. More problematic are the cases when such an information is requested by automated tools or when the 3rd party entities have a limited access to the operator's data with a physical location. In these cases, it may be necessary to estimate the clutter type based on the available datasets. To compare between two or more mobile networks, it seems crucial to compare the similar types of the networks or their parts. On the other hand, the examples commented above exhibit that each network may consist of parts with different characteristics. Due to the fact that the mobile network consists of up to 1 million nodes, each described by up to more than 10000 parameters and hundreds of performance time series, full node-by-node comparison is not manageable. The clutter-type identification provides a tool which is the enabler to divide a network into the comparable parts in a first step and to compare (and characterize) between these identified sections in the next stage. Such a benchmark is required to evaluate which operators use the most advanced networks and can be considered as a technology leader, which utilizes available resources to the greatest extend.

3 Mobile Network Data Characteristics

The data measured in the cellular network, typically used for network assessment and optimization, can be divided into two types: Configuration Management (CM) data and Performance Management (PM) data. CM data contains information on network node's configuration parameters and its values. Usually, it is modelled as a hierarchical structure reflecting the architecture of Base Transceiver Station (BTS) which is a physical element hosting from one up to tens of cells – logical elements, bounded to a certain range of frequency of radio waves, with physical anchor in form of antennas. Configuration data measured in single BTS may contain even more than 10000 parameters of different nature: categorical (single value from a closed set of allowed values), Boolean (true/false), text (e.g. human-readable identifiers), numerical – parameter values, integer - identifiers. It is worth noting that information on location is not a mandatory part of CM data, even though it conceptually may fit into configuration information. Operators usually use the separate systems to maintain information on physical equipment and its location. In this way, the location data storage within the CM is not a common. Performance Management (PM) data consists of the time series of the performance indicators associated to the certain network elements. For a single NE, hundreds of the performance indicators may be collected. The time aggregation may range from minute to day, with a typical value between 15 min and 1 h. Examples of PM data are, e.g.: the performance measures such as the throughput (transmission speed), the events counters (such as e.g. the number of radio connection requests), the quality measures (such as the percentage of the abnormally disconnected calls), the distribution of measurements of radio conditions reported by terminals connected to the certain cell.

4 Results

4.1 Experimental Protocol

Experiments were designed to answer 3 following questions:

Research Question 1: Are there any clear rules which allows to distinguish network element's clutter type based on expert-selected data?
Research Question 2: How to algorithmize the problem of clutter-type labelling of the network elements using a limited anount of input information?
Research Question 3: What is the performance of the selected machine learning algorithms in the clutter-type labelling of the network elements?

Experiment was divided into two parts. First, the data analysis has been performed. The subject of analysis was the CM data measured in real-world mobile network located in North America and surrounding the medium-size city (denoted as area 1, hereafter). A group of mobile communication experts selected CM parameters which were expected to carry information useful for clutter type estimation. CM data were accompanied with data on physical location of base stations along with the cell characteristics, e.g. the antenna tilts, the azimuths and the transmit power. Thanks to that, it was possible to calculate detailed coverage maps of each cell from the analyzed set. For each cell it was possible to assign it's dominating clutter type. Exploratory analysis of the resulting data set has been performed and preliminary models created to assess the ability to use such a dataset for prediction clutter type. The summary of analysis has been presented in Sect. 4.1.

Second part of analysis introduced changes in a way a dataset is formed (as explained in Sect. 4.3). Analyses were aimed at assessment of the performance of different machine learning algorithms in the task of prediction of the clutter type based on CM data.

4.2 Preliminary Studies - BTS Level

A complete characterization of the network element state requires huge number of parameters, as it was outlined in Sect. 2, and BTS location is one of the information useful for network optimization and management. The data science enables information retrieval using limited amount of data [7]. This case defines the prerequisite for the study reported here. Namely, the objective is to identify the clutter type with the use of a limited number of variables describing the BTS state. The advantage of proposed approach is that it can be used as the self-awareness component in cellular network, which corresponds to the conception of the network self-management and zero-touch networks.

In the study, the experimental configuration and performance data acquired in the LTE networks have been used, recorded in the area around one city in North American operator (denoted hereafter as the 'area 1').

Data need to be labelled so that for each cell a respective clutter type is assigned. Having knowledge on the cell location, the antenna parameters and the transmission power, the coverage area has been calculated for each cell using specialized coverage planning tool [8]. For each cell's coverage area, a land use classes were noted along with respective fraction of area occupied by this terrain type (similar to the ones presented in [5]). For a telecommunication network more important are the characteristics of propagation, traffic and user density than the actual land cover in the area of cell, thus the actual number of classes for estimation of clutter type may be reduced to 4: dense urban, urban, sub-urban and rural. In preliminary analysis, per-cell data was aggregated to the BTS level, following an assumption, that the BTS shall not serve two completely different environments. To assign the clutter type to BTS, the most represented clutter type among the cells in BTS was selected.

Distribution of the classes in the dataset available for training was as follows: Rural - 650, Suburban - 1092, Urban - 6881, and Dense Urban - 50.

The characteristics above exhibit the data imbalance in reported study. At start, the exploratory studies with the use of expert-augmented statistical approach applied to input-output data have enabled to propose a set of input variables limited to configuration data: *expected cell range* - median of expected cell range parameter values across cells in BTS; *number of relations* – total number of neighbor relations for all cells in BTS; *number of cells* – number of cell objects in a given BTS; *number of carrier components* – number of distinct carrier frequencies in a given BTS;

First two parameters provide estimates of network density, which is expected to be higher in urban environments. The latter ones describes BTS in terms of cells and carrier frequencies, which number is expected to be higher in urban areas as well. General statistics for these parameters have been shown in Fig. 1. Figures outline the distribution of values for each attribute per defined clutter type. This simple visualization shows that there is no clear features in selected input configuration parameters that could separate following types of clutters. Only dense urban clutter seems to have slightly different distribution of values. However, it is worth to notice here that we had very small sample of sites labeled as the dense urban clutter.

The next step was to check the inter-parameter relations, i.e. whether the attributes combined together create the separate classes corresponding to clutter types. Exemplary results of this action are presented in Fig. 2. One cannot clearly state on any dependence between those relations and clutter types. The other statistical properties for considered (configuration) input has been estimated, e.g. mutual information, chi-squared test, ANOVA F-value, but all observations obtained in this protocol have led to an observation on complex nature of the data interrelations and the same system under study [9,10]. To assess preliminarily the potential of the analytical approach, to distinguish between four states of clutter type with the use of selected configuration variables four algorithms have been tested: k-Nearest Neighbors (kNN), decision tree, random forest and support vector machines (SVM) [11]. Obtained results are presented in Table 1. The

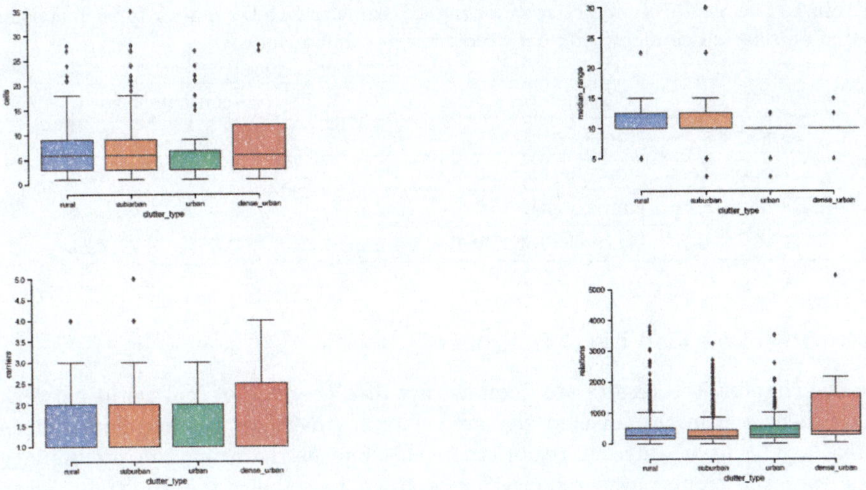

Fig. 1. Distributions of selected parameters per clutter type

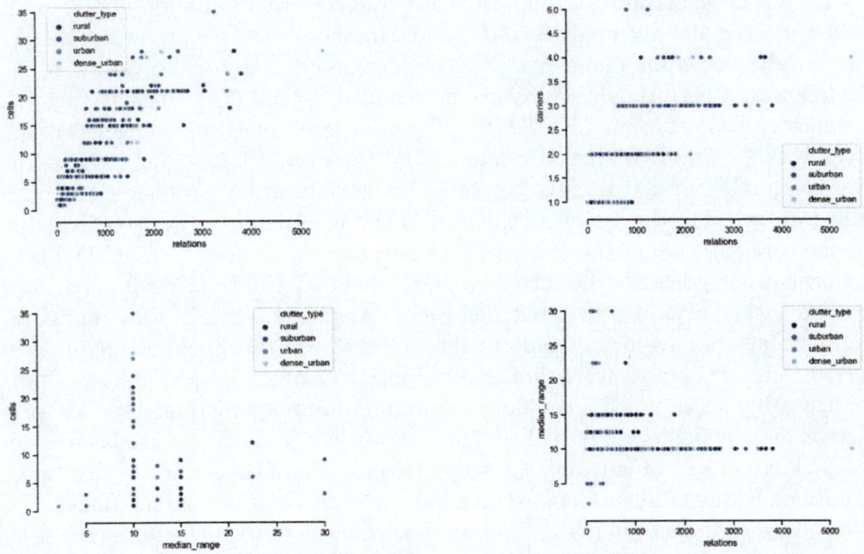

Fig. 2. Interrelations between configuration parameters recorded in operator network projected on possible differentiation directed towards the clutter type identification.

results obtained in this subsection allows to formulate answer to RQ1. Authors were not able to use rule-based approach to distinguish clutter types.

Table 1. The results of predictive modeling experiment with the clutter type clustering based on selected configuration data recorded in cellular network.

Model	Parameters	Accuracy	F1
kNN	k = 5	54.25%	46.80%
Decision Tree	max_depth = 10	61.13%	47.82%
Random Forest	max_depth = 10	61.94%	46.38%
SVM	kernel = rbf, C = 0.1, gamma = scale	63.56%	77.72%

4.3 Results - Cell Level Inference

In the chapter, the results are demonstrated for the case of real-world network data measured in the selected city area (North America), denoted hereafter as area 2. This is a different region than this one used during the preliminary analysis; the change was required since there have been the additional data (mentioned in the previous chapter) available for the authors only for this other location.

Multidimensional data exploration has triggered reformulation of the input vector, strategy for the analysis and the ML methods used for predictive modeling. Firstly, the input vector has been supplemented with 5 more variables, i.e. the following 4 performance data have been added: *LTE_1339A* - The average UE distance to base station, *LTE_806B* - The maximum of RRC connected users, *LTE_5242B* - The E-UTRAN Average RRC Connected UEs, *LTE_717B* - The average number of active users per cell. One configuration parameter has been added as well, i.e. the *Prach_size*, which is the maximum cell range calculated for the configuration of the Physical Random Access Channel ($PRACH$) Those performance metrics are the time series collected on a cell level. As it is outlined further, either mean or required percentile value is analyzed. Two configuration parameters have been removed: the expected cell range and the number of carriers since they presented similar distributions along analyzed classes.

Secondly, we assumed a top-down scenario for applied methodology, i.e. predictive binary clustering – with defined the urban and the rural classes – in the first stage, and broadening the scope to multiclassification in the next stage Additionally, the gradient boosted decision trees (XGBoost) classifier model have been added to the methodology, and the performance of the procedures has been assessed independently for two-, three- and four-class basketing. Finally, the further analyses have been focused on the cell level instead of the BTS level, as a consequence of employing per-cell PM data and with the accordance to the explanation (on the levels of concluding in telecommunication system) given above in Sect. 4.2. The number of cells described with available the CM and the PM data is as follows: Rural - 1062, Suburban - 1357, Urban - 397, Dense Urban - 41. The dataset was split on training and test dataset on ration 4:1 and parameters values has been normalized with L2 normalization before training.

Below, only the best results obtained in the study are presented for binary clustering, three and four baskets - see Table 2. Demonstrated results exhibit that

Table 2. The results of predictive modeling experiment with binary clustering based on the mean KPI aggregation recorded in cellular network elements deployed in area 2.

Classifier	Parameters	2 baskets		3 baskets		4 baskets	
		Accuracy	F1	Accuracy	F1	Accuracy	F1
kNN	n = 7	85.26%	84.00%	85.96%	85.37%	68.77%	44.49%
SVM	kernel = rbf, C = 2	89.47%	87.75%	85.26%	84.33%	71.05%	77.61%
Decision Tree	depth = 10	82.46%	81.00%	81.23%	59.52%	66.32%	52.89%
Random Forest	trees = 10, depth = 10	85.26%	83.19%	83.86%	60.93%	66.32%	52.89%
XGBoost	depth = 10, eta = 0.4, softprob	85.44%	83.46%	81.75%	81.67%	68.77%	44.02%

binary classification can provide a reliable (accuracy 90% and F1 score 88% for SVM model) prediction of the clutter types in the rural and combined the urban zones using a small input information on the network element configuration and performance. Regarding the practical, telecommunication aspects, three-baskets approach can also provide a valuable automation for network element identification, which can be further used, e.g. for comparisons between detailed configuration and performance in order to assure optimized network management. RQ2 and RQ3 were addressed in this part. An algorithmic approach was proposed to solve classification of network elements, and the performance of selected algorithms was examined. It is worth to note here that still there is the space for improvement of reported results. For example, the measurement action of input data should be extended to obtain more input samples, and to balance the representation of cells through following clutters (which have been unbalanced in reported study). Alternatively, methods to deal with data imbalance may be employed. In this study the focus was put on simple methods.

5 Conclusions

Synchronization between the network element parametrization, performance and surroundings (e.g. terrain conditions, population density, service preferences, etc.) contribute telecommunication vendors, service providers and users. In a telecommunication practice, it is not an isolated case when the BTS located in the urban area, due to the error in a manual clutter type labelling, is parametrized as the rural unit, which decreases the system efficiency, the users satisfaction and the business profit. Even the fundamental action for the network engineering, consisting in comparison between the network elements characteristics, can be significantly corrupted by their improper classification to clutter types. What is more, the forefront in wireless communication operates currently in the space of the system and the process digitalization and automation, moving towards the conception of zero-touch networks [6], thus expressed the manual mode need to be replaced with semi- or fully automated functionality, also for the case of clutter type identification.

The methodology of data processing with the use of machine learning applied to real-world configuration and performance datasets is introduced in the paper for the purpose of automation of computer recognition between the classes of the clutter types valid for cellular network elements. A limited amount of input information is used for this purpose. Namely, a bunch of several CM and PM variables has been selected analytically (with the support of domain knowledge) as sufficient to distinguish reliably (accuracy higher than 85%, F1 score higher than 84%) between three basic profiles of the clutters used in telecommunication practice, i.e. the rural, the urban (incl. suburban) and the dense urban. For binary (used categories: rural and urban) classification, SVM modeling can bring to even better accuracy (reaching almost 90%, estimated for data acquired in area 2). Moreover, the paper provides various angles and levels of view on the content of real-world CM and PM data, which triggers possible scenarios for prospective research, both for the objective defined in reported study and also for the other goals which require operation on network configuration and performance details. In reference to reported results, the crucial recommendation concerns supplementing input data with a greater number of samples for model training and also balancing a training dataset. Potentially, this will enable further extending/detailing in distinguishing the clutter types.

Demonstrated in the paper, the telco process studies can be extended to the field of cognitive networks – for example, the efficiency and the accuracy of an automated network optimization and management can be assessed in the next steps with the use of results depicted in this report.

Acknowledgement. Project co-financed by the European Union through the European Regional Development Fund under the Smart Growth Operational Programme 2014-2020. Project implemented as part of the competition of the National Center for Research and Development: Fast Track 6/1.1.1/2017 (Project number POIR.01.01.01-00-1009/17)

References

1. United Nations Department of Economic and Social Affairs, "Population Facts", no. 2020/2, December 2020
2. Dahlen, A., Johansson, A., Gunnarsson, F., Moe, J., Rimhagen, T, Kallin, H.: Evaluations of LTE automatic neighbor relations. In: Proceedings of the 2011 IEEE 73rd Vehicular Technology Conference (VTC Spring), Budapest, Hungary, 2011, pp. 1–5 (2011)
3. https://www.chinatelecom-h.com/en/ir/presentations/annpre190319.pdf. Accessed 15 March 2021
4. ITU - Facts and Figures 2020 - Interactive Report. https://www.itu.int/en/ITU-D/Statistics/Pages/ff2020interactive.aspx. Accessed 15 March 2021
5. International Telecommunication Union, "Telecom Network Planning for evolving Network Architectures. Reference Manual", January 2008
6. Ekudden, E.: Five technology trends augmenting the connected society, Ericsson Technology Review, pp. 2–12, September 2018

7. Dong, G., Liu, H. (eds.): Feature Engineering for Machine Learining and Data Analytics. CRC Press, Boca Raton, USA (2018)
8. https://www.forsk.com/sites/default/files/atoll_brochure.pdf . Accessed 15 March 2021
9. Deisenroth, M.P., Faisal, A.A., Ong, C.S.: Mathematics for Machine Learning. Cambridge University Press, Cambridge (2019)
10. Wackerly, D., Mendenhall, W., Scheaffer, R.L.: Mathematical statistics with applications. Thomson Brooks, Boston (2008)
11. Joshi, A.V.: Machine Learning and Artificial Intelligence. Springer, Cham, Switzerland (2020)

Cyber-Attack Detection from IoT Benchmark Considered as Data Streams

Paweł Zyblewski[1](✉), Marek Pawlicki[2], Rafał Kozik[2], and Michał Choraś[2]

[1] Department of Systems and Computer Networks, Faculty of Electronics, Wrocław University of Science and Technology, Wybrzeże Wyspiańskiego 27, 50-370 Wrocław, Poland
pawel.zyblewski@pwr.edu.pl

[2] Institute of Telecommunications and Computer Science, UTP University of Science and Technology, Bydgoszcz, Poland
chorasm@utp.edu.pl

Abstract. With the advancement of internet technologies, network traffic monitoring and cyber-attack detection are becoming more and more important for critical infrastructure. Unfortunately, there are still relatively few works in the literature that interpret the available benchmark data as data streams and take into account the dynamic characteristics of network packet tracking. The following work proposes an approach to generating data streams from the *IoT-23* dataset and evaluates the resulting streams based on their suitability for use in network intrusion detection research.

Keywords: Cyber-attack · Network intrusion detection · Data stream · Imbalanced data

1 Introduction

The constant advancement of technologies such as the *Internet of Things* (IoT), *Narrowband* IoT, or cloud computing brings many benefits, but also leads to the emergence of many new vulnerabilities and increased user concerns about cybersecurity [4]. Critical infrastructures such as *Sensitive Industrial Plants and Sites* (SIPS) and *Internet Industrial Control Systems* (ICS) are particularly affected, as they must be reliable and capable of operating even in unforeseen circumstances such as breakdowns or deliberate attacks targeting their infrastructure. Although critical structures have systems aimed at preventing such events from occurring and mitigating their effects, often these can become be a target of cyber-attack [11].

One of the currently widely researched tasks related to cyber-attack detection is network intrusion detection. Due to the ever-increasing network traffic in *Cyber-Physical Systems* (CPS), experts more and more often decide to employ machine learning algorithms to automate intrusion detection [6]. Often for this purpose, deep learning algorithms or various hybrid methods are used [9]. It is

M. Choraś et al. (Eds.): CORES 2021, IP&C 2021, ACS 2021, LNNS 255, pp. 230–239, 2022.
https://doi.org/10.1007/978-3-030-81523-3_23

worth noting, however, that despite the dynamic nature of network traffic observation, the current literature related to this issue often treats the available data as a completely static problem, disregarding the possibility of large amounts of data incoming over time [7].

It is this dynamic nature of the problem that makes *lifelong learning* approaches to network intrusion detection gain popularity recently. The notion of *lifelong learning* does not have a specific definition, but can be understood as *the ability to continually learn over time by accommodating new knowledge while retaining previously learned experience* [16]. As its special case, the data stream classification task can be considered, as it is characterized not only by the influx of large amounts of data over time but also by the possibility of the *concept drift* phenomenon occurring [12]. *Concept drift* is the most recognizable characteristic of a data stream, and its occurrence may lead to a shift in the problem's decision boundary, and thus to the degradation of the classification model used. Therefore, data stream classification algorithms must be designed to deal with the potentially fluctuating characteristics of the problem. The most common are batch-based or online ensemble methods that process the stream based on separate windows or incrementally, instance by instance. In addition, the subject of imbalanced data stream classification [21], which is rarely discussed in the literature, should be highlighted, as it combines data stream classification and classification of data with the skewed class distribution [10,19]. In this case, apart from classical imbalance, in which in a binary classification the cardinality of one (minority) class significantly exceeds the cardinality of the other (majority) class, the so-called dynamic imbalance may occur. This phenomenon can be described as a drift in the *prior* class probability. This work tries to treat the task of network intrusion detection as a binary data classification, using the following *stat-of-the-art* methods:

- *Online Bagging* (OB) [15], which incrementally updates base classifiers based on the *Poisson*($\lambda = 1$) distribution,
- *Oversampling-Based Online Bagging* (OOB) and *Undersampling-Based Online Bagging* (UOB) [19], which integrate OB with resampling,
- *Learn ++*.CDS (*Concept Drift with* SMOTE) and *Learn++*.NIE (*Nonstationary and Imbalanced Environments*) [5], extending popular *Learn++*.NSE in order to classify imbalanced stream data,
- *Weighted Aging Ensemble* (WAE) [20], which employs various weights calculation and classifier selection methods.

A very serious problem when classifying imbalanced data streams is access to real-life data. There are overviews of the available datasets [2,14,18], but they usually turn out to be useless due to the problem being too easy, the sequential arrangement of classes, or too low imbalance ratio. The same applies to testing algorithms designed for the needs of network intrusion detection from the data stream. There are, admittedly, relatively recent benchmark datasets, e.g., *IoT-23* [17], *LINET-2020* [3], or *NetML-2020* [1], but as mentioned earlier, they are usually used as static datasets and no attempt is made to use them as data streams. Therefore, this work attempts to generate data streams based on the *IoT-23* dataset, and check whether they are suitable for the evaluation of solutions for cyber-attack detection.

In short, the main contributions of the following work are as follows:

- The proposition of approach to the generation of data streams based on the *IoT-23* dataset,
- Conducting experiments on generated streams using *state-of-the-art online* and *chunk-based* data stream classification methods,
- Determining the possibility of using data streams generated on the basis of the *IoT-23* dataset for the evaluation of *lifelong learning* cyber-attack detection methods.

2 *IoT-23* Dataset as Data Stream

The *IoT-23* dataset consists of 23 subsets representing the network traffic. Twenty of these subsets contain real captures of both malware and benign traffic, and three of them (*Honeypot*) consist entirely of benign packets. The dataset was made available at the beginning of 2020 by Stratosphere Laboratory in Czechia, with the intention of using it for the purposes of designing and evaluating intrusion methods based on machine learning. It contains 20 features with various characteristics, e.g. time-stamp, nominal and numeric values.

As the data from *IoT-23* is to be transformed into streams for the imbalanced data stream classification task, only 20 subsets containing malware traffic were used, as honeypot subsets contain only one class and have relatively small cardinality. First, each of the 20 subsets was treated as a data stream by sorting its instances by timestamp. Then, multiple substreams were created from each data stream by the process of binarization. Substreams were created for the case of benign packets versus all malware packets and benign packets versus each of the malware classes separately. Additionally, a small amount of noise from the normal distribution was introduced to smooth out any changes in the imbalance ratio. In the case of larger datasets (>1 GB), initial sampling was performed (different for each subset). Thanks to this, the characteristics of the stream were preserved, while allowing for its faster processing for the purposes of conducting experiments. All problem features have been encoded to discrete values.

Despite the fact that this process was carried out for all 20 scenarios, only some of them were suitable for the evaluation of imbalanced data stream classification algorithms. The selection was made by examining the imbalance ratio (prior class probabilities) displayed by each of the generated streams. A large number of streams received had an unusually high imbalance ratio, with most chunks containing only the majority class instances. It is an interesting problem and these streams may be used in the future to evaluate anomaly or outlier detection methods, but these tasks are not the subject of this article. Therefore, the 3 scenarios presented in Table 1 were used to carry out the experiments. For each of these scenarios, one of the streams resulting from the binarization was selected:

- *CTU-IoT-Malware-Capture-1-1* – in this case, the stream resulting from the benign vs all approach was chosen. Although it has a low Imbalance Ratio, it was chosen due to the diversification of the streams used in the experiments. Here, no sampling was performed.
- *CTU-IoT-Malware-Capture-33-1* – the stream obtained by comparing benign packets to packets in class 2 Was selected. This stream displays the highest imbalance ratio while still keeping samples from both classes in each data chunk. In this scenario, every seventieth flow was sampled.
- *CTU-IoT-Malware-Capture-43-1* – the stream obtained by comparing benign class to class 3, displaying moderate imbalance ratio, was selected. In this scenario, one out of eighty flows was sampled.

Table 1. Summary of the selected malicious IoT scenarios.

Dataset name	Duration (hrs)	#Packets	Malware name
CTU-IoT-Malware-Capture-1-1	112	1.686.000	Hide and Seek
CTU-IoT-Malware-Capture-33-1	24	54.000.000	Kenjiro
CTU-IoT-Malware-Capture-43-1	1	82.000.000	Mirai

Additionally, in order to obtain a data stream displaying dynamic imbalance - i.e. a drift in prior class probabilities - the *33-1-2–43-1-3* stream was prepared by combining the *CTU-IoT-Malware-Capture-33-1* and *CTU-IoT-Malware-Capture-43-1* scenarios. Figure 1 shows the prior class probabilities in relation to time for each data stream obtained.

3 Experimental Evaluation

This section presents the experiment's motivation and set-up.

Research Questions

The experiment was designed to answer the following questions:

Q1. Can the tested ensemble methods, designed for the imbalanced data stream classification task, be successfully used for the network intrusion detection form data stream?

Q2. Is it possible to use the streams obtained from the *IoT-23* dataset for the evaluation of advanced intrusion detection methods (e.g. lifelong deep learning algorithms)?

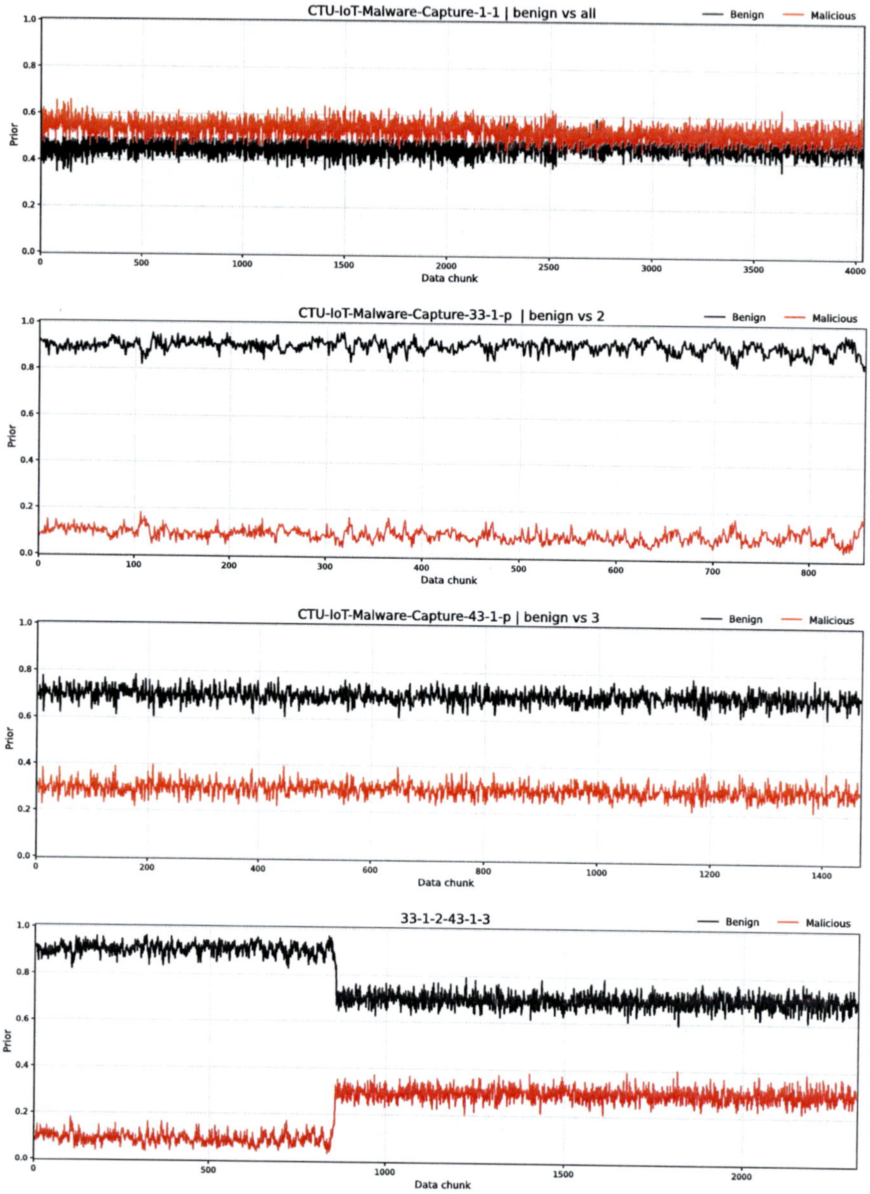

Fig. 1. Prior probabilities of generated data streams.

The evaluation of the conducted experiment was conducted in accordance with *Test-Then-Train* procedure [12], and based on 6 metrics widely used in the problems of imbalanced data classification, namely BAC, $Gmean_s$, F_1 *score*, *precision*, *recall*, and *specificity*. The 6 algorithms described in Sect. 1 were used as reference methods, in which two different base classifiers were used, namely *Gaussian Naïve Bayes* (GNB) and *Hoeffding Tree* (HT). Experiments were conducted using *stream-learn* package [13] and can be replicated according to the *Python* source code which is available on the *Github* repository[1].

3.1 Results of the Experiment

Figures 2, 3, and 4 show successively the results of the conducted experiments for each of the analyzed scenarios. Radar charts show the average values for all six evaluation metrics used, while the runs display the $Gmean_s$ value over the entire stream length.

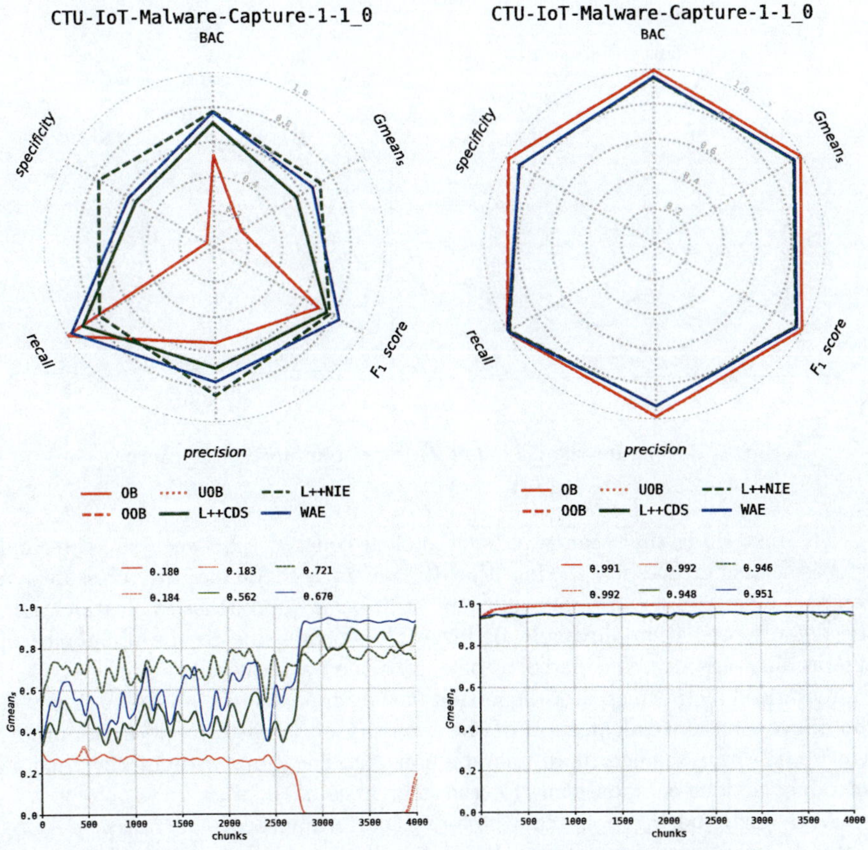

Fig. 2. Results for the *CTU-IoT-Malware-Capture-1-1_0* capture.

[1] https://github.com/w4k2/cores21-ll-iot32.

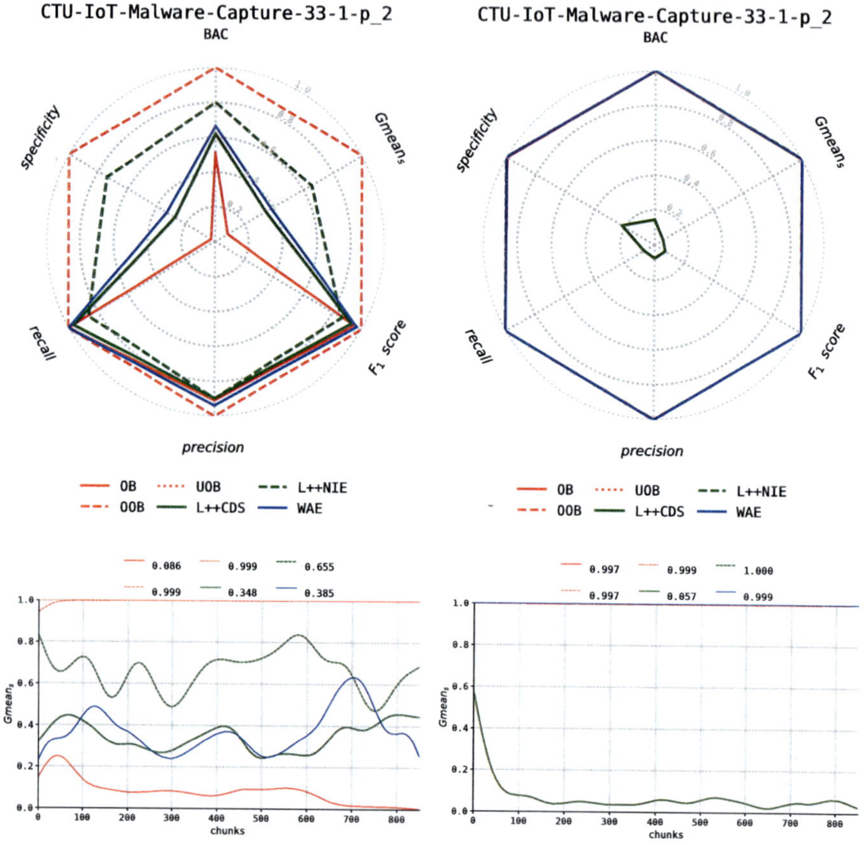

Fig. 3. Results for the *CTU-IoT-Malware-Capture-33-1_2* capture.

The first thing that stands out is the clear difference between the GNB and HT base classifier. Despite the fact that the patterns of the majority class form a compact cluster in the support space for both basic approaches to classification, the default prediction threshold of fuzzy classifiers - due to the imbalance of the problem - is still considered to belong to the patterns of the minority class characterized by much greater dispersion in the support space. In the case of HT, the above-mentioned dispersion of the minority class patterns does not occur, which allows for a much more unambiguous classification, without the need to introduce actions compensating for the imbalance of the sets.

Also worth noting is by far the worst performance of the *Learn ++*.CDS method, which even when using *Hoeffding Trees* achieves metric values much lower than the other approaches, which classify the stream almost perfectly. This can be caused by class overlapping that is worsened by using the SMOTE algorithm [8], which leads to a blurry decision boundary and makes the problem more difficult.

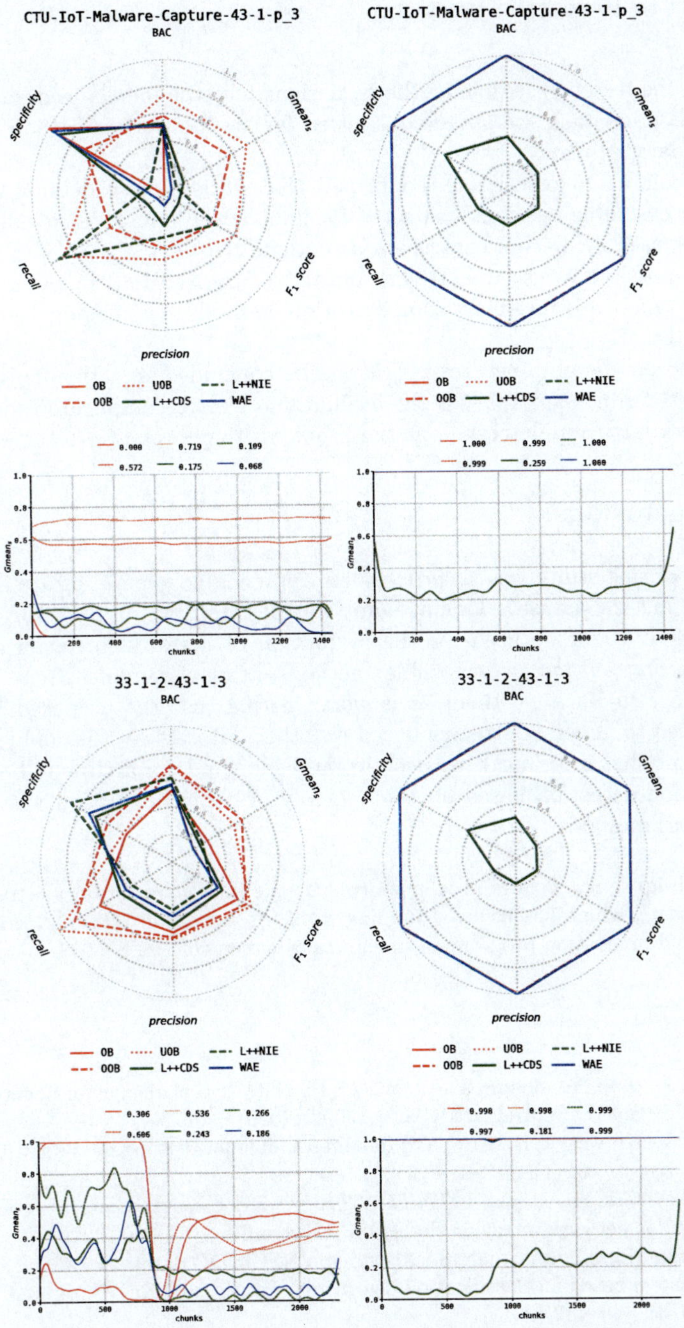

Fig. 4. Results for the *CTU-IoT-Malware-Capture-43-1_p* capture and *33-1-2-43-1-3* merged stream.

Answers to Research Questions

Q1. Can the tested ensemble methods, designed for the imbalanced data stream classification task, be successfully used for the network intrusion detection form data stream?

A1. The conducted experiments confirmed, that *state-of-the-art* chunk-based and online ensemble methods dedicated for imbalanced data stream classification can be used for network intrusion detection.

Q2. Is it possible to use the streams obtained from the *IoT-23* dataset for the evaluation of advanced intrusion detection methods (e.g. lifelong deep learning algorithms)?

A2. Based on the obtained results, it can be concluded that the analyzed data streams are not suitable for the evaluation of more complicated methods of network intrusion detection, as the problems they present are too easy.

4 Conclusions

The aim of this study was to propose an approach to generating data streams from the *IoT-23* dataset. This attempt was dictated by the need for real data streams that could be used for the evaluation of network intrusion detection methods. *State-of-the-art* ensemble methods using *Hoeffding Trees* obtained results close to 100% on them. It is worth noting that the proposed approach can be used to generate streams based on other datasets. Additionally, some of the streams that have not been used in the conducted research may be suitable for intrusion detection based on the anomaly or outlier detection with the use of one-class methods.

Acknowledgment. This work is supported by the InfraStress project https://www.infrastress.eu/, which has received funding from the European Union's Horizon 2020 research and innovation programme under grant agreement No. 833088.

References

1. Barut, O., Luo, Y., Zhang, T., Li, W., Li, P.: NetML: a challenge for network traffic analytics. arXiv preprint arXiv:2004.13006 (2020)
2. Cano, A., Krawczyk, B.: Kappa updated ensemble for drifting data stream mining. Mach. Learn. **109**(1), 175–218 (2020)
3. Damasevicius, R., et al.: LITNET-2020: an annotated real-world network flow dataset for network intrusion detection. Electronics **9**(5), 800 (2020)
4. Dhanabal, L., Shantharajah, S.: A study on NSL-KDD dataset for intrusion detection system based on classification algorithms. Int. J. Adv. Res. Comput. Commun. Eng. **4**(6), 446–452 (2015)
5. Ditzler, G., Polikar, R.: Incremental learning of concept drift from streaming imbalanced data. IEEE Trans. Knowl. Data Eng. **25**(10), 2283–2301 (2013)

6. Dong, B., Wang, X.: Comparison deep learning method to traditional methods using for network intrusion detection. In: 2016 8th IEEE International Conference on Communication Software and Networks (ICCSN), pp. 581–585. IEEE (2016)
7. Dutta, V., Choraś, M., Pawlicki, M., Kozik, R.: A deep learning ensemble for network anomaly and cyber-attack detection. Sensors **20**(16), 4583 (2020)
8. Fernández, A., Garcia, S., Herrera, F., Chawla, N.V.: SMOTE for learning from imbalanced data: progress and challenges, marking the 15-year anniversary. J. Artif. Intell. Res. **61**, 863–905 (2018)
9. Garg, S., Kaur, K., Kumar, N., Kaddoum, G., Zomaya, A.Y., Ranjan, R.: A hybrid deep learning-based model for anomaly detection in cloud datacenter networks. IEEE Trans. Netw. Serv. Manag. **16**(3), 924–935 (2019)
10. Klikowski, J., Woźniak, M.: Employing one-class SVM classifier ensemble for imbalanced data stream classification. In: International Conference on Computational Science, pp. 117–127. Springer (2020)
11. Kozik, R., Choraś, M., Flizikowski, A., Theocharidou, M., Rosato, V., Rome, E.: Advanced services for critical infrastructures protection. J. Ambient. Intell. Humaniz. Comput. **6**(6), 783–795 (2015)
12. Krawczyk, B., Minku, L.L., Gama, J., Stefanowski, J., Woźniak, M.: Ensemble learning for data stream analysis: a survey. Inf. Fus. **37**(Supplement C), 132–156 (2017)
13. Ksieniewicz, P., Zyblewski, P.: stream-learn-open-source python library for difficult data stream batch analysis. arXiv preprint arXiv:2001.11077 (2020)
14. Lu, J., Liu, A., Dong, F., Gu, F., Gama, J., Zhang, G.: Learning under concept drift: a review. IEEE Trans. Knowl. Data Eng. **31**(12), 2346–2363 (2018)
15. Oza, N.C.: Online bagging and boosting. In: 2005 IEEE International Conference on Systems, Man and Cybernetics, vol. 3, pp. 2340–2345 (2005)
16. Parisi, G.I., Kemker, R., Part, J.L., Kanan, C., Wermter, S.: Continual lifelong learning with neural networks: a review. Neural Netw. **113**, 54–71 (2019)
17. Parmisano, A., Garcia, S., Erquiaga, M.: Stratosphere laboratory. A labeled dataset with malicious and benign IoT network traffic (2020)
18. de Souza, V.M.A., Silva, D.F., Batista, G.E.A.P.A.: Classification of data streams applied to insect recognition: initial results. In: 2013 Brazilian Conference on Intelligent Systems, pp. 76–81 (2013)
19. Wang, S., Minku, L.L., Yao, X.: Resampling-based ensemble methods for online class imbalance learning. IEEE Trans. Knowl. Data Eng. **27**(5), 1356–1368 (2015)
20. Woźniak, M., Kasprzak, A., Cal, P.: Weighted aging classifier ensemble for the incremental drifted data streams. In: Flexible Query Answering Systems, pp. 579–588. Springer, Heidelberg (2013)
21. Zyblewski, P., Sabourin, R., Woźniak, M.: Preprocessed dynamic classifier ensemble selection for highly imbalanced drifted data streams. Inf. Fus. **66**, 138–154 (2021)

Check for updates

Some Principles of Building Steganographic Communication Networks

Jurii Ryabinin(✉) and Oleg Finko

North-Caucasus Federal University, 355017 Stavropol, Russia

Abstract. This paper proposes a model of interaction of steganographic systems, which formalizes the description of the principles of construction of steganographic telecommunications. This model is based on the principle of decomposition of the OSI model and includes seven levels: application level, representation level, stegosession level, stegotransport level, stegonetwork level, stegochannel level and stegocontainer level. The main functions of all levels are determined. The article presents the order of interaction of steganographic systems at all levels. A method of redundant data encoding is proposed for the stegochannel level. It is based on an unconventional method of applying codes of the Residue Number System (RNS). The data represented by the RNS code is embedded in steganographic containers. The advantages of the proposed method of coding are increasing the stability of the steganographic channel to the destructive effects of interference and the attackers. An example of the functioning of a steganographic communication network is given. The process of encapsulation and decapsulation of steganographic data at seven levels of the interaction model of steganographic systems is presented. The order of addressing steganographic packets is formulated (in general outline.

Keywords: Steganography · Network · Secure data transmission · Steganographic communication network · Residue number system · Chinese reminder theorem

1 Introduction

Modern communication networks are characterized by regular media exchange (for example, pics, sound, video). Media content can be used as steganographic containers for storing or secure transmission of sensitive data by steganography methods. By ways of embedding payload into a container and types of containers, steganography is classified into: classical, digital, computer (Network Steganography), radio-electronic [1–3]. In this text, "steganography" term will be applied only to digital steganography. In digital steganography, containers mean objects that have analogue nature (that is, the embedding of payload is done in digitized

The authors express their gratitude to S. Makarenko, Doctor of Engineering Science, for valuable comments and suggestions.

M. Choraś et al. (Eds.): CORES 2021, IP&C 2021, ACS 2021, LNNS 255, pp. 240–251, 2022.
https://doi.org/10.1007/978-3-030-81523-3_24

continuous signals). Digital steganography methods solve the following tasks: copy protection, secure document profile, authentication, secure communication, access differentiation [1,2]. Literature analysis showed that most research in the field of steganography is aimed at the implementation of steganographic communication channels (steganographic channels/stegochannels). At the same time, the theory of construction of steganographic communication networks (SCNs) is not sufficiently developed.

Thus, the purpose of this article is to substantiate the possibility of construction SCNs in the interests of formation of hidden communication networks.

2 Problem Solving

2.1 General Statements

SCNs can be built on the basis of and within existing communication networks, which operate according to the general principle set out in the OSI model [4]. It is known [1,2], that most methods of steganography perform embedding of container redundancy in spatial, spectral and/or temporal regions. This redundancy is also used by well-known standards for compressing digital media files (for example, jpeg, jpeg 2000, etc.). Therefore, we can argue that steganographic systems function at the level of OSI model data representation (Fig. 1).

One of the first papers aimed at formalizing data transfer processes in the SCNs is [5]. The author proposed a general reference model of interaction of steganographic systems. In article [6] the authors proposed a conceptual model of interaction of steganographic systems. It reflects the overall approach to the formation of the SCNs and includes seven levels.

In continuation of the presented works, we use the principle of decomposition of the communication network levels of the OSI model to describe the proposed interaction model of steganographic systems for the formation of SCNs (Fig. 1). The model includes 7 levels.

2.2 Application Level

The Application layer (analogue of the OSI Model Application Layer) provides access to common resources—messages and containers. Here, the types of container (photo, video, audio, etc.) and types of message (text, photo, video data, or others) are determined.

2.3 Presentation Level

The Presentation level (analogue of the OSI Model Data Presentation level) is responsible for converting the message (part of the message) and the container (part of the container) to the form needed for execution of data embedding by the steganographic algorithm.

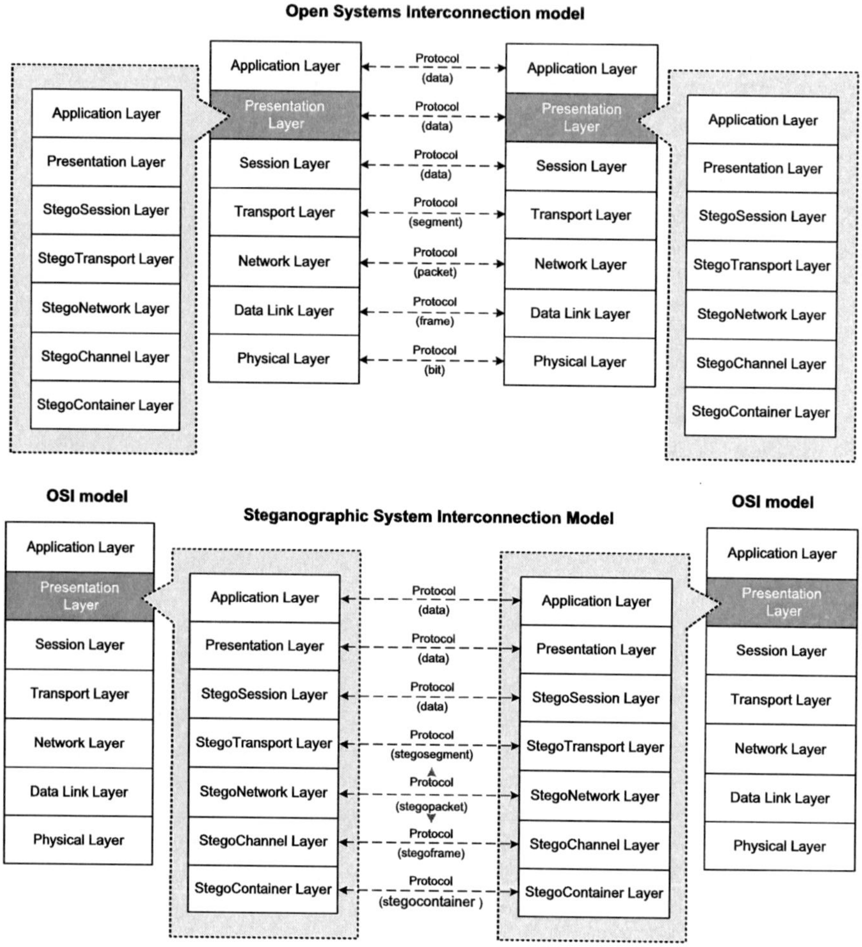

Fig. 1. Different ways to represent the interactivity diagram of logical levels of the interaction model of steganographic systems

2.4 Stegosession Level

Stegosession level functions (analogue of OSI model Session Data Level) are relevant, primarily for steganographic systems using stream containers and embedding data in near-real time. Session synchronization tools must be implemented at this level. This will allow data transfer to maintain information about the status of this transfer as checkpoints and to get back to the last of them in case the session breaks.

2.5 Stegotransport Level

At the *level of stegotransport* (analogue of the Transport layer of the OSI model), the transmitted message is divided into stegosegments, if necessary. At this level, the issues of managing the resources of the SCNs and ensuring a given level of communication quality should be implemented. So, the requirements: urgency, importance of information, capabilities of lower levels to detect and correct errors, etc., should determine the types of service provided. For example: the retention period (buffering) of the information transmitted by the origin source, the procedure for confirming (acknowledgment) of the receipt of the data at the destination or the need for retransmission.

2.6 Stegonetwork Level

The main tasks of the *stegonetwork level* (analogue of the OSI model Network layer) are: to determine the routes of transfer of stegopackets from a source to a recipient. This task requires defining network addresses for SCNs (let's call them stegoaddresses). Stegoaddresses are specified in the stegopacket overhead (Fig. 2). Also, network data encoding operations can be performed here to increase the rate of data transfer over the SCNs. At the stegonetwork level, routers for SCNs operate.

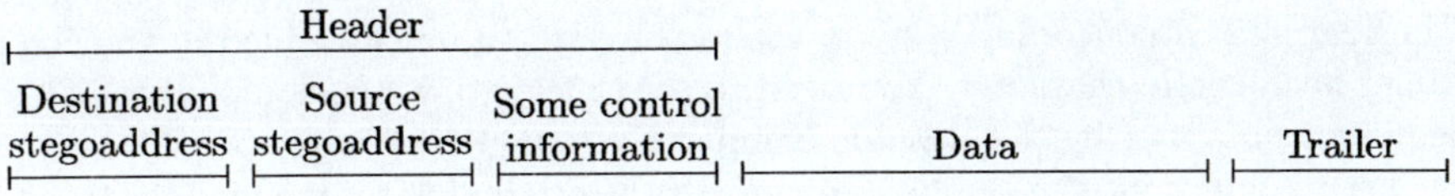

Fig. 2. Stegopacket structure (in general outline)

2.7 Stegochannel Level

The stegochannel level (analogue of the Data Link Layer of OSI model) establishes a logical connection between the interacting Nodes of the SCNs. It prepares the data for the first level, that is, if necessary, splits the stegopacket into stegoframes (the processes of encapsulation and decapsulation of data are represented on Fig. 6). Fragmentation into stegoframes is caused by the steganographic transformation algorithm (stegoalgorithm) used in a particular SCN. A stegoframe consists of a given level overhead field, higher level overhead fields, additional service information (for example, conditioned by a stegoalgorithm), and a data field.

To increase the reliability of the transmitted data, the processes of redundant data encoding must be implemented at the stegochannel level. At the same time, it is known [1,2], that different container formats allow to embed a different amount of payload with a given level of steganographic resistance.

Definition 1. Steganographic resistance—the ability of a steganographic system to hide from stegoanalytics the fact of transferring confidential data, as well as the ability to stealthily resist attempts by an attacker to destroy, distort, delete transmitted data.

Definition 2. Container capacity—the ability of a container to ensure the embedding of payload with a given level of steganographic resistance (calculation of steganographic capacity is developed, for example, in work [7]).

The container utilization factor (efficiency factor) is determined by rule [1]:

$$K_{ef}(T) = \frac{L(C_{\max}(T))}{L(T)}, \tag{1}$$

where C—is the hidden message, T—is the container, $L(C)$—gives the message size, $L(T)$—is the container volume, $L(C_{\max}(T))$—is the container capacity.

The given value of the $K_{ef}(T)$ steganographic system is achieved at $L(C) \leq\leq L(C_{\max}(T))$. Therefore, in the process of redundant data coding, it is important to consider $L(C_{\max}(T))$. In general, different containers can be characterized by different values $L(C_{\max}(T))$.

Way to Redundant Coding. To achieve a given level of steganographic resistance, the amount of embedded data must not exceed the container capacity. Consider such an encoding system, where the data to be embedded in the container is one code character. You need to select the appropriate alphabet. This requirement is met by the Residue Number System (RNS) code.

Let us determine the symbol sequence of the original data intended for transmission over the stegochannel:

$$C_j = (\underbrace{a_{j,0}\, a_{j,1} \cdots a_{j,i_1-1}}_{c_{j,1}} \parallel \underbrace{a_{j,i_1} \cdots a_{j,i_2-1}}_{c_{j,2}} \parallel \cdots \parallel \underbrace{a_{j,i_{k-1}} \cdots a_{j,i_k-1}}_{c_{j,k}}), \tag{2}$$

for $\sum_{s=0}^{k-1} i_s$ digits $a_{j,i_s} \in \{0, 1\}$, where "$\parallel$"—is the concatenation symbol. Then, C_j one could assume: $C_j = c_{j,1} \parallel c_{j,2} \parallel \cdots \parallel c_{j,k}$.

Let $c_{j,1}, c_{j,2}, \ldots, c_{j,k}$; $c_{j,1} = \sum_{s=0}^{i_1-1} a_{j,s}2^s$, $c_{j,2} = \sum_{s=0}^{i_2-1} a_{j,s}2^s$, ..., ..., $c_{j,k} = \sum_{s=0}^{i_k-1} a_{j,s}2^s$—integer non-negative numbers, such that:

$$0 \leq c_{j,1} < 2^{i_1}, \quad 0 \leq c_{j,2} < 2^{i_2}, \quad \ldots, \quad 0 \leq c_{j,k} < 2^{i_k}. \tag{3}$$

Let's take a set of pairwise simple modules $p_{j,1}, p_{j,2}, \ldots, p_{j,k}$:

$$2^{i_1} < p_{j,1}, \quad 2^{i_2} < p_{j,2}, \quad \ldots, \quad 2^{i_k} < p_{j,k}. \tag{4}$$

To minimize module values, we also introduce an optional condition:

$$p_{j,1} < 2^{i_1+1}, \quad p_{j,2} < 2^{i_2+1}, \quad \ldots, \quad p_{j,k} < 2^{i_k+1}. \tag{5}$$

Then the numbers $c_{j,1}$, $c_{j,2}$, ..., $c_{j,k}$ can be declared the smallest non-negative remainders of some number X_j. According to [8], any integer $0 \le X_j << P$, where $P = \prod_{i=1}^{k} p_{j,i}$ uniquely appears to be a sequence $(c_{j,1}, c_{j,2}, \ldots, c_{j,k})_{\text{RNS}}$, where $c_{j,i} = X_j \bmod p_{j,i}$; $i = 1, 2, \ldots, k$; $p_{j,1}, p_{j,2}, \ldots, p_{j,k}$—are pairwise primes. Then the system:

$$X_j = c_{j,1} \bmod p_{j,1}, \quad X_j = c_{j,2} \bmod p_{j,2}, \quad \ldots, \quad X_j = c_{j,k} \bmod p_{j,k}, \tag{6}$$

has a single solution X_j:

$$X_j = \left(\sum_{i=1}^{k} P_i y_i c_{j,i} \right) \bmod P, \tag{7}$$

where $P_i = p_{j,i}^{-1} P$; $y_i = P_i^{-1} \bmod p_{j,i}$; $P = \prod_{i=1}^{k} p_{j,i}$, if the above conditions are met. A non-negative integer X_j can be unambiguously recovered in various ways [9,10].

To protect the transmitted data from errors, the RNS code extension operation (Redundant RNS—RRNS) is performed. Choose redundant grounds $p_{k+1}, \ldots, p_{k+r}$ according to the rule:

$$p_{j,1}, p_{j,2}, \ldots, p_{j,k} < p_{j,k+1} < \cdots < p_{j,k+r}. \tag{8}$$

Get excess remainders $c_{j,k+1} = X_j \bmod p_{j,k+1}; \ldots; c_{j,k+r} = X_j \bmod p_{j,k+r}$.

The proposed mathematical model can be represented as a tree graph isomorphic transformation (Fig. 3). The message (data) C_j is split into k parts. Parts of the message are treated as RNS remainders of some number X_j on a system of pairwise simple modules $p_{j,1}, p_{j,2}, \ldots, p_{j,k}$. An RRNS code extension operation is performed: redundant grounds are entered $p_{k+1}, \ldots, p_{k+r}$ and redundant remainders are obtained $c_{j,k+1}, \ldots, c_{j,k+r}$. The resulting extended system of residues is transmitted over the stegochannel. As shown below, the length $c_{j,i}$, $i = 1, 2, \ldots, k+r$ is determined by the container capacity.

Errors may occur during the data transfer process. Then a sequence will be obtained on the receiving side $(c^*_{j,1}, c^*_{j,2}, \ldots, c^*_{j,k+r})_{\text{RRNS}}$:

$$\begin{cases} c^*_{j,1} = (c_{j,1} + e_{j,1}) \bmod p_{j,1}; \\ c^*_{j,2} = (c_{j,2} + e_{j,2}) \bmod p_{j,2}; \\ \ldots\ldots\ldots\ldots\ldots; \\ c^*_{j,k} = (c_{j,k} + e_{j,k}) \bmod p_{j,k}, \end{cases} \tag{9}$$

where the sequence $(e_{j,1}, e_{j,2}, \ldots, e_{j,k+r})$ corresponds to the error vector.

Next, restore the number changed in the communication channel X^*_j, for example, by rule (7) and perform a check for errors. Error detection is performed following the rule:

$$X^*_j \ge P. \tag{10}$$

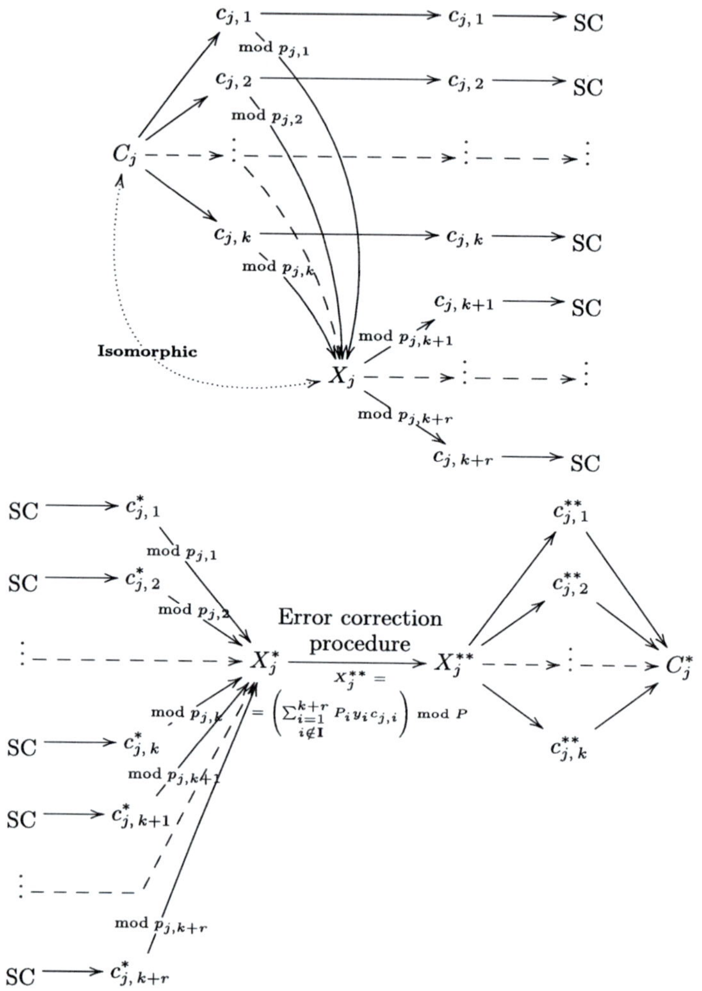

Fig. 3. Structural diagram of isomorphic transformation principle (SC—Steganographic Channel; **I**—set numbers of erroneous residues)

If the condition is not met, the error is detected, the invalid remainder is determined and corrected [8,11]. The q-multiple error means arbitrary invalidation of q characters $c_{j,i}$. Then restore the transmitted message.

The presented mathematical model and method of data encoding for the stegochannel in work [12] are described in more detail. The method allows you to limit the amount of embedded data in the container from above $T_{j,i}$, $i = 1, 2, \ldots, k+r$. This is achieved by selecting modules according to (8) by rules:

$$p_{j,i} > \lceil \text{antlog} \hat{L}_{j,i}(C_{\max}(T_{j,i})) \rceil, \tag{11}$$

$$c_{j,i} < p_{j,i}, \tag{12}$$

$\lceil \cdot \rceil$—brackets mean the largest integer, antlog—anti-logarithm, $\hat{L}_{j,i}(C_{\max}(T_{j,i})) = L_{j,i}(C_{\max}(T_{j,i})) - N_j$; N_j—the amount of service information $\chi_{j,i}$ about the RRNS code modules used (explained below) and the length of the stegoframe overhead $h_{j,i}$. Then according to the proposed model $c_{j,i} = h_{j,i} \parallel \hat{c}_{j,i}$, $\hat{c}_{j,i}$—information data. Therefore, the container capacity determines the size of the information part, taking into account the size of the stegoframe overhead $h_{j,i}$ and the amount of service information $\chi_{j,i}$. The length of the overhead and service information are fixed.

Data $\alpha_{j,i} = c_{j,i} \parallel \chi_{j,i}$ where $\chi_{j,i}$—service information, which is the numbers of the modules used $p_{j,i}$ (selected from tables, in advance known to correspondents) is subject to embedding into the $T_{j,i}$ container. container $c_{j,i}$ to their respective modules $p_{j,i}$.

For the service information to be protected $\chi_{j,i}$ against errors, it is encoded by another method of coding, for example [12], on the principle of majority data coding. Thus, one stegoframe is embedded in one container. Figure 4 presents an example of the stegoframe structure of the $\alpha_{j,i}$ stegochannel level.

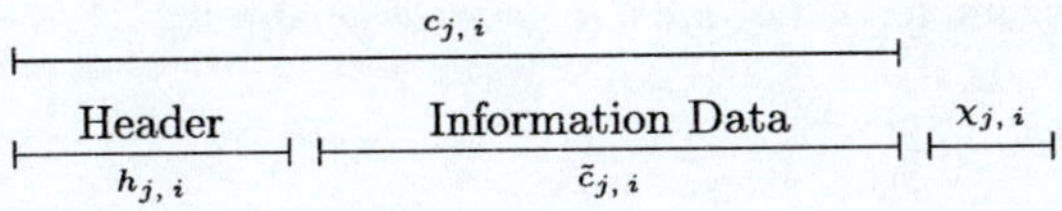

Fig. 4. Stegoframe structure (in general)

Figure 5 presents an example of the implementation of the proposed method of encoding data at the stegochannel level. The example is made for the transmitting side—for three informational and two verification code characters.

Figure: $T_i[0]$—empty container, $T_i[\alpha_i]$—container with built-in payload. The proposed method of coding allows the receiving side to fully recover no more than $r/2$ remainders accepted with error (if all the conditions described in the mathematical model were met).

2.8 Stegocontainer Level

Stegocontainer level is the first level (analogue of the OSI Model Physical Layer). The physical communication channel is characterized by the transmission of signals by means of various types of modulation and methods of communication channel compaction [13]. By analogy, various ways of embedding data are considered at the stegocontainer level. Thus, the stegocontainer level is responsible for embedding stegoframes into containers by digital steganography methods. A number of ways of embedding data are described, for example, in [1,2]. Stegocoder, stegodetector and stegodecoder operate at this level.

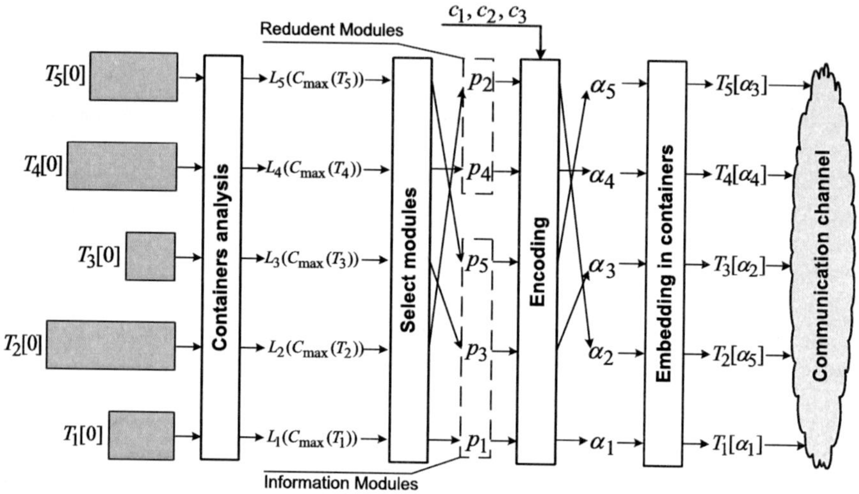

Fig. 5. Implementation of RRNS encoding method for three informational and two verification code characters (transmitting network node)

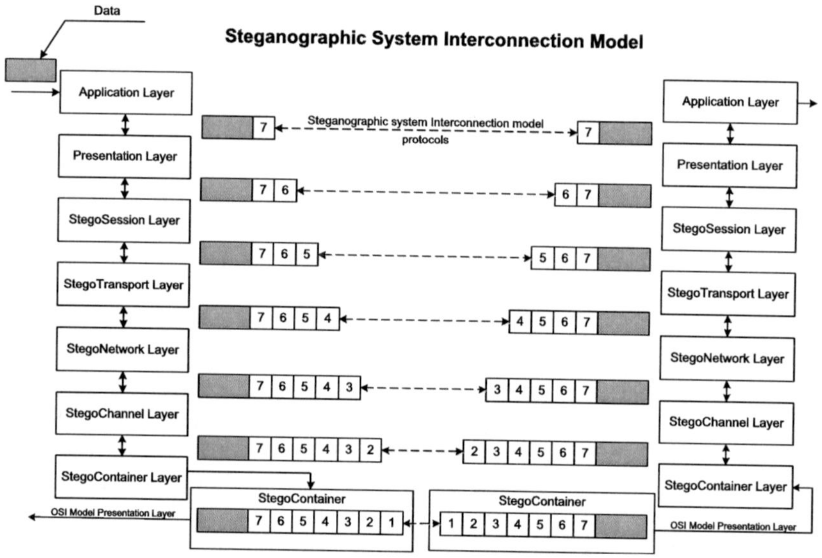

Fig. 6. Processes of data encapsulation and decapsulation in the model of interaction of steganographic systems

2.9 Example of Steganographic Communication Network Implementation

Let us consider an example of an abstract communication network. On the basis of this communication network, an SCN can be formed (an example is presented as an unweighted graph on Fig. 7).

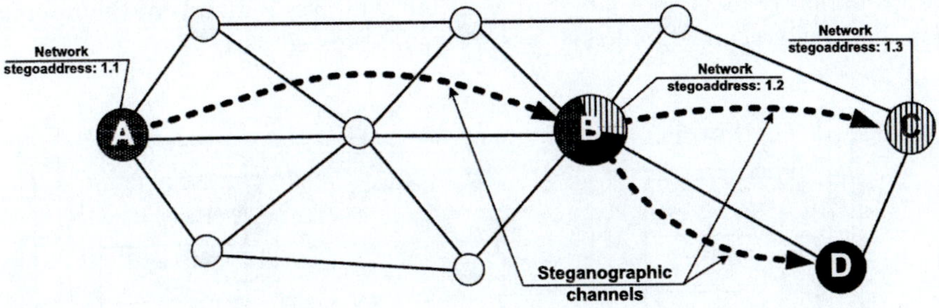

Fig. 7. Example of elementary steganographic communication network

Consider the process of transferring data over SCN from origin source "A" to the recipient "C". The application level of the interaction model of steganographic systems receives a task for transmitting a steganographic message. Through the OSI model representation level, this level executes a request to provide it with a container and message. It should be noted here that the SCN operates on the basis of an existing communication network. Therefore, the data embedding must be carried out in containers currently available at the node, which are to be transferred to some (in advance unknown to which) node. In our case (Fig. 7) Node "A" through the basic communication network on which the SCN is formed, transfers its media data to node "B".

Having got the message and the container, the level initiates encapsulation of the data and transmits it to the lower levels of the interaction model of steganographic systems. Stegopackets are formed at the stegonetwork level (Fig. 2). The stegonetwork level overhead contains the origin node's network stagaaddress and the recipient node's network stagaaddress. In our example (Fig. 7) these are stegoaddresses (conditionally) "1.1" and "1.3" respectively. Next, the stegopacket is transferred to the level of the stegochannel. Here, the stegopacket (in general) can be defragmented into stegoframes. Stegoframes are subjected to a redundant data encoding procedure. Then stegocoder embeds them into containers. The filled container (stego) is transmitted by transport of the OSI reference model to the addressee. Let us recall that for the base communication network, the destination address of the recipient of the stego is the address of "B" node. However, node "B" is an intermediate node for the SCN.

Node "B", having received a filled container, transfers it to the first level of the interaction model of steganographic systems (Fig. 8). Here begins the decapsulation process, which is performed in reverse order. However, after reaching the stegonetwork level and forming a packet (from stegoframes), node "B" determines what is not the end user of a given stegopacket. Its network stagoaddress in this network is "1.2". Therefore, this stegopacket is re-encapsulated and embeded in a (possibly another) container, which is planned, for example, to go to node "C".

The newly formed stego is transferred further by transports of the OSI reference model to the "C" node. This node is the end recipient. Therefore, the decapsulation process does not stop here, but will pass all levels of the interaction model of steganographic systems, reaching the end user.

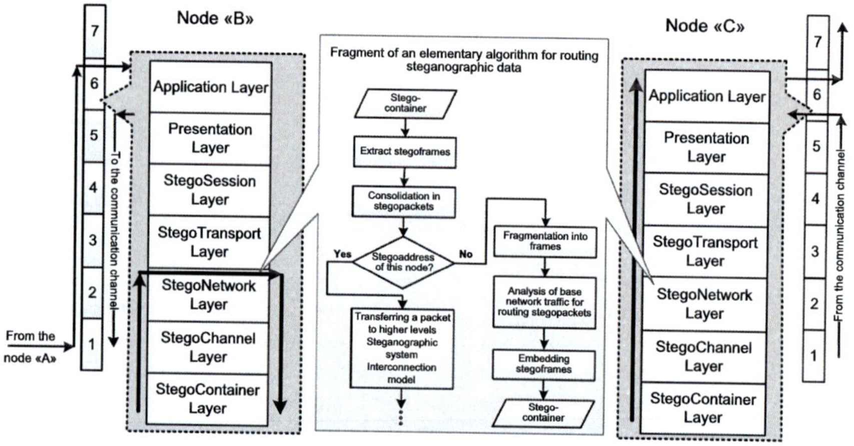

Fig. 8. Processes of encapsulation and decapsulation of a steganographic message at intermediate and end nodes

3 Conclusion

The article describes the general principles of SCN construction. A method of redundant data encoding for the stegochannel level is proposed. The need to decapsulate information to the presentation level of the OSI model on each of the SCN routers leads to additional delays in the transmission of data over the SCN. Also, the input service information, in the form of overheads, additionally limits the hidden carrier capacity of containers [1,2].

References

1. Agranovski, A.V., Balakin, A.V., Gribunin, V.G., Sapozhnikov, S.A.: Steganografiya, cifrovie vodyanie znaki I steganaliz: Monografiya. M.: Vuzovskaya kniga (2009). ISBN 987-5-9502-0401-2 (in Russian)

2. Fridrich, J.: Steganography in Digital Media. Principles, Algorithms, and Applications. Cambridge University Press, Cambridge (2010)
3. Lubacz, J., Mazurczyk, W., Szczypiorski, K.: Principles and overview of network steganography. IEEE Commun. Mag. **52**(5), 225–229 (2014). https://doi.org/10.1109/MCOM.2014.6815916
4. Tanenbaum Andrew, S., Wehterall David, J.: Computer Networks, 5th edn. Prentice Hall, Hoboken (2011). ISBN-13 978-0-13-212695-3
5. Makarenko, S.I.: The steganographic system interconnection basic reference model and the justification of new areas of steganography theory's development: Voprosy kiberbezopasnosti **2**(3), 24–32 (2014). https://www.cyberrus.com/wp-content/uploads/2014/07/vkb_03_04.pdf. Accessed 8 Feb 2020. (in Russian)
6. Abazina, E.S., Cvetkov, K.J.: Konceptualnaya model vzaimodeistviya stegosistem peredachi dannih v sostave etalonnoi modeli vzaimideistviya otkritih system: Trudi Voenno-kosmicheskoi akademii imeni A.F. Mozhaiskogo, No 668. pp. 70–80 (2019). http://trudvka.ru/download/2019/vipusk%20668.pdf. Accessed 8 Feb 2020. (in Russian)
7. Jeremiah, J.H., William, A.P: Capacity of steganographic channels. IEEE Trans. Inf. Theory **55**, 1775–1792 (2009). https://doi.org/10.1145/1073170.1073175. Accessed 8 Feb 2020
8. Omondi, A., Premkumar, B.: Residue Number System. Theory and Implementation. Imperial College Press, London (2007)
9. Finko, O.A.: Large systems of Boolean functions: realization by modular arithmetic methods. Autom. Remote Control **65**(6), 871–892 (2004). https://doi.org/10.1023/B:AURC.0000030901.74901.44
10. Finko, O.A.: Number restoration in the system of residual classes with a minimum number of radices. Eng. Simul. **16**, 329–334 (1999)
11. Mandelbaum, D.M.: On a class of arithmetic codes and a decoding algorithm. IEEE Trans. Inf. Theory **24**, 643–644 (1978)
12. Patent RU 2649753, Int. Cl.H04L 9/00 Method of safe coding information for its transmission over open communication channels using steganography techniques: J.E. Ryabinin, O.A. Finko, D.V. Samoilenko. Date of publication: 04.04.2018. Bull. No 10. https://patents.google.com/patent/RU2649753C2/ru?oq=Patent+RU+2649753
13. Sklar, B.: Digital Communications. Fundamentals and Applications, 2nd edn. Prentice Hall, Hoboken (2001). ISBN 0-13-084788-7

SAT-Based Cryptanalysis of Salsa20 Cipher

Sylwia Stachowiak[1(✉)], Mirosław Kurkowski[2], and Artur Soboń[3]

[1] Department of Computer Science, SWPS University of Social Sciences and Humanities, Warsaw, Poland
sstachowiak@swps.edu.pl

[2] Institute of Computer Sciences, Cardinal Stefan Wyszyński University, Warsaw, Poland
m.kurkowski@uksw.edu.pl

[3] Developers Division, HSBC Service Delivery, Kraków, Poland

Abstract. Modeling through Boolean encoding is often used for the investigation of various kinds of algorithms or other computer systems. Such an approach can be used successfully to cryptanalysis of symmetric ciphers. In this case, cipher testing begins with encoding the cipher algorithm into a Boolean propositional formula. Then the randomly selected bits representing plaintext and cryptographic key are encoded as formulas too. Specially dedicated programs, called SAT solvers, which can solve an SAT problem, can then compute ciphertext values for such inputs. The final cryptanalysis is an attempt at computing a key value from plaintext and ciphertext. In this way, we perform cryptanalysis with plaintext and ciphertext. In this paper, we show how SAT techniques behave for the Salsa20 cipher which is one of a stream cipher widely used in many security systems of computer networks. In our work, we compared a number of selected SAT solvers. Some are relatively old but still very efficient and some are modern and popular.

Keywords: Symmetric ciphers · Satisfiability · SAT-based cryptanalysis · Salsa20

1 Introduction

Boolean encoding can be used for investigating algorithms properties in many areas of computer science [2,14,15,23]. Using this approach, often a whole investigated system is encoded as a boolean propositional formula. The encoding is prepared to express the desired property as a satisfiability (SAT) problem for the formula obtained. The SAT problem results from the hardness of computing whether exists a boolean valuation that satisfies a given logical formula [2,23]. It is well known that SAT is an NP-complete problem. It was stated in the famous Cook's Theorem. NP-completeness states that every problem in the NP class can be reduced to this problem in polynomial time.

M. Choraś et al. (Eds.): CORES 2021, IP&C 2021, ACS 2021, LNNS 255, pp. 252–266, 2022.
https://doi.org/10.1007/978-3-030-81523-3_25

Paradoxically in many cases, despite the hard computational complexity of SAT solving, there are programs that can often compute a solution (variables propositional valuation) for rather large and complex logical formulas, consisting sometimes of hundreds of thousands of variables. A good example of a scientific field where the SAT can be successfully applied is cryptanalysis.

Described above procedures, among other techniques, can be successfully used for proving the security properties of many kinds of security systems including ciphers and their applications. A good example is cryptanalysis of symmetric ciphers or their parts/modifications. This idea well supplements other classic and successfully used cryptanalysis methods [5,6,13,20,21]. In the literature can be still found new papers devoted to this topic [7,10,11,16–19,22,24–27].

In this paper, we investigate how SAT techniques behave for the Salsa20 cipher which is one of a stream cipher widely used in many security systems of computer networks. In particular, we apply direct boolean encoding to encode the Salsa20 and investigating how known SAT solvers deal with cryptanalysis of this cipher. In our research, we investigate cryptanalysis with plain and ciphertexts.

The rest of this article is organized as follows. In Sect. 2, we present the basics of the Salsa20 algorithm, presenting all parts of the cipher from simple 4-byte word operations, to the Salsa20 hash function, to the Salsa20 encryption function. We do this to the extent necessary to explain our ideas and Boolean encoding. Section 3 contains information about the cryptanalysis procedure and the direct, boolean encoding algorithm of the analyzed cipher. In Sect. 4 we present the obtained experimental results. At the end of the article, we will present conclusions and future directions related to our research.

2 Salsa20

Salsa20 is a symmetric streaming cipher that was developed in 2005 by Bernstein. This cipher was submitted to the eSTREAM project organized by the EU ECRYPT network in 2004–2008, which aimed to propose new stream ciphers suitable for common use. Salsa20 is built on a pseudorandom function based on the operations: add-rotate-xor (ARX) - adding words, the XOR operation, and the rotation operation. The internal state consists of sixteen 4-byte words from 0 to 15 placed in a simple square matrix (Fig. 1).

Salsa20 consists of eight keywords, two-stream position words, two nonce words (additional bits of stream position), and four constant words that depend on the key length. In the case of a 32-bit key, the constant words are "expand 32-byte k". For a 16-byte key, the constant words mean "expand 16-byte k". With the 64-byte input data string constructed in this way, the algorithm uses the expansion function Salsa20, appropriate depending on the key length, which using the hash function Salsa20 and returns the 64-byte sequence of output data. In the last step, the data returned by the extension function is XORed with a 64-byte plaintext string. In this way, we obtain a ciphertext with a length of 64-bytes.

"expa"	key	key	key
key	"nd 3"	nonce	nonce
pos.	pos.	"2-by"	key
key	key	key	"te k"

Fig. 1. Initial state of Salsa20

The hash function works by breaking the 64-byte string received from the expansion function into 4-byte words, then subjecting each word to the so-called *liitleendian* function to rearrange the bytes in the word. In the next step, the algorithm transforms the word sequence using 10 rounds of the *doubleround* function, using the functions: column rounds and row rounds, which modify the columns and rows of the word matrix, respectively, using the *quarterrouround* function transforming words using the addition and rotation. The following operations are performed with 4-byte words which can be written hexadecimally (Fig. 2).

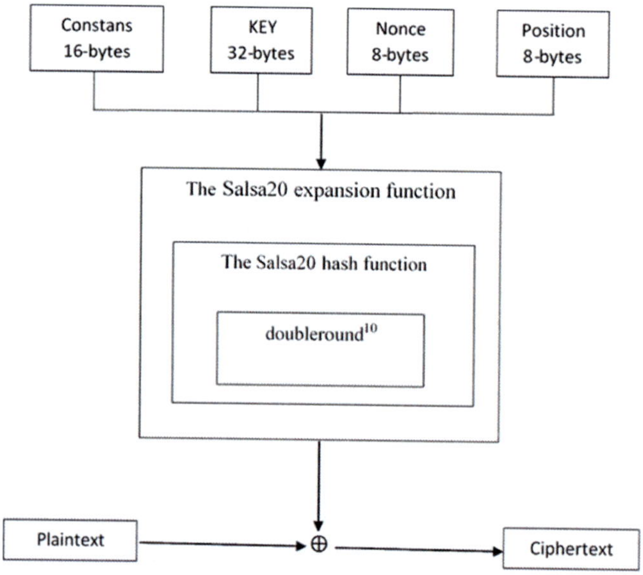

Fig. 2. A scheme of the Salsa20 cipher

2.1 Operations on 4-Byte Words Used in Salsa20

The Sum of Two Words. We calculate the addition of two 4-byte words according to the formula: $(u + v) mod\ 2^{32}$. The result of this action is 4 bytes long. For example, $0x1234abcd + 0x56789ef0 = 0x68ad4abd$.

The Exclusive-or of Two Words. The XOR operation of the two words u and v, we write as $u \oplus v$, is the sum of u and v without carry. In other words, if $u = \sum_i 2^i u_i$ and $v = \sum_i 2^i v_i$ then $u \oplus v = \sum_i 2^i (u_i + v_i - 2u_i v_i)$. For example, $0x1234abcd \oplus 0x56789ef0 = 0x444c353d$.

The w-Bits Left Rotation of a Word u. For $w \in \{0, 1, 2, 3, \ldots\}$ we denote the w-bits left rotation of a word u $u <<< w$ and its result is a unique, non-zero word, equal to $(2^w \cdot u)\, mod\ 2^{32} - 1$, except $0 <<< w = 0$. In other words, if $u = \sum_i 2^i u_i$ then $u <<< w = \sum_i 2^{i+w\ mod\ 32} u_i$. For example, $0x1234abcd <<< 4 = 0x234abcd1$.

The Quarterround Function. The quarterround function takes 4 words as input and returns a 4 word string. If y denote a sequence of 4 words and $y = (y_0, y_1, y_2, y_3)$, then: $quarterround(y) = (z_0, z_1, z_2, z_3)$, where

$$\begin{aligned} z_1 &= y_1 \oplus ((y_0 + y_3) <<< 7), \\ z_2 &= y_2 \oplus ((z_1 + y_0) <<< 9), \\ z_3 &= y_3 \oplus ((z_2 + z_1) <<< 13), \\ z_0 &= y_0 \oplus ((z_3 + z_2) <<< 18). \end{aligned}$$

The quarterround function is executed directly, i.e. without the need for additional memory allocation: first y_1 becomes z_1, then y_2 becomes z_2, then y_3 becomes z_3, and finally y_0 becomes z_0. Each modification used in the quarterround function is invertible, so the entire function is invertible.

The Rowround Function. Let y is a sequence of 16 words. Then $rowround(y)$ is also a sequence of 16 words. Let $y = (y_0, \ldots, y_{15})$. Then $rowround\,(y) = (z_0, \ldots, z_{15})$, where:

$$\begin{aligned} (z_0, z_1, z_2, z_3) &= quarterround\,(y_0, y_1, y_2, y_3), \\ (z_5, z_6, z_7, z_4) &= quarterround\,(y_5, y_6, y_7, y_4), \\ (z_{10}, z_{11}, z_8, z_9) &= quarterround\,(y_{10}, y_{11}, y_8, y_9), \\ (z_{15}, z_{12}, z_{13}, z_{14}) &= quarterround\,(y_{15}, y_{12}, y_{13}, y_{14}). \end{aligned}$$

It is easy to see that the input data can be represented as a $4x4$ simple matrix:

$$\begin{pmatrix} y_0 & y_1 & y_2 & y_3 \\ y_4 & y_5 & y_6 & y_7 \\ y_8 & y_9 & y_{10} & y_{11} \\ y_{12} & y_{13} & y_{14} & y_{15} \end{pmatrix}.$$

The rowround function modifies the rows of this matrix using the quarterround function. In the first row, the rowround function modifies y_1 first, then y_2, then y_3, and finally y_0; on the second row, the rowround function modifies y_6 first, then y_7, then y_4, then y_5; in the third row, the rowround function modifies y_{11} first, then y_8, then y_9, and finally y_{10}; in the fourth row, the rowround function transforms y_{12}, then y_{13}, then y_{14}, and finally y_{15}.

The Columnround Function. The column round function proceeds in the same way as the row round function. If we denote a sequence of 16 words by x then *columnround*(x) is also a sequence of 16 words. If $x = (x_0, \dots, x_{15})$ then $columnround\,(x) = y = (y_0, \dots, y_{15})$, where:

$$\begin{aligned}(y_0, y_4, y_8, y_{12}) &= quarterround(x_0, x_4, x_8, x_{12}),\\ (y_5, y_9, y_{13}, y_1) &= quarterround\,(x_5, x_9, x_{13}, x_1)\,,\\ (y_{10}, y_{14}, y_2, y_6) &= quarterround\,(x_{10}, x_{14}, x_2, x_6)\,,\\ (y_{15}, y_3, y_7, y_{11}) &= quarterround\,(x_{15}, x_3, x_7, x_{11})\,.\end{aligned}$$

We can write a formula equivalent to it. Such formula has the following form:

$$\begin{aligned}&(y_0, y_4, y_8, y_{12}, y_1, y_5, y_9, y_{13}, y_2, y_6, y_{10}, y_{14}, y_3, y_7, y_{11}, y_{15})\\ &= rowround\,(x_0, x_4, x_8, x_{12}, x_1, x_5, x_9, x_{13}, x_2, x_6, x_{10}, x_{14}, x_3, x_7, x_{11}, x_{15})\,.\end{aligned}$$

Like the rowround function, the inputs $(x_0, \dots, x_{15})$ can be represented as a square simple matrix:

$$\begin{pmatrix} x_0 & x_1 & x_2 & x_3\\ x_4 & x_5 & x_6 & x_7\\ x_8 & x_9 & x_{10} & x_{11}\\ x_{12} & x_{13} & x_{14} & x_{15}\end{pmatrix}.$$

The columnround function is nothing more than a transposition of the rowround function: it modifies the columns of the matrix by introducing a permutation of each column through the quarterround function. Thus, in the first column, this function modifies x_4 first, then x_8, then x_{12}, and finally x_0; in the second column, the columnround function modifies x_9 first, then x_{13}, then x_1, and finally x_5; in the third column, the columnround function modifies x_{14} first, then x_2, then x_6, then x_{10}; in the fourth column, columnround transforms x_3 first, then x_7, then x_{11}, and finally x_{15}.

The Doubleround Function. The argument of this function is a sequence of 16 words marked as x. The function returns a sequence of 16 words as defined:

$$doubleround\,(x) = rowround(columnround\,(x)).$$

The *doubleround* function first modifies parallel columns of input data and then performs transformations on the parallel rows of data resulting from the previous operation. Each word is modified twice.

The Littleendian Function. Let b be a 4 byte sequence. Then *littleendian*(b) is the word. If $b = (b_0, b_1, b_2, b_3)$ then $littleedian\,(b) = b_0 + 2^8 b_1 + 2^{16} b_2 + 2^{24} b_3$. Note that the littleendian function is invertible.

The Salsa20 Hash Function. Let x be a 64-byte string. Then $Salsa20(x)$ is a 64-byte sequence, which we get from the formula:

$$Salsa20\,(x) = x + doubleround^{10}\,(x)\,.$$

Looking at the Salsa20 hash function, we can see that the input 64-byte string x transforms into a sequence of 16 words $(x_0, x_1, \ldots, x_{15})$ such that: if $x = (x[0], x[1], x[2], \ldots, x[63])$, then:

$$\begin{aligned} x_0 &= \mathit{littleendian}\,(x[0], x[1], x[2], x[3]), \\ x_1 &= \mathit{littleendian}(x[4], x[5], x[6], x[7]), \\ &\vdots \\ x_{15} &= \mathit{littleendian}(x[60], x[61], x[62], x[63]). \end{aligned}$$

Then the string $(x_0, x_1, \ldots, x_{15})$ is modified during the 10 iterations of the doubleround function. Hence:

$$(z_0, z_1, \ldots, z_{15}) = \mathit{doubleround}^{10}(x_0, x_1, \ldots, x_{15})$$

In the final step of the Salsa20 hash function, the 16 input words and the 16 words obtained as output from 10 iterations of the doubleround function are added, and this result is converted into a 64-byte string using the inverse of the littleendian function. The output from the Salsa20 hash function is 64-byte word concatenation:

$$\begin{aligned} &\mathit{littleendian}^{-1}\,(z_0 + x_0), \\ &\mathit{littleendian}^{-1}\,(z_1 + x_1), \\ &\qquad\vdots \\ &\mathit{littleendian}^{-1}\,(z_{15} + x_{15}). \end{aligned}$$

The Salsa20 Expansion Function. Let us denote a 32-byte or 16-byte sequence by k, and a sequence of 16 bytes by n. Then $Salsa20_k(n)$ is a sequence of 64-bytes. If the string k is 32-byte then it is split into two 16-byte substrings k_0 i k_1, and the ASCII text "expand 32-byte k" is represented as: $\sigma_0 = (101, 120, 112, 97)$, $\sigma_1 = (110, 100, 32, 51)$, $\sigma_2 = (50, 45, 98, 121)$, $\sigma_3 = (116, 101, 32, 107)$. Then the Salsa20 expansion function is defined using the Salsa20 hash function as written below:

$$Salsa20_{k_0,k_1}\,(n) = Salsa20(\sigma_0, k_0, \sigma_1, n, \sigma_2, k_1, \sigma_3).$$

However, when the string k is 16-byte and the ASCII text "expand 16-byte k" is $\tau_0 = (101, 120, 112, 97)$, $\tau_1 = (110, 100, 32, 49)$, $\tau_2 = (54, 45, 98, 121)$, $\tau_3 = (116, 101, 32, 107)$. Then the Salsa20 expansion function is defined using the Salsa20 hash function as follows:

$$Salsa20_k\,(n) = Salsa20(\tau_0, k, \tau_1, n, \tau_2, k, \tau_3).$$

The Salsa20 Encryption Function. Let k be a 32-byte or 16-byte string.Let $v$be an 8-byte sequence and let m be an l-byte sequence for $l \in \{0, 1, \ldots, 2^{70}\}$.

Then the function Salsa20 encrypting the data m with the key k and nonce v we denote $Salsa20_k(v) \oplus m$ is a sequence of l-bytes long. Then k is a secret key, ideally, 32-bit; m is the plaintext; v is a nonce, which is the unique message number, and $Salsa20_k(v) \oplus m$ is the ciphertext. It could also be that m is a ciphertext, in which case $Salsa20_k(v) \oplus m$ s the original plaintext. $Salsa20_k(v)$ is a 2^{70}-byte sequence

$$Salsa20_k(v,0), Salsa20_k(v,1), Salsa20_k(v,2), \ldots, Salsa20_k(v, 2^{64}-1),$$

where i is the unique 8-byte sequence $(i_0, i_1, \ldots, i_7)$ such that:

$$i = i_0 + 2^8 i_1 + 2^{16} i_2 + ... + 2^{56} i_7.$$

The formula $Salsa20_k(v) \oplus m$ implicitly truncates the $Salsa20_k(v)$ to the same length as m. That is:

$$Salsa20_k(v) \oplus (m[0], m[1], ..., m[l-1]) = (c[0], c[1], ..., c[l-1]),$$

where $c[i] = m[i] \oplus Salsa20_k(v, [i/64])[i \ mod \ 64]$.

3 Boolean Encoding and SAT Based Cryptoanalysis

After presenting what the Salsa20 cipher is, now we can show, proposed in [10], the method of direct, boolean encoding of the Salsa20 cipher. Firstly we show encoding three simple operations on 4-byte words and the littleendian function. Then, we present the encoding of the quarterround function.

The Sum of Two Words. Let $(u_1, u_2, \ldots, u_{32})$ and $(v_1, v_2, \ldots, v_{32})$ be vectors representing words with a length of 4-bytes, where u_i and $v_i \in \{0,1\}$, for $i = 1, 2, \ldots, 32$, and c_i is the carry bit, $c_i \in \{0,1\}$, for $i = 0, 1, \ldots, 31$. The result of this operation is the vector $(q_1, q_2, \ldots, q_{32})$, where $q_i \in \{0,1\}$, for $i = 1, 2, \ldots, 32$. The adopted markings Boolean encoding for summing two words is of the form:

$$\bigwedge_{i=1,\ldots,\ 32} q_i \Leftrightarrow (u_i \oplus v_i) \oplus c_{i-1},$$

where $c_0 = 0$ and $c_i \Leftrightarrow (c_{i-1} \wedge u_i) \vee (c_{i-1} \wedge v_i) \vee (u_i \wedge v_i)$.

The Exclusive-or of Two Words. XOR operation on two 4-byte words u and v, where $(u_1, u_2, \ldots, u_{32})$ and $(v_1, v_2, \ldots, v_{32})$ will be vectors of words u and v, and $u_i \in \{0,1\}$ and $v_i \in \{0,1\}$, for $i = 1, 2, \ldots, 32$. Vector $(q_1, q_2, \ldots, q_{32})$, where $q_i \in \{0,1\}$, for $i = 1, 2, \ldots, 32$, will represent the xoring result. Then encoding Boolean XOR operation of two 4-byte words u, v looks like this:

$$\bigwedge_{i=1,\ldots,32} q_i \Leftrightarrow (u_i \oplus v_i).$$

The W-Bit Left Rotation of a Word. Let $(u_1, u_2, \ldots, u_{32})$, where $u_i \in \{0,1\}$, for $i = 1, 2, \ldots, 32$ will be a vector representing a 4-byte word, and w will be a number in the set $\{0, 1, 2, 3, \ldots, 31\}$ and let the vector $(q_1, q_2, \ldots, q_{32})$, where $q_i \in \{0,1\}$, for $i = 1, 2, \ldots, 32$, will represent the word u after the rotation operation. Then w-bits left rotation of word u written with a boolean formula has the form:

$$\bigwedge_{w=0,1,2,\ldots,31} \left(\bigwedge_{i=1,\ldots,32-w} (q_i \Leftrightarrow u_{i+w}) \wedge \bigwedge_{i=32-w+1,\ldots,32} (q_i \Leftrightarrow u_{i-(32-w)}) \right).$$

The Littleendian Function. Let $(u_1, u_2, \ldots, u_{32})$, where $u_i \in \{0,1\}$, for $i = 1, 2, \ldots, 32$ will be a vector representing a 4-byte word and let the vector $(q_1, q_2, \ldots, q_{32})$, where $q_i \in \{0,1\}$, for $i = 1, 2, \ldots, 32$, will represent the word u after modification introduced by the littleendian functions. Then you can introduce Boolean encoding for the littleendian function as follows:

$$\bigwedge_{i=1,\ldots,8} (q_i \Leftrightarrow u_{i+24}) \wedge \bigwedge_{i=9,\ldots,16} (q_i \Leftrightarrow u_{i+8})$$

$$\wedge \bigwedge_{i=17,\ldots,24} (q_i \Leftrightarrow u_{i-8}) \wedge \bigwedge_{i=25,\ldots,32} (q_i \Leftrightarrow u_{i-24}).$$

Transforming the Quarterround Function into a Boolean Formula. Let $(y_{0_1}, y_{0_2}, \ldots, y_{0_i})$, $(y_{1_1}, y_{1_2}, \ldots, y_{1_i})$, $(y_{2_1}, y_{2_2}, \ldots, y_{2_i})$, $(y_{3_1}, y_{3_2}, \ldots, y_{3_i})$, where $y_{0_i} \in \{0,1\}$, $y_{1_i} \in \{0,1\}$, $y_{2_i} \in \{0,1\}$, $y_{3_i} \in \{0,1\}$ for $i = 1, 2, \ldots, 32$ will be vectors representing the input to the quarterround function and let $(z_{0_1}, z_{0_2}, \ldots, z_{0_i})$, $(z_{1_1}, z_{1_2}, \ldots, z_{1_i})$, $(z_{2_1}, z_{2_2}, \ldots, z_{2_i})$, $(z_{3_1}, z_{3_2}, \ldots, z_{3_i})$, where $z_{0_i} \in \{0,1\}$, $z_{1_i} \in \{0,1\}$, $z_{2_i} \in \{0,1\}$, $z_{3_i} \in \{0,1\}$ for $i = 1, 2, \ldots, 32$, they will be vectors representing the output data. Hence the quarterround function in the form of boolean encoding it has the following notation:

$$\bigwedge_{i=1,\ldots,25} z_{1_i} \Leftrightarrow y_{1_i} \oplus \left(\left(y_{0_{i+7}} \oplus y_{3_{i+7}} \right) \oplus c_{i+6} \right)$$

$$\wedge \bigwedge_{i=26,\ldots,32} z_{1_i} \Leftrightarrow y_{1_i} \oplus \left(\left(y_{0_{i-25}} \oplus y_{3_{i-25}} \right) \oplus c_{i-26} \right),$$

$$\bigwedge_{i=1,\ldots,23} z_{2_i} \Leftrightarrow y_{2_i} \oplus \left(\left(z_{1_{i+9}} \oplus y_{0_{i+9}} \right) \oplus d_{i+8} \right)$$

$$\wedge \bigwedge_{i=24,\ldots,32} z_{2_i} \Leftrightarrow y_{2_i} \oplus \left(\left(z_{1_{i-23}} \oplus y_{0_{i-23}} \right) \oplus d_{i-24} \right),$$

$$\bigwedge_{i=1,\ldots,19} z_{3_i} \Leftrightarrow y_{3_i} \oplus \left(\left(z_{2_{i+13}} \oplus z_{1_{i+13}} \right) \oplus e_{i+12} \right)$$

$$\wedge \bigwedge_{i=20,\ldots,32} z_{3_i} \Leftrightarrow y_{3_i} \oplus \left(\left(z_{2_{i-19}} \oplus z_{1_{i-19}} \right) \oplus e_{i-20} \right),$$

$$\bigwedge_{i=1,\ldots,14} z0_i \Leftrightarrow y0_i \oplus \left(\left(z3_{i+18} \oplus z2_{i+18}\right) \oplus f_{i+17}\right)$$

$$\wedge \bigwedge_{i=15,\ldots,32} z0_i \Leftrightarrow y0_i \oplus \left(\left(z3_{i-14} \oplus z2_{i-14}\right) \oplus f_{i-15}\right).$$

where c_i, d_i, e_i, f_i denote the carry bits in the corresponding operations adding two words and $c_i \in \{0,1\}$, $d_i \in \{0,1\}$, $e_i \in \{0,1\}$, $f_i \in \{0,1\}$ for $i = 0, 1, \ldots, 31$.

SAT Solvers. The SAT solver is a program that looks for a variables valuation that satisfies the given formula if any exists. Many used today's SAT solvers are based on the DPLL algorithm, which is now one of the most popular methods for automatically checking the satisfiability of Boolean formulas [8,9]. DPLL's name comes from the names of the authors of the work where a procedure has been proposed: Davis, Putnam, Logemann, and Loveland. DPLL requires the use of conjunctive normal form (CNF). The algorithm is based on relapses, i.e. it generates valuation gradually until the value found makes the formula unsatisfiable. The algorithm then returns until its value is set and changes it to the opposite.

The DPLL algorithm proceeds as follows:

- First, a decision on a chosen randomly variable valuation is done.
- Next, a deduction of possible values of other variables, in cases where it is possible, is performed.
- Then, a diagnosis procedure of whether there is a conflict that makes it impossible satisfiability of the formula is executed. If a conflict occurring, the relapse procedure is applied. If there is no possibility of recurrence, it means that there is no valuation that satisfies this formula and the UNSAT message should be returned.
- If the formula is satisfiable, return the SAT message. Otherwise, the algorithm continues by returning to the first point and deciding to evaluate the next variable.

3.1 Cryptoanalysis Procedure

In this paper we deal with cryptanalysis with plaintext and ciphertext. The cryptanalysis procedure used in this work consists of several steps. The first one involves encoding a single round of the investigated cipher into a propositional Boolean formula. Then the formula encoding the given number of rounds (or the entire cipher) is automatically generated. In the next stage, the formula obtained is converted into the CNF format. Here together with randomly selected bits of the plaintext and the key, we use the SAT-solver for calculating the ciphertext. The last step is for the SAT-solver to find the value of the variables representing the key bits. For this task, we use a formula encoding the appropriate number of cipher rounds, bits of the plaintext, and the previously obtained ciphertext. In our investigations, we use and compare several SAT-solvers: a few new ones and a few rather old but still efficient and popular.

4 Experimental Results

In our research, we use formulas that encode the use of rounds Salsa20 algorithm with a 16-byte key. For our experiments, we have used several SAT solvers [1,3,4,12]. A good example is glu_vc that is the solver's successor glu_alt. It is based on the solutions used in Glucose 3.0 and improved in aspects such as reducing the base of learned clauses and the choice of so-called decision variables. glu_vc is a solution from the CDCL family (Conflict-Driven, Clause-Learning), which uses VSIDS heuristics, indicating the sentence variables preferred for further branching the tree. In addition, the authors optimized the solution for the formulas they meet. glu_vc was presented at the SAT Competition in 2017.

Another example of SAT-solvers is PicoSat, which is a SAT Solver created by Johannes Kepler University. It's an idea is based on the MiniSAT 1.14 algorithm. In this solution, the developers focused on minimizing the use of memory to increase the efficiency of the tool. Additionally for our research, we used: Painless MapleCOMSPS, MiniSAT - ver. 2.2 and Sat4J.

Parallel solutions were also used for further testing. Both pLingeling and Glucose-Syrup are parallel versions of sequential solvers, respectively Lingeling and Glucose 4.0. The input parameter is the number of threads and for each of them (so-called worker threads) a separate instance of the sequential Solver is started.

We should here also notice that, in addition to the working threads, the main thread is started, which reads a file with the formula in the form of DIMACS and communicates with the other threads to coordinate their work - the working threads do not communicate with each other.

Table 1 shows the number of variables used and the generated clauses for encoding formulas.

Table 1. Variables and clauses in encoded formula

Rounds	Variables	Clauses
1	6144	29376
2	9216	47424
3	12288	65472
4	15360	83520
5	18432	101568
6	21504	119616
7	24576	137664
8	27648	155712
9	30720	173760
10	33792	191808

The first of our experiment shows how quickly we obtain the ciphertext using the indicated number of rounds of the Salsa20 algorithm with a 16-byte key (see Tables 2 and 3).

Table 2. Searching for a ciphertext - Sequential SAT solvers.

Rounds	Minisat	Sat4J	PicoSAT	glu_vc
	Time [ms]	Time [ms]	Time [ms]	Time [s]
1	22	732	41	19
2	15	326	37	14
3	28	1141	36	16
4	22	413	58	20
5	25	379	69	22
6	29	377	78	26
7	34	412	81	29
8	36	434	98	32
9	43	461	91	35
10	44	481	108	38

Analyzing the attack on one double round of the Salsa20 cipher (see Tables 4 and 5), we can see that the MiniSat achieved the best time, although the other sat-solvers, both sequential and parallel, achieved similar times.

Attempts to attack two double-rounds of the Salsa20 algorithm were unsuccessful, so we tested the attack on two double-rounds of Salsa20 with the evaluations of variables representing the key bits added to the encoded formula.

We conducted this experiment in two variants: the first assumes adding the initial evaluations of the variables representing the key to the encoding formula (see Tables 4 and 5), the second - adding the final evaluations of the variables representing the key (see Tables 6 and 7).

The obtained results show that SAT solvers calculate the evaluations of the variables that satisfiable the formula faster when they have the values of the initial variables representing the key added than when they have the evaluations of the final variables representing the key, even when we add the same number.

When looking at the work of programs that test the satisfiability of Boolean formulas, we notice that SAT sequence solvers are stable, while parallel SAT solvers such as Glucose Syrup or PalnleSS communicated errors or memory overflow. The best times were achieved by MiniSat, and from Sat parallel solvers: Glucose Syrup.

All experimental results have been done using the machine with the following parameters: Architecture: x86_64, CPU op- mode(s): 32-bit, 64-bit, Byte Order: Little Endian, CPU: 4, On-line CPU(s) list: 0–3 memory, Thread(s) per core: 1, Core(s) per socket: 4, Socket(s): 1, Producent ID: GenuineIntel, family CPU: 6,

Table 3. Searching for a ciphertext - Parallel SAT solvers

Rounds	PaInLeSS	Glucose Syrup	Plingeling
	Time [ms]	Time [ms]	Time [ms]
1	1263	ERROR	57
2	1340	ERROR	39
3	2149	ERROR	54
4	2168	ERROR	56
5	1646	ERROR	63
6	1804	ERROR	75
7	1889	ERROR	83
8	2148	ERROR	122
9	2065	ERROR	101
10	2377	ERROR	110

Table 4. Attack on one and two double rounds of Salsa20 with initial key values added - Sequential SAT solvers

Rounds	Added nonce bits	Added key bitskey bits	Minisat	Sat4J	PicoSAT	glu_vc
			Time [ms]	Time [ms]	Time [ms]	Time [ms]
1	128	–	28	453	33	27
2	128	96	251	1630	231	91
2	128	80	9548	39018	66305	13038
2	128	72	18048	23228791	1418996	1789619
2	128	68	14970605	–	–	–

Table 5. Attack on one and two double rounds of Salsa20 with initial key values added - Parallel SAT solvers.

Rounds	Added nonce bits	Added key bits	PaInLeSS	Glucose Syrup	Plingeling
			Time [ms]	Time [ms]	Time [ms]
1	128	–	1579	37	59
2	128	96	4879	174	311
2	128	80	4578	5308	6126
2	128	72	ERROR	1934703	2544064
2	128	68	ERROR		

Table 6. Attack on the one and two double Salsa20 rounds with added final key values - Sequential SAT solvers

Rounds	Added nonce bits	Added last key bits	Minisat	Sat4J	PicoSAT	glu_vc
			Time [ms]	Time [ms]	Time [ms]	Time [ms]
1	128	–	28	453	33	27
2	128	96	101	642	68	49
2	128	80	119285	126680	186984	282699
2	128	72	13707609	–	29711494	–

Table 7. Attack on the one and two double Salsa20 rounds with added final key values - Parallel SAT solvers.

Rounds	Added nonce bits	Added last key bits	PaInLeSS	Glucose Syrup	Plingeling
			Time [ms]	Time [ms]	Time [ms]
1	128	–	158	37	59
2	128	96	1810	70	165
2	128	80	ERROR	61535	196726
2	128	72	ERROR		

model: 142, model name: Intel(R) Core(TM) i7-8565U CPU @ 1.80 GHz, version: 11. Linux VirtualBox 4.15.0-88-generic #88-Ubuntu SMP x86_64 GNU/Linux.

5 Conclusion and Future Work

In this paper, we showed our results obtained in the case of the SAT-based cryptanalysis with plaintext and ciphertext of the Salsa20 cipher in a version with a 16-byte key. We presented in detail all cipher's parts and the procedure of encoding the Salsa20 algorithm into a Boolean formula. We converted this formula to CNF and DIMACS format. Then, using this transformed formula and randomly generated plaintext and key bits, also presented in DIMACS format, we searched for the ciphertext values using SAT programs. The next step in the Salsa20 cryptanalysis was to find cryptographic key values by the SAT solver using known plaintext and ciphertext bits. We tested the possibility of the attack with one double-round and two double-rounds of the Salsa20 algorithm with 96, 80, 72, 68 initial 128-bit key values and 96, 80, 72 final key values added. We used different SAT solvers to get the best results. Among them are SAT single-threaded (sequential) and multi-threaded (parallel) solvers.

In the next study, we will try to examine the ChaCha cipher which is a modification of the Salsa20. We also plan to consider several types of formula optimization, which should reduce the size of the formulas and the time needed for cryptanalysis. We will use our experience to SAT cryptanalysis of several other ciphers.

References

1. Biere, A.: PicoSAT essentials. J. Satisfiabil. Boolean Model. Comput. (JSAT) **4**, 75–97 (2008)
2. Biere, A., Heule, M., van Maaren, H., Walsh, T. (eds.): Handbook of Satisfiability, vol. 185 of Frontiers in Artificial Intelligence and Applications. IOS Press (2009)
3. Biere, A.: Lingeling, Plingeling, Picosat and Precosat at SAT race 2010. Technical report. FMV 10/1, Institute for Formal Models and Verification, Johannes Kepler University (2010)
4. Biere, A.: Lingeling, Plingeling and Treengeling entering the SAT competition 2013. In: Proceedings of SAT Competition 2013, vol. B-2013-1 of Department of Computer Science Series of Publications B, pp. 51–52. University of Helsinki (2013)
5. Biham, E., Shamir, A.: Differential cryptanalysis of DES-like cryptosystems. J. Cryptol. **4**(1), 3–72 (1991)
6. Courtois, N., Bard, G.V.: Algebraic cryptanalysis of the data encryption standard. In: Galbraith, S.D. (ed.) IMA International Conference. Lecture Notes in Computer Science, vol. 4887, pp. 152–169. Springer, Heidelberg (2007)
7. Chowaniec, M., Kurkowski, M., Mazur, M.: New results in direct SAT-based cryptanalysis of DES-like ciphers. In: Proceedings of ACS 2018. AISC, vol. 889, pp. 282–294. Springer, Cham (2018)
8. Davis, M., Putnam, H.: A computing procedure for quantification theory. J. ACM **7**(3), 201–215 (1960)
9. Davis, M., Logemann, G., Loveland, D.W.: A machine program for theorem-proving. Commun. ACM **5**(7), 394–397 (1962)
10. Dudek, P., Kurkowski, M., Srebrny, M.: Towards parallel direct SAT-based cryptanalysis. In: PPAM 2011 Proceedings. LNCS, vol. 7203, pp. 266–275. Springer (2012)
11. Dwivedi, A.D., et al.: SAT-based cryptanalysis of authenticated ciphers from the CAESAR competition. In: Proceedings of the 14th International Joint Conference on on e-Business and Telecommunications (ICETE 2017) - vol. 4: SECRYPT, pp. 237–246 (2017)
12. https://github.com/arminbiere/cadical. Accessed 15 Jan 2020
13. Kościelny, C., Kurkowski, M., Srebrny, M.: Modern Cryptography Primer. Springer, Heidelberg (2013). https://doi.org/10.1007/978-3-642-41386-5
14. Kurkowski, M., Penczek, W.: Applying timed automata to model checking of security protocols. In: Wang, J. (ed.) Handbook of Finite State Based Models and Applications, pp. 223–254. Chapman and Hall/CRC Press (2013)
15. Kurkowski, M., Penczek, W.: Verifying timed security protocols via translation to timed automata. Fundam. Inform. **93**(1–3), 245–259 (2009)
16. Lafitte, F., Lerman, L., Markowitch, O., van Heule, D.: SAT-based cryptanalysis of ACORN. IACR Cryptology ePrint Archive **2016**, 521 (2016)
17. Lafitte, F., et al.: Applications of SAT Solversin cryptanalysis: finding weak keys and preimages. JSAT **9**, 1–25 (2014)
18. Massacci, F.: Using Walk-SAT and Rel-SAT for cryptographic key search. In: Dean, T. (ed.) IJCAI, pp. 290–295. Morgan Kaufmann (1999)
19. Massacci, F., Marraro, L.: Logical cryptanalysis as a SAT problem. J. Autom. Reason. **24**(165), 165–203 (2000)
20. Matsui, M.: The first experimental cryptanalysis of the data encryption standard. In: Desmedt, Y. (ed.) CRYPTO. LNCS, vol. 839, pp. 1–11. Springer, Heidelberg (1994). https://doi.org/10.1007/3-540-48658-5_1

21. Menezes, A., van Oorschot, P.C., Vanstone, S.A.: Handbook of Applied Cryptography. CRC Press, Boca Raton (1996)
22. Morawiecki, P., Srebrny, M.: A SAT-based preimage analysis of reduced Keccak hash functions. Inf. Process. Lett. **113**(10–11), 392–397 (2013)
23. Schoening, U., Toran, J.: The Satisfiability Problem: Algorithms and Analyses. Lehmanns Fachbuchhandlung GmbH (2013)
24. Soboń, A., Kurkowski, M., Stachowiak, S.: Towards complete SAT-based cryptanalysis of RC5 cipher. In: Proceedings of 2019 IEEE 15th International Scientific Conference on Informatics, pp. 369–374. IEEE Press (2019)
25. Soos, M., Nohl, K., Castelluccia, C.: Extending SAT solvers to cryptographic problems. In: Proceedings of 12th International Conference on Theory and Applications of Satisfiability Testing, SAT 2009, pp. 244–257 (2009)
26. Szymoniak, S., Siedlecka-Lamch, O., Kurkowski, M.: Timed analysis of security protocols. In: Proceedings of 37th ISAT Conference, Advances in Intelligent Systems Computing, vol. 522, pp. 53–63. Springer (2016)
27. Zbrzezny, A.M., Siedlecka-Lamch, O., Szymoniak, S., Kurkowski, M.: SMT solvers as efficient tools for automatic time properties verification of security protocols. In: 20th International Conference on Parallel and Distributed Computing, Applications and Technologies, PDCAT 2019, pp. 320–327. IEEE (2019)

Smart Meters Reading Process Improvement by UAV Utilization

Piotr Kiedrowski(✉), Łukasz Zabłudowski, and Beata Marciniak

Telecommunication and Computer Science Institute,
UTP University of Science and Technology, Bydgoszcz, Poland
{piotr.kiedrowski,lukasz.zabludowski,beata.marciniak}@utp.edu.pl

Abstract. The paper describes UAV application in automatic meter reading systems. The proposal increases the reliability of communications between SMs and TSC. The reading process must be improved because the communication capacity of the last mile RF networks has deteriorated significantly, as a result of the process of continuous replacement of RF smart meters to PLC units. The UAV is equipped with a communication node. The UAV communication node has a firmware dedicated for repeaters which allows it to operate as the mobile repeater. The additional moving node changes the topology of the wireless network with more connectivity introduced by the UAV. In this paper the algorithm of space track determination for the UAV is proposed. This algorithm is implemented on the UAV changing its trajectory according to the received sequences of packets types. Performance of the algorithm was tested in practice and compared to the efficiency of SMs reading process using fixed or random trajectories of the UAV. The application of the proposed algorithm makes the independence from the last mile RF network topology and changes in the propagation conditions.

Keywords: Dron · Mobile data collecting · Smart metering · UAV

1 Introduction

In the first decade of the 21st century many different solutions have been implemented regarding the smart meters (SMs) which allow for remote readings volume of the consumed energy by the households. SMs, paired with proper communication and IT systems form the Smart Metering system which is a part of the Smart Grid.

The main issue during the time which has been mentioned above was the selection of the technology that would be able to fulfil the requirements of the last mile communication systems. In that time 3 types of technologies were considered and utilized: General Packet Radio Service (GPRS), Radio Frequency (RF) and Power Line Communications (PLC). The main disadvantage of the GPRS technology was the Distributed System Operators' (DSOs) dependence from the mobile operators, high production cost and additional exploitation

M. Choraś et al. (Eds.): CORES 2021, IP&C 2021, ACS 2021, LNNS 255, pp. 267–275, 2022.
https://doi.org/10.1007/978-3-030-81523-3_26

costs during service to cover the mobile communication expenses. Despite of these disadvantages this solution found its common usage in Denmark [1]. In different countries around the Europe this solution was used as a last resort where all other could not be implemented because of big distances between the households [2]. While the use of GPRS in SMs can be considered as a niche solution, its widespread use considers the equipment of the Transformer Station Data Concentrators (TSCs) with the GPRS module for the purposes of the IP network access. The TSC acts as a gateway between IP network and a last mile network where connections between SMs and a TSC are implemented in RF or PLC.

In the first years of this century, both DSOs and meter manufacturers considered the choice of technology for the implementation of the last mile network. Those technologies are very different from each other. The main differences are: interference resistance, the number of available bands, the number of channels available in the band, the ability to detect the source of interference or the cost of production of communication modules. Many smart metering systems have been implemented where the last mile network was based on RF technology. The main drawback for this solution is low resistance to intentional and unintentional interference. This is crucial and for this reason, along with the standardization of PLC interfaces [3–5] in 2012, the PLC technology replaces the RF technology. All new installations of smart metering systems are based on PLC technology. In existing last mile networks based on RF technology, if the meter needs to be replaced, it is replaced with a new PLC unit. The most common reasons for replacing meters are: the need to legalize them or an increasing number of solar installations [6] and thus the need to enable prosumer settlements [7]. As a result of migration to PLC technology RF nodes and RF links are successively eliminated, which causes communication problems. These problems are manifested by the following phenomena: increasing the number of re-queries, preventing the transmission of long frames. These problems are getting worse with the elimination of next RF meters. In order to improve the operation of the network, the authors propose the use of Unmanned Aerial Vehicle (UAV) acting as repeater. In the proposed solution the use of UAV as a repeater is a much better solution than using it as a traffic concentrator (TC) in a meaning of replacing an existing, stationary TSC. The important benefits of using the proposed solution are following: saving time (procedures of replacing TSC to mobile TCs are quite time consuming), no encryption key and TCS address are not required, no alarms associated with the switching off the TSC are reported.

2 State of the Art

Nowadays UAVs are widely used for delivering different types of services. Using UAVs different types of data can be captured from the air which can be widely used in construction, telecommunication, railway, excavations, mines, environment monitoring etc.

Substantial research has been done in the area of UAVs deployments for wireless communication improvement but most of the research focus on routing packets to a mobile sink. In [8], authors consider the problem of prolonging network lifetime by employing a mobile sink for data gathering. To achieve this they propose the heuristic algorithm for finding a trajectory of the sink. In this paper, we propose an algorithm for finding a trajectory for UAV which acts as a data repeater. Utilizing the UAV as a repeater instead of a sink does not cause steady connectivity disruptions, this problem is described in [9] where the authors tried to solve it by proposing a multiple sink network architecture. In [10] the authors proposed an extra mobile sink for the WSN (Wireless Sensor Network) supervision and monitoring application. The described problem in the introductory section of this work, namely the presence of gaps in the network, was solved by the authors of [11], in which they also proposed to use an UAV but in this case its role was a mobile gateway. Our research in area of smart metering is focusing on the practical usage of mobile repeater mounted on the UAV instead of a mobile sink [12] or data collector [13,14]. Proposed solution provides a flying smart grid communication node gap filler. In the contrast to the methods described in [15] our UAVs utilize a specially formulated algorithm which detects the sequences sent by individual SMs and TC so the hovering UAV is able to change its trajectory along the path.

Also in [15], the problem of UAV trajectories description, assessment or optimization is signalized. Numerous factors may be used for the comparison of trajectories [16]. In our research we applied Quality of Service (QoS) approach. To describe the QoS we used the same parameters as those used by DSOs or SMs vendors.

3 Communication Scheme Used in RF Last Mile Networks for SMs Reading

The architecture of the RF last mile network dedicated for communication with SMs is not complicated and it is similar to the WSN architecture [17]. It consists of components like a TSC (which plays the same role as the sink in WSN) and other network elements in particular SMs. The TSC address is also the last mile network address. SM acts as destination node or repeater.

The communication protocol utilizes three types of packets: command, response and ACK/Cancel. SM reading process scenario is repetitive and presents as follow: 1) The TSC sends a command packet with an appropriate message type. 2a) If destination SM is in the direct range of the TSC, it sends response packet containing the proper message type and requested data. 2b) If the destination SM is not in the direct range of the TSC, command packet is relayed by other SMs after random time in a range from 60 ms up to 700 ms per each hop, this allows the packet to reach the destination node and send response packet containing the proper message type and the requested data. 3)After response packet is received by the TSC it sends ACK/Cancel packet immediately. This kind of packet is the acknowledge message for the queried SM and cancels relaying process of response packet in SMs, which have not relayed it yet.

The described procedure for forwarding packets via SMs or repeaters has to be changed if a repeater is mounted on the UAV. The change is a consequence of using packet encryption. The following rule was used to forward the packet: the packet is forwarded if its length is different from the one previously forwarded. Although due to encryption there is no access to the packet control field, and thus no access to the packet type identifier. It is possible to specify its type by the length. This property will be used in the packet sequence detection algorithm used to change the position of the UAV described later.

4 Research Process and Methods

The research method was an experiment carried out in the actual last mile network, implemented in technology identical to those operated by DSOs. This was possible because one of the Authors took part in the construction works focusing on creating a smart metering system based on the RF technology. As a result of this research, the smart metering system was implemented in 2011, and the experimental SMs network is still working on the UTP campus. The network consists of 70 SMs and additionally 30 RF communication modules, which are seen in the network as SMs nodes. The load is not attached to the smart meters, which allows desired SMs to be turned off without damaging the daily operation of the university.

To assess the influence of SM removing, two metrics of the effectiveness of the reading process may be used [2]: the first is Reading Rate Indicator (RRI) and the second is Reading Speed Indicator (RSI). The *RRI* is a percentage value which indicates how many of all the nodes were readout successfully; *RRI* is described as following:

$$RRI = \frac{R}{R+U} \cdot 100\% \tag{1}$$

where: R is the number of the SMs read successfully, U is the number of unread SMs.

The *RSI* is the total read time of all nodes to the number of them in the tested network;*RSI* is described as following:

$$RSI = \frac{\sum_{i=1}^{R} t_i + \tau U}{R+U} \tag{2}$$

where: t_i is the ith SM read time and τ is the TSC constant value for the response time from the SM.

We should notice that if the *RRI* is smaller the *RSI* is bigger and vice versa. This relation is easy to explain, namely if the communication conditions do not allow to read the SM then: 1) the *RRI* decreases, 2) the read attempt of unreachable SM takes the most time, because t_i is always smaller than τ. The *RSI* describes the average value of the readout time of a single node. The unit of *RSI* is seconds per node.

The UAV engagement as an additional node in the network always improves and results in reducing the *RSI* value and increasing the *RRI* value if it was less than 100%. To evaluate the effectiveness of improving communication in the last mile network it is proposed to compare the *RRI* and *RSI* values before using the UAV in the network with those obtained after its use.

5 Space Track Determination Algorithm

The starting point for the UAV is near the TSC. UAV flies on a fixed altitude "trying" to move in a circle or arc, which center is the point where the TSC is located. The term "tries" results from the fact that the radius of the circle changes, as the effect of the algorithm. The altitude of the UAV determines the highest obstacle in the area where the UAV performs its mission. Any interference by the UAV operator disables the mission until it is resumed by the UAV operator manually. When UAV is deployed the operator and the UAV are informed about detecting the packet sequence in the following order: Command, Response and ACK/Cancel (CRA) and goes into the state: Looking for Command. From the safety reasons the take off procedure is done manually by the operator.

The UAV position change algorithm uses packet type sequences received by the UAV. For the detection of packet sequences by UAV an algorithm has been proposed and is presented in Fig. 1. This algorithm is a consequence of the communication scheme used in the last mile network for reading electricity meters in RF technology and knowledge of network behavior in conditions of significant reduction in network performance due to the elimination of a large number of SMs. The presented algorithm is implemented on the microcontroller of the UAV. The repeater is mounted to the UAV gear (Fig. 2) and connected with two wires i.e. ground and repeater's serial transmitter, which transmits what was received by repeater's radio module, it allows the UAV to find packets sequences.

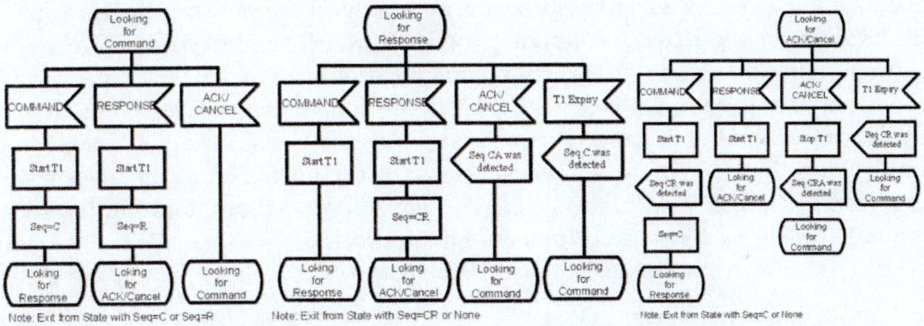

Fig. 1. The algorithm of detection of packet sequence for the following states: a) Looking for Command, b) Looking for Response, c) Looking for ACK/Cancel.

Fig. 2. Repeater mounted on the UAV gear during operation.

Table 1. Generated primitives and UAV's reaction tables.

Primitives	UAV's reactione
Seq C was detected	Decrease the radius(move closer to TSC), Start T2
Seq CA was detected	Move side, Start T2
Seq CR was detected	Stop UAV, Start T2 twice
Seq CRA was detected	Increase the radius(move away from TSC), Start T2

Primitives generated as a result of the algorithm operation are an argument of the UAV traffic control function according to the principle set out in Table 1.

According to the presented algorithm the UAV can change its position every 2 s or 4 s, the next flight parameters are: distance between the old and new position –s and speed – V, these parameters are set independently for the individual last mile network. The default values adopted result from the typical number of nodes in the network and the surface area in which the network operates and the capacity of the battery supplying the UAV. With the increase of V and/or s, the number of nodes with which communication cannot be established or it is not possible to perform a read though flying time over takes less time. After conducting a number of experiments it can be stated that increasing V and/or s and leaving the UAV in motion consistent with the operation of the algorithm, as a result of which it will travel a similar route several times does not improve communication parameters in the last mile network.

6 Experimental Results

We experimented with UAV which flew over the campus for checking the proposed algorithm and for optimizing its parameters. As a measure of the algorithm effectiveness we used RRI, RSI and the UAV flight time, during the experiment the network topology was changed 10 times. The UAV is on air until RRI is smaller than 100% or its battery discharge i.e. around 25 min. The changes in the network topology depended on turning off SMs in the different localizations. The number of active SMs was changed too. In Fig. 3 we present an area with the exemplary network. The tested area consists of three kinds of subareas, they are: subareas where communication with the SMs exists (green bordered in Fig. 3), subareas where SMs were turned off (yellow bordered in Fig. 3) and subareas where SMs are turned on but there is no communication with them.

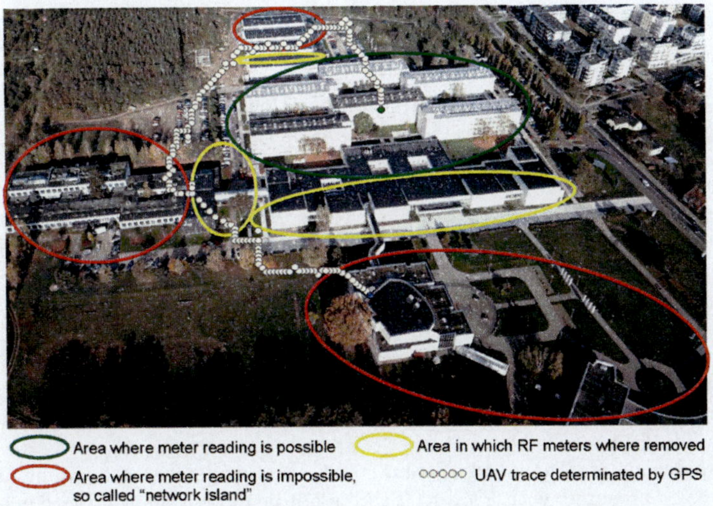

Fig. 3. Photo picture (taken by the UAV) of the campus where tests were conducted.

There were 100 SMs installed in the area presented in Fig. 3 but only 50 of them were online so that $R + U = 50$ and R was between 22 and 25 depending on radio propagation conditions so that RRI was between 44% and 50%. After the use of an UAV the results were as follow: $RRI = 100\%$, RSI around 4 s/node and the UAV flight time was almost 5 min.

7 Conclusion

The use of the UAV as a mobile communication node in the RF network for SMs reading improves this process. In the presence of large communication gaps in the network, the use of UAVs seems to be the only effective method of ensuring communication between SMs and the TSC. The use of many stationary repeaters is pointless due to: their number, costs and lack of space for their installation. The effectiveness of the use of UAVs is specified in a measurable way. In this work authors explain why it is better to use the UAV as the repeater rather than the TC. The method of employment the UAV as the mobile repeater that change its position according the algorithm based on received packet sequences detection may be applied to improve communication in other wireless networks e.g. WSNs, assuming that the communication protocol is known.

A promising solution for autonomous UAV flight over the wireless smart metering network is the use of received packets decryption facility. Thanks to decryption capabilities it will be possible to access to additional information like destination and source addresses etc. Nevertheless such an approach has its drawbacks. In particular the necessity of providing authentication for particular last mile network which require security service key to be known.

References

1. Stefan, M.: Automation of Smart Grid operations through spatio-temporal data-driven systems. Ph.D. Series, Aalborg University Press, p. 152 (2019)
2. Polak, M.: Experiences of smart grid implementation in Sweden. Rynek Energii **IX**(1), 252–256 (2014)
3. Recommendation ITU-T G.9902, Narrowband orthogonal frequency division multiplexing power line communication transceivers for ITU-T G.hnem networks (2012)
4. Recommendation ITU-T G.9903, Narrowband orthogonal frequency division multiplexing power line communication transceivers for G3-PLC networks (2017)
5. Recommendation ITU-T G.9904, Narrowband orthogonal frequency division multiplexing power line communication transceivers for PRIME networks (2012)
6. Bessa, R.J., Trindade, A., Miranda, V.: Spatial-temporal solar power forecasting for smart grids. IEEE Trans. Ind. Inf. **11**(1), 232–241 (2015). https://doi.org/10.1109/TII.2014.2365703
7. Agathokleous, C., Tuan, L.A., Steen, D.: Stochastic operation scheduling model for a Swedish prosumer with PV and BESS in nordic day-ahead electricity market. In: 2019 IEEE Milan PowerTech, pp. 1–6. (2019).https://doi.org/10.1109/PTC.2019.8810651
8. Bassam, N.A., Jerew, O.D.: Energy aware and delay-tolerant data gathering in sensor networks with a mobile sink. In: 3rd MEC International Conference on Big Data and Smart City (ICBDSC), pp. 1–5. (2016). https://doi.org/10.1109/ICBDSC.2016.7460372
9. Abuarqoub, A.: Optimal mobile sink selection scheme for multiple sink mobile wireless sensor networks. In: European Intelligence and Security Informatics Conference, Manchester, pp. 169–172 (2015). https://doi.org/10.1109/EISIC.2015.19
10. Derdour, Y., Kechar, B., Khelfi, F.: Data collection in WSNs with a mobile sink for a supervision application. In: 2013 Science and Information Conference, London, pp. 856–863 (2013)
11. Deaconu, I., Voinescu, A.: Mobile gateway for wireless sensor networks utilizing drones. In: RoEduNet Conference 13th Edition: Networking in Education and Research Joint Event RENAM 8th Conference, Chisinau, pp. 1–5 (2014) https://doi.org/10.1109/RoEduNet-RENAM.2014.6955319
12. Shenoy, R., Keshavan, B.K.: Hybrid drone for data transaction. In: 2017 IEEE International Conference on Smart Energy Grid Engineering (SEGE), Oshawa, pp. 238–241 (2017) https://doi.org/10.1109/SEGE.2017.8052805
13. Tkachov, V., Tokariev, V., Dukh, Y., Volotka, V.: Method of data collection in wireless sensor networks using flying ad hoc network. In: 2018 International Scientific-Practical Conference Problems of Infocommunications. Science and Technology, pp. 197–201 (2018) https://doi.org/10.1109/INFOCOMMST.2018.8632048

14. Cardei, I., Cardei, M., Papa, R.: UAV-enabled data gathering in wireless sensor networks. In: IEEE 37th International Performance Computing and Communications Conference, pp. 1–8 (2018). https://doi.org/10.1109/PCCC.2018.8711145
15. Martinez-Caro, J.-M., Cano, M.-D.: IoT system integrating unmanned aerial vehicles and lora technology: a performance evaluation study. Wirel. Commun. Mob. Comput. **11**, 1–12 (2019). https://doi.org/10.1155/2019/4307925
16. Mozaffari, M., Saad, W., Bennis, M., Nam, Y.-H., Debbah, M.: A tutorial on UAVs for wireless networks: applications, challenges, and open problems. IEEE Commun. Surv. Tutor. **21**(3), 2334–2360 (2019). https://doi.org/10.1109/COMST.2019.2902862
17. Elhabyan, R., Shi, W., St-Hilaire, M.: Coverage protocols for wireless sensor networks: review and future directions. J. Commun. Netw. **21**(1), 45–60 (2019). https://doi.org/10.1109/JCN.2019.000005

Defuzzification Method Comparison in Real Data Packets Control Method for Network Protection

Łukasz Apiecionek(✉)

Institute of Computer Science, Kazimierz Wielki University, Bydgoszcz, Poland
lukasz.apiecionek@ukw.edu.pl

Abstract. Unfortunately still one of huge problem for IT systems are Distributed Denial of Service attacks. Many servers which are accessible from public network were a victim of such attacks. This article presents the method for using fuzzy logic and defuzzification method for protecting servers resources. Using different defuzzification method let the administrators to react faster or slower on the situation with his server conditions. Presented comparison let to check how different method could be used in different control mechanism.

Keywords: DDoS · Network security · Fuzzy logic

1 Introduction

Nowadays, during Covid-19 pandemic, everybody needs a fast access to information from every part of the network. The systems have to be accessible online, while Distributed Denial of Service (DDoS in short) attacks have become a problem as they cause network unavailability by blocking services via seizing system resources in computers in the network until they stop working. Administrators are afraid of DDoS attacks on they systems. Lot of users could describe the situation, when they have been starting some work and lost the connection. In such situation they cannot even log out of the system. Usually papers present methods of fighting the DDoS attack problems [1–6] by using the Intrusion Detection System and Intrusion Prevention System (IDS/IPS in short) solutions. Such systems are efficient provided that they have a description of wellknown attacks. Sometimes IDS/IPS solutions are using some kind of Artificial Intelligence. Such solutions could learn the actions in some specific scenarios of attack. Other solutions suggest using a firewall mounted on the network edge. However, this firewall will only block the incoming traffic on specific ports or IP address ranges, which is not sufficient. The authors presented some method which could solve this problem – intelligent FTBint method [8]. This method could be implemented in all possible places of the networks, especially on firewalls and routers. Presented method could eliminate the effects of DDoS attacks. The idea of developing this mechanism was to prepare method for dealing with them. The main algorithm

M. Choraś et al. (Eds.): CORES 2021, IP&C 2021, ACS 2021, LNNS 255, pp. 276–283, 2022.
https://doi.org/10.1007/978-3-030-81523-3_27

should be easy implemented in already used network devices and could help to solve the problem of DDoS attacks. But using this method let the administrators to decide how fast they will block the traffic to the server. This decision could be made using fuzzy logic and the some defuzzification method. Using different methods provide different possibilities which are presented in this paper. The structure of this paper is as follows: Sect. 2 shortly describes the issue of the DDoS attacks and introduces the proposed method. Section 3 provides a 5 defuzzification method and the results which were achieved with them. Finally Sect. 4 provides a conclusion and discussion over the developed method.

2 The Idea of Proposed Method

The proposed idea consist of some existing algorithms and some new proposition, author followed the methods of Apiecionek et al. 2016 [8]. All together provide ready to use solutions. The aim of the attack is a server, so the server know the best that it is under attack. If the routers could get the information, that they have to block some specific traffic, or traffic for specific server, they could do it using already implemented mechanism e.g. mentioned access lists.

The proposition of the method could works on servers, routers, and firewalls and has got a following steps:

- Step (1) - server is collecting information about their traffic statistic,
- Step (2) - server is recognizing that there is under attack using their traffic statistic and their work status,
- Step (3) - server is establishing IPSec channel to network provider's router,
- Step (4) - server is passing information about type of traffic which has to be blocked by the network using SNMP over IPSec channel,
- Step (5) - router or firewall starts to block specific traffic,
- Step (6) - router or firewall is spreading the information about specific traffic which has to be blocked via SNMP trap message to other routers and firewalls.
- Step (7) – the specific traffic of the attacker is blocked via the network devices.

These steps are presented on Fig. 1.

2.1 Algorithm for Dropping Packets on Routers and Firewalls

The algorithms for dropping packets on routers and firewalls are using well known QoS method idea with some additional modules. What should be noticed is that during normal network activity, data streams are transmitted to the receiver over a router. When an attack occurs, the number of streams grows rapidly up to the maximum number supported by the network resources. When the number of packets exceeds the given limits, the proposed method has to remove them from the queue by using a wellknown mechanism called Random Early Detection (RED). In a standard method, the RED mechanism is executed on incoming packets with data streams categorization. In the proposed method, the RED mechanism should not be executed on the data streams that had been transmitted earlier. Thus, the method has to:

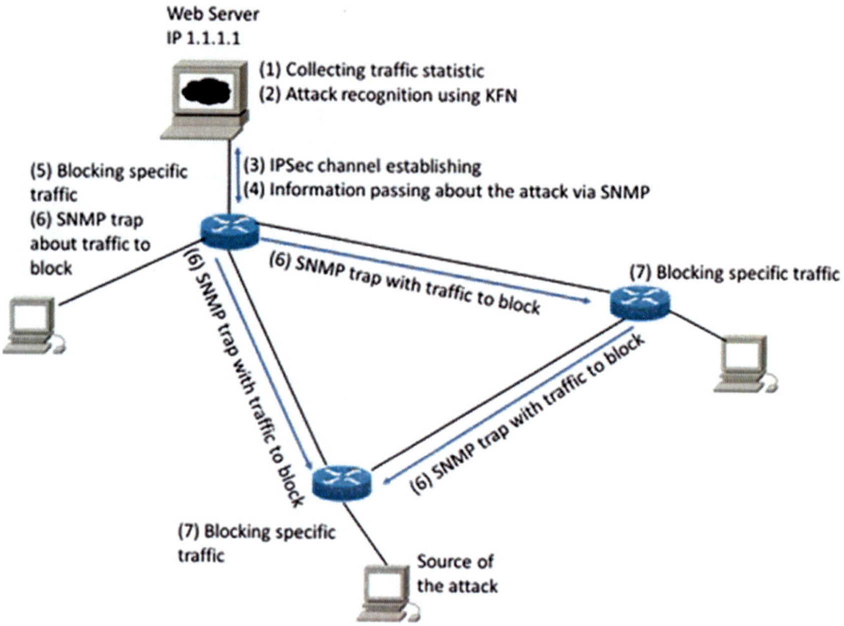

Fig. 1. The algorithm concept

- keep the data streams history from the moment before the attack and this period should not be neither too long, as otherwise it will not capture the appropriate traffic, nor too short, in order to protect the device from heavy load,
- after detecting the attack, set the traffic filtering rules to privilege historical packets, i.e. by selecting them with the use of filtering rules on a router module,
- block all other malicious traffic, i.e. by using RED mechanism [9] or firewall rules [10,11] but in intelligent way.

2.2 The Intelligent Dropping Module

A special firewall dropping module was also developed, the role of which is to filter the traffic on special server's open port and to limit it according to the determined policy in intelligent way. During the server's regular work all packets are passed through the module. The regular work means that the network is not under any attack. When the attack is recognized and the filtering module is informed about it, the intelligent dropping module starts his work.

At the beginning of the filtration a list of the IP addresses which communicate with the server correctly, i.e. which are not a part of the DDoS attack – $listIP$ – is taken from the server. During the filtration each packet is checked whether it is on the list of the valid IP addresses $listIP$; if so, the packet is sent to the network, if no, a counter of the passed packets *packet_counter* is checked whether its value is greater than the allowable packet limit *packet_limit* in a time slot t1. If the limit of the packets is exceeded, the packet is dropped - DROP. In the following time slot the number of current packets *packet_counter* is zeroed and the above mentioned filtration process is restarted. Then the dropping module has to regulate the opening connection limit in intelligent way. It is made using fuzzy logic in such way that:

- the incoming packets are counted and give us a value,
- fhe four following values let us to prepare the Oriented Fuzzy Number (OFN in short) $A = [value1, value2, value3, value4]$, when the $value1 < value2 < value3 < value4$we get the positive number, in other situation we get the negative number, after 4 timeslots, we get new OFN during each timeslots, consequently we get $OFN_{t1}, OFN_{t2}, \ldots, OFN_{tn}$,
- it we get the positive OFN number, the number of connection is increasing, so the $packet_limit$ has to be decreased, in other situation we could increase it,
- the amount of how the packet number should be changed we could calculate using the formula:

$$new_packet_limit = packet_limit \times \frac{OFN_{t+1}^{-1}}{OFN_t^{-1}}$$

  where OFN_{t+1}^{-1}, OFN_t^{-1} are the OFN after defuzzification process,
- when the *new_packet_limit* is lower than minimum, it will stay at minimum, while he *new_packet_limit* is greater than maximum, it will stay at maximum level.

The process of decreasing the packet limit should depend on the server type. Different server type and services could require different method and limits. This should be also provided by the servers to the filtering module in future solutions. Moreover, the limit values may require experimental determination or setting them basing on the server's resources, its operating system, the amount of memory and processors type.

During the test there were five defuzzification method tested: Golden ratio, Mandala, FOM, MOM, LOM.

3 The Method Test Results

3.1 Test Environment

In order to verify presented idea, the method was implemented and tested. The test had got to prove that the server will indeed work continuously. The implementation consisted of described method for the firewall IPTables module on Debian Linux system based on kernel 2.6.32. There were a simulation made on a simple network presented in Fig. 2. Those test had to simulate most common types of attacks on the servers. There were Asterix FreePBX distribution with configuration web page as a server side, working under CentOS with kernel 2.6.32 and an Apache 2.2.15 http server, equipped with 1 GB of RAM. There were a machine as an hacker side – based on Debian Linux with kernel 2.6.32 (equipped with 512 MB of RAM) and DDOSIM software (Layer 7 DDoS Simulator v0.2). The server was connected to the real network through the firewall with proposed method implemented, which was running on Linux Debian with kernel 2.6.32 with 512 MB of RAM.

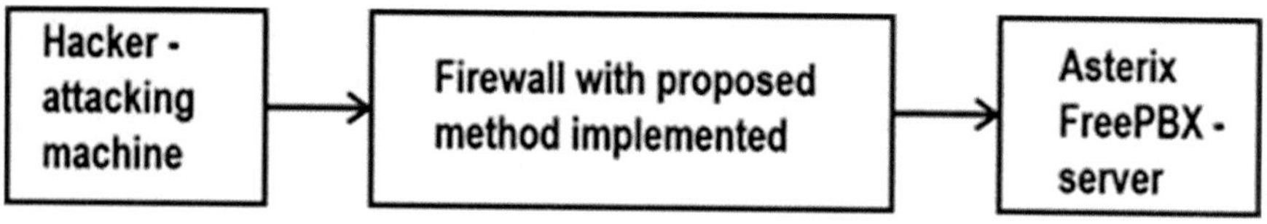

Fig. 2. Diagram of the test network

During the test, the memory usage and http server response times were observed. The attack consisted in sending a large amount of HTTP Get requests to the server. Without the proposed method, the server's memory usage raised up to 100% and finally the server stopped responding. After launching the proposed method as an IPTables module, RAM memory usage was observed.

The proposed method was configured with the following limits:

- time slot for analyzing the amount of transmitted $datatime\_slots = 1$ s;
- allowable limit of $packetspacket\_limit = 30$;
- minimal limit of packets per timeslots = 100; maximal limit of packet = 1000.

The test of the attack on the server were performed and the conditions were as follows: 2000 HTTP GET requests sent every 10 s, there were 5 mentioned defuzzification method used.

3.2 Test Results

As it was mentioned, the implemented method was installed on firewall at server side. When the attack starts, the server were passed the information about their connection ($listIP$) to the implemented method working on firewall. Then, the method started its work and were limiting the impact of the attack.

None of the fifth cases resulted in server overload. Prior to the attack a connection with the server through the firewall was established and it remained active because it was started before the attack and the computer was recognized as allowed to communicate with the server. The information about this computer were on the list $listIP$.

RAM usage on the http server is presented in Fig. 3.

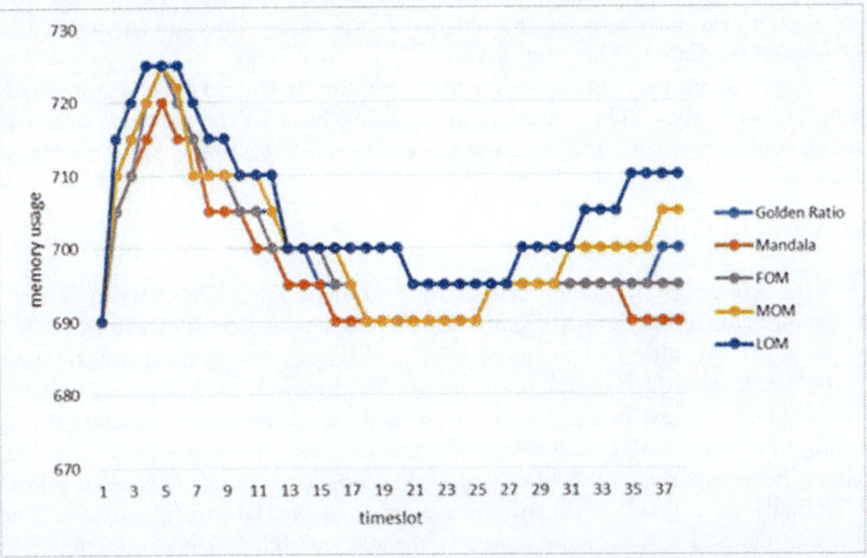

Fig. 3. RAM usage on http server.

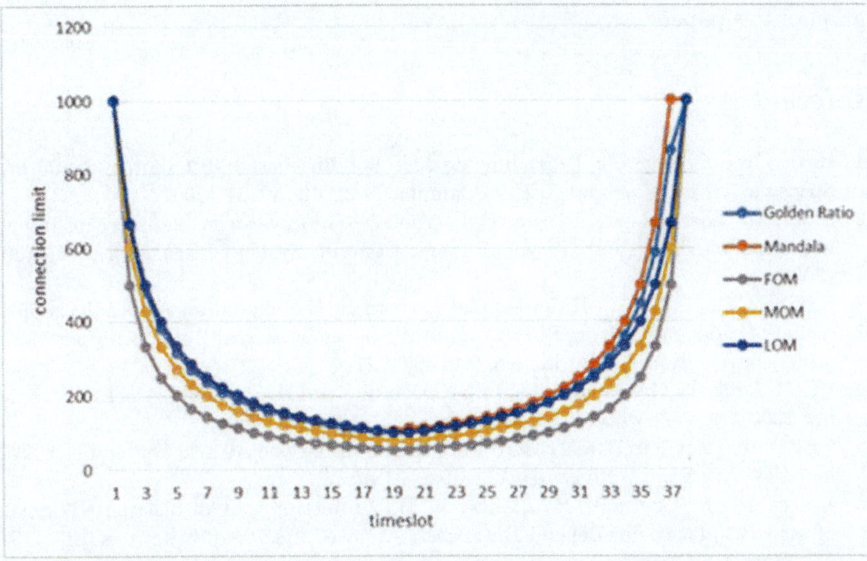

Fig. 4. Packet limit on time slots regulated by intelligent dropping module.

In the previous section the description of the mechanism for intelligent narrowing the amount of packets in a single time slot was mentioned. Figure 4 shows how different defuzzification method changes the packet limit. This process should take into account the nature of the server which are under attack and should be done by the administrator.

What is worth to mention is that after the attack, the network is going back to its ordinary state. The packet limits is going back to the previous numbers and the network will be still working without requiring administrator's work.

4 Conclusions

The article presents the comparison of application of five different types defuzzificators for controlling with the use of method against DDoS type attack. Based on the fuzzy logic after estimation of growth or drop of new connections to server the defuzzification process follows. The applied method has an impact on outcome value. If the method generates bigger outcomes the process of connections closing to server is faster. However it also means the opening process will be also faster. The comparison of defuzzification methods is necessary because it allows to estimate an impact of the results achieved on the controlled process. The choice of method will allow for faster or slower reaction. In the conducted tests it has an impact on the server resources consumption (RAM). In other cases it can have an impact e.g. on furnace control or air conditioning.

The author declares that there is no conflict of interest regarding the publication of this paper.

References

1. Rocky, K., Chang, C.: Defending against flooding-based distributed denial-of-service at-tacks: a tutorial. IEEE Commun. Mag. **40**, 42–51 (2002)
2. Moor, D., Shannon, C., Brown, D.J., Voelker, G.M., Savage, S.: Inferring internet denial-of-service activity. ACM Trans. Comput. Syst. (TOCS) **24**(2), 115–139 (2006)
3. Schuba C. L., Krsul, I., Huhn M. G., Spafford E. H., Sundaram A.: Analysis of a denial of service attack on TCP. Computer Science Technical Reports. Paper 1327 (1996). http://docs.lib.purdue.edu/cstech/1327
4. CERT Advisory CA-1996-21 TCP SYN Flooding and IP Spoofing Attacks, November 2000. http://www.cert.org/advisories/CA-1996-21.html
5. CERT Advisory CA-1996-01 UDP Port Denial-of-Service Attack, September 1997. http://www.cert.org/advisories/CA-1996-01.html
6. Apiecionek, Ł., Czerniak, J.M., Zarzycki, H.: Protection tool for distributed denial of ser-vices attack. In: Beyond Databases, Architectures, and Structures, pp. 405-414. Springer (2014)
7. Piechowiak, M., Zwierzykowski, P.: the evaluation of unconstrained multicast routing algorithms in ad-hoc networks. In: The International Science Conference: Computer Networks CN2012. Szczyrk, Poland (2012)

8. Apiecionek, Ł., Makowski, W.: Intelligent FTBint method for server resources protection, beyond databases, architectures and structures. In: Advanced Technologies for Data Mining and Knowledge Discovery. Series Communications in Computer and Information Science, vol. 613, pp. 683–691 (2016)
9. Changwang, Z., Jianping, Y., Zhiping, C., Weifeng, Ch.: RRED: robust RED algorithm to counter low-rate denial-of-service attacks. IEEE Commun. Lett. **14**(5), 489–491 (2010)
10. Cheswick, W.R., Bellovin, S.M.: Firewalls and Internet Security: Repelling the Wily Hacker. Addison-Wesley Publishing Company, Boston (1994).ISBN 0-201-63357
11. Chapman, B., Zwicky, E.D.: Building Internet Firewalls. O'Reilly & Associates Inc., Sebastopol (1995).ISBN 1-56592-124-0

Extending Machine Learning-Based Intrusion Detection with the Imputation Method

Mikołaj Komisarek[1], Marek Pawlicki[1(✉)], Piotr Soboński[3], Aleksandra Pawlicka[2], Rafał Kozik[1], and Michał Choraś[1]

[1] UTP University of Science and Technology, Bydgoszcz, Poland
marek.pawlicki@utp.edu.pl
[2] ITTI Sp z o.o., Poznań, Poland
[3] United Technologies Research Centre Ireland Limited, a Part of Collins Aerospace, Cork, Ireland

Abstract. Cybersecurity is relevant to everyone, as cyberthreat concerns individuals and whole societies, and a precise cyberattack targeted at critical infrastructure may pose danger to millions of citizens. At a European level, several initiatives have aimed at protecting CI, one of them being InfraStress. This paper presents a part of the InfraStress architecture, the ST.CD.2 component, which is dedicated to cyberthreat detection. The question of handling missing data, the used dataset and experimental setup have been discussed. The ST.CD.2 component has been demonstrated in industrially relevant environment; the results of this experiment have been provided. When compared with the results achieved by means of the traditional ways of handling missing data, the proposed method proves superior. Then, the final conclusions are given and future work is briefly described.

Keywords: Machine learning · Stream processing · Handling missing data

1 Introduction

It is an inescapable fact that the world has become globally connected. Digitalization makes it easier to access and spread any kind of information, enables communicating with people all around the globe, allows landing oneself a remote job from virtually any place in the world, fosters commercial competition, to name just a few advantages [17]. The breakout of the COVID-19 global pandemic has accelerated the pace of digitization even more. Almost overnight, millions of people had to switch to working, learning, purchasing and socializing in an online manner [11].

Yet, despite the fact digitalization brings a plethora of advantages, it has also encouraged the emergence of a new type of threats, i.e. cyberthreats. Thus, the fact that people have been increasingly reliant on various digital services means

M. Choraś et al. (Eds.): CORES 2021, IP&C 2021, ACS 2021, LNNS 255, pp. 284–292, 2022.
https://doi.org/10.1007/978-3-030-81523-3_28

they may be more and more vulnerable to falling victim to cyberthreats. As a matter of fact, cyberattacks have been the crime which grows the fastest, and their scope, size, level of sophistication and the funds allocated to performing them, all have been on the increase [8,13]. So much so, that as of December 2020, the European Union, when stating its strategy for the Digital Decade, deemed cybersecurity "an integral part of European's security", a key element of the process of creating "a resilient, green and digital Europe" [11].

Nowadays, it is fairly commonly known that inadequate levels of cybersecurity lead to individuals' money and data being stolen. However, the threat is likely to escalate to a massive scale. This is especially true for the attacks on the so-called critical infrastructures (CI), i.e., the systems which are crucial to keeping vital societal functions. Sensitive Industrial Plants and Sites (SIPS) are an example of such infrastructures. In the light of the growing dependence of the vital services on digital systems, and the sharp spike in the prevalence of cyberthreats, cyber- and cyber-physical attacks on SIPS have become as serious a potential threat as natural disasters or technical failure. The efforts to guarantee CI stays secure and resilient yet again emphasize the fundamental role cybersecurity occupies in todays' Europe [7].

The number of cyberthreats is on the rise, but so are the efforts of researchers and scientists aimed at finding the best cybersecurity solutions and being one step ahead of the wrongdoers. Regarding CI, a number of actions have been carried out at a European level, in order to actively help protect it against cyberthreats. One of such initiatives has been InfraStress (the full name being "Improving resilience of sensitive industrial plants & infrastructures exposed to cyber-physical threats, by means of an open testbed stress-testing system"), funded by the European Union's Horizon 2020 research and innovation programme. InfraStress started on June 1st, 2019, and will last two years and four months. The objectives of InfraStress encompass providing an advanced, cutting-edge framework for detecting threats of cyber/physical nature and an innovative methodology for assessing resiliency, at the same time integrating C/P Situational Awareness and Threat Intelligence. In the project, 27 partners from 11 European countries have joined forces. The InfraStress solution is now assessed in 5 pilot locations. This paper concentrates on Pilot2, involving DePuy Synthes (DPS), i.e., a plant within a medical device manufacturing site, located on the south coast of Ireland. The industrial processes taking place at the site entail dangerous elements, such as volatile and toxic materials, high voltage, pressure and temperature, etc., making it a Sensitive Industrial Plant. The interdependencies with the other CIs in DPS's vicinity, as well as possible cascading effects in the face of C/P threats, and their influence over employers and the supply chain will be evaluated during InfraStress.

The InfraStress architecture comprises of several state-of-the-art components, each of them being crucial to the effective functioning of the system. One of those components, called ST.CD.2, has been developed by the University of Science and Technology in Bydgoszcz, Poland, which is one of the InfraStress partners. This particular component is going to be the focus of this paper. Firstly, the testbed and stream processing are going to be presented. Then, the experimental setup along with the data used are going to be described, followed by the conclusions and a brief discussion of the future work.

2 Infrastress Cyber Detection Solution

2.1 Leveraging Apache Kafka, Elacticsearch and TensorFlow

The ST.CD.2 component environment is distributed and scalable. A container environment is used to run all the required modules. This allows for a quick implementation process and possible scaling of solutions. A container is a standard unit of software that is capable of packing the source code files of a program together with all the necessary dependencies required to transform that code into the resulting program. A container runs fast and works independently of other system environments. The ecosystem components consist of the following pieces of software:

Apache Kafka [2] is a distributed platform handling stream data. The software enables the creation of streaming data pipelines in real time. The project is available under the Open-Source license. A great advantage of using Kafka is the guarantee of message delivery. In a distributed environment resistance to failures is of utmost importance. A distributed system, running on many devices simultaneously, should work without any disturbances even if several servers stop working. Kafka by default guarantees that a message is delivered at least once, this means that a message can be received again, but none will be lost. In Kafka, messages sent by a sender under a particular subject will be stored in the order in which they are sent. Individual recipients have access to messages in the order in which they are stored.

Elasticsearch and Kibana
Elasticsearch [6] is a search and data analysis engine that also allows for data warehousing. It is characterised by scalability and high speed. Communication with the engine is done using REST queries, which makes it very easy to implement in our system. Some of the most important Elasticsearch characteristics are:

- Data is stored in indexes, or collections of documents
- Data can be mined using Elasticsearch DSL queries sent through a REST API.
- It is a distributed system and scales very well horizontally - one can add more nodes, while Elasticsearch handles processes like appropriate cluster extension and data division between servers.
  The use of this component in ST.CD.2 was dictated by the storage of large amounts of data and fast search. The second important aspect was to combine the Elasticsearch database with the Kibana tool. This combination allows for fast browsing of the stored information and creation of advanced charts and visualisations.

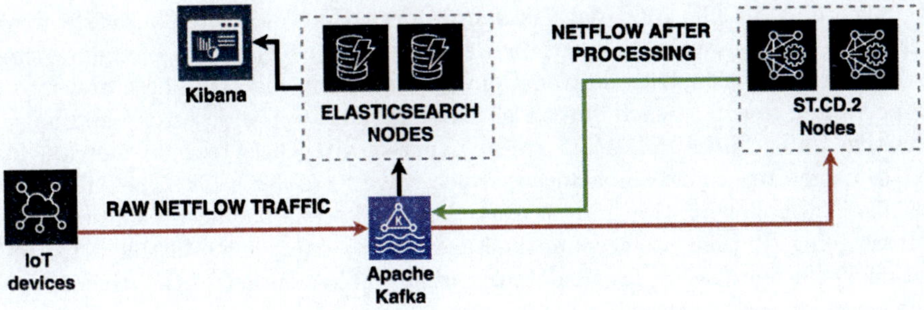

Fig. 1. ST.CD.2 enviroment

Machine Learning

The last and most important element in ST.CD.2 is the part which enables traffic analysis and anomaly detection. The software works completely with Kafka streams and the Elasticsearch database. The entire component is written in Python, opening the doors to the utilisation of state-of-the-art ML libraries like Tensorflow [19].

The component feeds network traffic to ML-based models and streams the classification results back to another Kafka topic.

The solution is displayed in Fig. 1. The ST.CD.2 component consists of several main modules that work closely together. The software contains the following modules:

1. Producers - Module responsible for sending processed data directly to the Elasticsearch database or to the Kafka stream. Data is usually stored as JSON or as comma-separated records.
2. Connectors - A mechanism that allows to connect to a Kafka stream and define the appropriate topic, time interval, and how many new data samples to load.
3. PreProcessor - capable of processing any object scheme. The data from a collector can be passed through a preprocessor and properly prepared before being fed to a ML algorithm.
4. Classifier - A module that loads the appropriate files and feeds it to an ML model, collecting the classification result.

3 Handling Missing Data for Network Intrusion Detection

The module leveraging ML for classification of network traffic lies at the heart of the ST.CD.2 component. Systems leveraging AI are only as good as the data they are trained with. This reliance on inputs opens up the question of data quality.

According to [9], good data is simply data that facilitates a satisfactory result. In case of Network Intrusion Detection, the effectiveness of the system is measured using metrics derived from the confusion matrix. These are, most importantly, Recall - which illustrates the sensitivity of the detector - and False Positive Rate. Those indicators are augmented with the Precision metrics. In NIDS data often comes imbalanced, which causes problems for some ML algorithms [12]. Following [15], in this work the dataset was balanced using random subsampling. To take into account the issues of data balancing, Matthiews Correlation Coefficient (MCC) is calculated in addition to Accuracy (ACC) to describe the classifier performance.

Following [9] data completeness has to be taken into account. In cybersecurity-related data, some values are often missing. This is reflected in many benchmark datasets. To formulate a well-performing ML model, the issue of missing data has to be tackled.

Regardless of the pattern of missingness, in cybersecurity research the method most often used for dealing with missing data is either Complete Case Analysis (CCA) [4], mean substitution [20], or substitution with constant or most frequent values [21].

3.1 Data Imputation

CCA simply discards the datapoints where any values are missing. This can lead severe loss of information - in [10] an example can be found where values absent from just 16% of cases result in discarding 80% of the dataset, due to their rather unfortunate distribution.

All substitution methods, on the other hand, influence the statistical characteristics of the feature, potentially causing bias of the ML model. The research into handling of missing data lead to the formulation of 'multiple-imputation methods'. Those approaches, like MICE [3], KNN-Imputation [22], or MissForest [18] circumvent the issues caused by substitution or deletion by taking into account the whole dataset. Both [18] and the preliminary research of the authors of this work indicate that MissForest outperforms KNN-Imputation and MICE in the cybersecurity use case, thus it was employed for dealing with missing values in this work. **MissForest** is a reinterpretation of the RandomForest algorithm.

3.2 Dataset and Experimental Setup

The data used in this work comes from the Aposemat IoT-23 dataset [1]. This novel IoT benchmark was formulated by the Stratosphere Lab to include the most relevant malicious scenarios for IoT - by capturing real network traffic on physical (not simulated) hardware. The research utilises randomly sampled 10% of the set stratified to reflect similar class distributions as the entire dataset.

The ST.CD.2 component allows for the utilisation of any machine learning algorithm provided by sk-learn [16] or TensorFlow [19] libraries or any innovative modifications of those. In this work, the RandomForest (RF) algorithm was used as the classifier to put emphasis on the influence of data imputation and assure

Table 1. The class-level performance metrics for RF classifier trained using the dataset after discarding incomplete features

	Precision	Recall	F1-score
Attack	0.72	0.98	0.83
Benign	0.98	0.29	0.45
C& C	0.94	0.15	0.26
C2HearBeat	0.72	0.08	0.15
DDoS	0.69	1.00	0.82
Okiru	0.35	0.77	0.48
PortScan	0.94	0.02	0.03
Macro avg	0.76	0.47	0.43
Weighted avg	0.74	0.52	0.44

reproducibility of results. RF is also excellent for fast prototyping, as it gives high performance with much less optimisation than neural networks [5].

3.3 The Experiment and the Results

Using the dataset as-is yields sub-par results. Large parts of some of the features included in the dataset are missing - for example, in the '*duration*' column 23550433 of all 31973855 samples contained only the '-' mark, indicating a missing value. That constitutes over 73% of all the values. Using Complete Case Analysis would cripple the dataset severely. One of the approaches to such large percentage of missing values in one feature could be to just discard the feature. The performance of the RF classifier trained with the dataset where the features containing missing values were not taken into account is showcased in Table 1. As a frame of reference, the performance of the RF classifier trained with the dataset with the missing features substituted by zeros is displayed in Table 2.

The two approaches yielded very similar results - just as if the classifier algorithm was not capable of using the features containing such high number of missing values. The performance of the RF classifier trained with the dataset imputed using the MissForest algorithm is illustrated in Table 3. The increase in the effectiveness of classification is striking. The classifier-level metrics in Table 4 - the Matthews Correlation Coefficient (MCC), False Positive Rate (FPR) and Accuracy - are provided to complete the picture of the difference in classification results.

Table 2. The class-level performance metrics for RF classifier trained using the dataset after missing value substitution

	Precision	Recall	F1-score
Attack	0.95	0.88	0.92
Benign	0.98	0.29	0.45
C& C	0.83	0.14	0.23
C2HearBeat	0.65	0.13	0.22
DDoS	0.70	0.99	0.82
Okiru	0.35	0.78	0.48
PortScan	0.81	0.04	0.07
Macro avg	0.75	0.46	0.46
Weighted avg	0.71	0.52	0.45

Table 3. The class-level performance metrics for RF classifier trained using the dataset after MissForest

	Precision	Recall	F1-score
Attack	0.95	0.89	0.92
Benign	1.00	1.00	1.00
C& C	0.89	0.59	0.71
C2HearBeat	0.87	0.44	0.58
DDoS	0.69	1.00	0.82
Okiru	1.00	0.78	0.87
PortScan	1.00	0.77	0.87
Macro avg	0.91	0.78	0.82
Weighted avg	0.92	0.89	0.89

Table 4. Comparison of classifier-level metrics

Classifier - level metrics	Feature deletion	CCA	MissForest
Accuracy	0.52	0.52	**0.89**
MCC	0.42	0.42	**0.86**
FPR	0.24	0.24	**0.0008**

4 Conclusions and Future Work

To realise the objectives of the Infrastress project the ST.CD.2 component was designed and developed. It is an innovative solution which allows handling streams of data while scaling well, but most importantly allows to use the latest advancements in machine learning to realise network intrusion detection. In this work a part of the pipeline which handles missing data and the effect it has on the effectiveness of ML algorithms is presented. The MissForest imputation method improved the performance metrics of the RF algorithm used for intrusion detection by a substantial margin. To facilitate reproducibility of the research results the IoT-23 benchmark dataset was used, along with a standard implementation of RandomForest. In the near future the research will be conducted on real-world data coming directly from the DPS infrastructure in a manner that is compliant with the GDPR [14].

Acknowledgement. This work is funded under INFRASTRESS project, which has received funding from the European Union's Horizon 2020 research and innovation programme under grant agreement No. 833088.

References

1. Agustin, P., Sebastian, G., Maria Jose, E.: Stratosphere laboratory. A labeled dataset with malicious and benign IoT network traffic (2020). https://www.stratosphereips.org/datasets-iot23
2. ApacheKafka: https://kafka.apache.org
3. Azur, M.J., Stuart, E.A., Frangakis, C., Leaf, P.J.: Multiple imputation by chained equations: what is it and how does it work? IJMPR **20**(1), 40–49 (2011). https://doi.org/10.1002/mpr.329
4. Baio, G., Leurent, B.: An introduction to handling missing data in health economic evaluations. In: Care at the End of Life (2016). tinyurl.com/19hh58ba
5. Choraś, M., Pawlicki, M.: Intrusion detection approach based on optimised artificial neural network. Neurocomputing (2020)
6. Elasticsearch: elasticsearch/elasticsearch (2015). https://tinyurl.com/ltzbptc5
7. EU Science Hub: Cybersecurity (2021). https://tinyurl.com/2c8telmz
8. European Council: Cybersecurity: How the EU tackles cyber threats (2021). https://www.consilium.europa.eu/en/policies/cybersecurity/
9. Fan, W., Geerts, F.: Foundations of data quality management. Synthesis Lect. Data Manage. **4**(5), 1–217 (2012). https://tinyurl.com/183828pu
10. Graham, J.W.: Missing Data. Springer, New York (2012). https://doi.org/10.1007/978-1-4614-4018-5
11. Joint Communication to the European Parliament and the Council: The EU's Cybersecurity Strategy for the Digital Decade (2020)
12. Ksieniewicz, P., Woźniak, M.: Imbalanced data classification based on feature selection techniques. In: IDEAL, pp. 296–303. Springer (2018)
13. Nec: Why is cyber security important? – Why we need cyber security (2020). https://tinyurl.com/3h52993g
14. Pawlicka, A., Jaroszewska-Choras, D., Choras, M., Pawlicki, M.: Guidelines for stego/malware detection tools: achieving GDPR compliance. IEEE Technol. Soc. Mag. **39**(4), 60–70 (2020)
15. Pawlicki, M., Choraś, M., Kozik, R., Hołubowicz, W.: On the impact of network data balancing in cybersecurity applications. In: ICCS, pp. 196–210. Springer (2020)
16. Scikit-learn. https://scikit-learn.org/
17. StartSmarter: The Advantages and Disadvantages of Digitalisation (2019). https://startsmarter.co.uk/the-advantages-and-disadvantages-of-digitalisation/

18. Stekhoven, D.J., Buhlmann, P.: MissForest–non-parametric missing value imputation for mixed-type data. Bioinformatics **28**(1), 112–118 (2012). https://tinyurl.com/3h74tza2
19. TensorFlow. https://www.tensorflow.org
20. Tripathi, A.K., Rathee, G., Saini, H.: Taxonomy of missing data along with their handling methods. In: ICIIP, pp. 463–468 (2019). https://doi.org/10.1109/ICIIP47207.2019.8985715
21. Zeng, D., Xie, D., Liu, R., Li, X.: Missing value imputation methods for TCM medical data and its effect in the classifier accuracy. In: IEEE Healthcom, pp. 1–4 (2017). https://doi.org/10.1109/HealthCom.2017.8210844
22. Zhang, Y., Kambhampati, C., Davis, D.N., Goode, K., Cleland, J.G.F.: A comparative study of missing value imputation with multiclass classification for clinical heart failure data. In: 2012 9th ICNC-FSKD, pp. 2840–2844 (2012). https://doi.org/10.1109/FSKD.2012.6233805

Progress in Advanced Computer Systems

Green IT: Energy Efficient Constructions and Applications for Data Centers and Clusters

Andriy Luntovskyy(✉) and Tenshi Hara

Saxon University of Cooperative Education – State Study Academy Dresden, Hans-Grundig-Straße 25, 01307 Dresden, Germany
{andriy.luntovskyy,tenshi.hara}@ba-sachsen.de

Abstract. This paper is devoted to an actual concept of "Green IT" with the background of the "hardcore" resource consuming applications. The given work discusses the energy efficient constructions for computing centres and clusters, which provide optimisation of performance as well as PW, PUE and ERE factors. The following both approaches are examined: 1) workload consolidation and intelligent VM migration; 2) optimisation of the cooling process (guaranteeing PUE) and energy recycling (guaranteeing ERE and CO_2 footprint reduction). The optimal ERE within the data centres and computing clusters provides advanced resource-consuming distributed applications like pharmacology simulations, Big Data analytics, Blockchain, crypto-currency mining.

Keywords: Green IT · Performance · Speedup · PW · PUE · ERE · Scalability · Blockchain · Big data analytics · Machine learning

1 Green IT Problems and Challenges: Motivation

The "Green IT" concept stems from topical problems [1–4]:

1. resources are limited and renewable energy is too expensive,
2. electricity costs are rising permanently – partly because of IT – developing into a significant cost factor; individual devices indeed consume less energy;however, the overall count of consuming devices steadily increases due to clustering, cloud computing and IoT, and
3. growth of "greenhouse gas" emissions and, especially, CO_2 footprints for IT, are to be mentioned.

Therefore, providing a reasonable and sustainable Green IT concept is only possible under consideration of the following two groups of aspects [5–7]:

1. economic aspects, including
   - efficient use of hardware resources, energy efficient hardware, software and network protocols, energy efficient cooling processes in IT,
   - recycling of waste heat,

M. Choraś et al. (Eds.): CORES 2021, IP&C 2021, ACS 2021, LNNS 255, pp. 295–309, 2022.
https://doi.org/10.1007/978-3-030-81523-3_29

as well as

2. environmental and social aspects, including
   - increased people awareness, reduction of electronic waste and electric smog,
   - UN and EU climate programs, and environment-friendly IT placement.

Surely, there too are many politically motivated statements, such as the following:

- "*Not the CO2 emissions of factories, aeroplanes or cars, but the Internet is the climate killer*" (translated from a 2017 article in Die Welt),
- "*A single Google search query uses just as much electricity as a 10WCFL needs to light up for 108 s*" (according to Google themselves[1]), and
- "*The data centres worldwide can generate so much CO2 emissions as international air traffic*" (claimed in a 2015 article in The Guardian).

Besides these publicity claims and whether one believes them or not, the statistics are inexorable [1–11]:

- in 2017, IT services consume reached 3% of world's energy overall consumption,
- in 2025, IT services will consume 20% of the world's available energy!

IBH IT-Service GmbH, as an Internet service provider and system integrator, has many years of experience in the professional equipment and service-oriented operation of data centres. The following questions are often discussed between the provider and users [2]:

- What standards must be applied to the construction of high-performance data centres?
- How can standardised 19"-racks be optimally used?
- What role do lightning and fire protection as well as energy efficiency play?

IBH represents the most modern design principles and important technical details on fail-safe power supply, expansion of redundant communication channels and highly secure services [2]. Nowadays however, massively energy-efficient computing power is required due to the deployment of new applications [4,5] such as data mining, AI, or Blockchain. This can be illustrated under use of the example of CLOUD&HEAT Technologies GmbH [5]:

- What is the PUE (Power Usage Effectiveness) factor and how is it calculated?
- What is the ERE (Energy Reuse Efficiency) factor and how is it calculated?
- How do you achieve a higher level of CPU and GPU integration?
- How is the server waste heat recycled to useful heat, e.g., in a building's hot water and heating supply?

---

[1] The data was provided by Google in 2009. Since then, the efficiency of data centres surely has increased. Nevertheless, it definitely debunks a often stated "1 h per search"-claim to be found in the Internet and many publishing media.

CLOUD&HEAT delivers fully integrated solutions for secure data centres and computing infrastructures with higher energy efficiency. The projects enable real transition to a digital future through the deployment of powerful racks and highly integrated CPU containers with efficient water- and air-cooling as well as energy recycling.

There is a further well-known fact, too: cluster CPU under-utilisation is a common problem in IT [4–10] (cf. Fig. 1a). Of their available CPU time, under-utilisation can be quantified as Twitter Clusters: 40% to 50%, Google Clusters: 25% to 35%, and Amazon EC2 Clusters: approx. 20%. It often leads to sub-optimal values of PUE: ca 1.12 to 1.89 in this area. Many solutions have no pre-designed energy recycling; often, excess heat is simply cooled away, however at least an efficient cooling process is used. Figure 1a provides a scenario for VM migration with three virtual machines: VM1, VM2, and VM3. It is aimed at load balancing, dynamic deployment, and efficient utilisation of the physical servers S_1, S_2, ..., S_n via live time migration[2]. Therefore, the next question might be formulated like this: What is energy efficient service execution? Current approaches to Green IT are herewith as follows [1–12]:

- design of multi-dimensional, highly-scalable, energy-efficient multi-processor servers with multiple CPUs per board and multiple boards per server,
- efficient inter-processor communication via powerful links in radio and light frequency ranges for interaction and load balancing between the jobs, and
- flexible virtualisation and VM migration processes taking into account the power, load, network, storage usage as well as jobs structure in order to increase QoS.

Furthermore, the following challenges for Green IT can now be expressed:

1. Advanced efficient constructions for computing centres are required.
2. Workload consolidation and intelligent VM migration are necessary.
3. Optimisation of the cooling processes (guaranteeing PUE) and energy recycling (guaranteeing ERE and CO_2 footprint reduction) gain on the meaning.
4. The optimal ERE within the data centres and computing clusters provides advanced and—simultaneously—resource consuming distributed applications such as pharmacology simulations, Big Data analytics, Blockchain, or crypto-currencies mining.

The remainder of this paper's content follows as:

§2: State-of-the-art and energy efficient constructions and approaches to performance and energy consumption, as well as QoS optimisation are discussed.
§3: Performance parameters and a generalised law for clusters speedup are examined under considering of the amount of the available processors or VMs and the workload.
§4: CLOUD&HEAT efficiency parameters are discussed.
§5: The evaluation of the solutions is offered and proposed approaches are given.

---

[2] Live Time Migration: the VM remains active while it is migrated during run-time. Service interruptions are minimised, ideally not occurring at all.

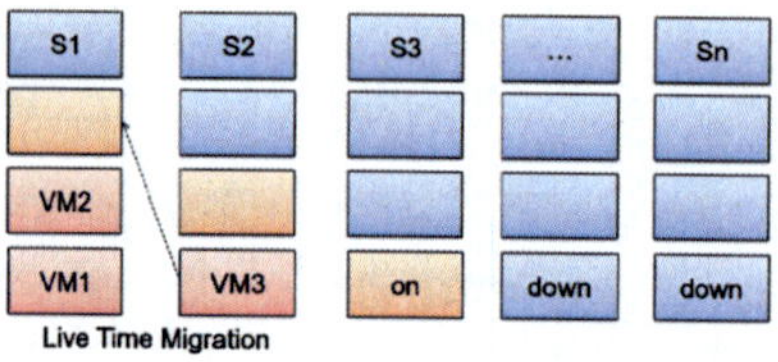

(a) Cluster-CPU under-utilisation as a known problem

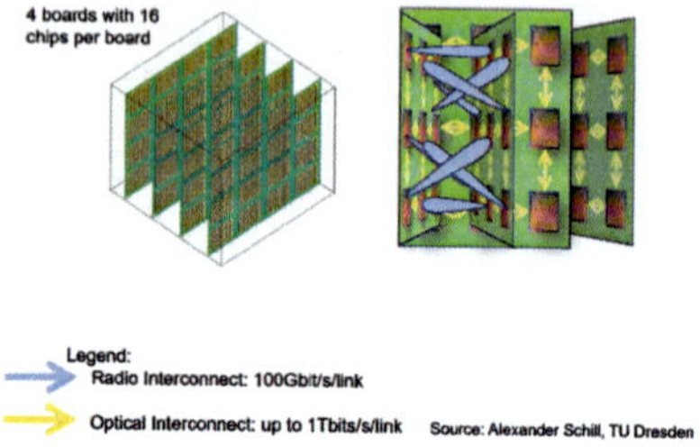

(b) High-Rate Inter-Chip Comm. with n Nodes and $C = 0.5n\,(n-1)$ interconnections [3, 6]

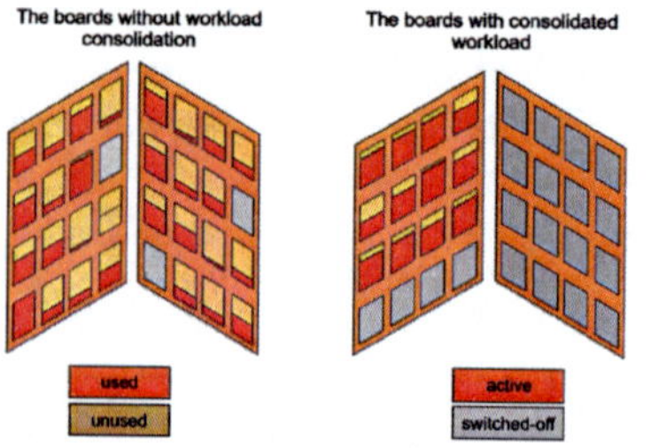

(c) Workload Consolidation by Distributed Computing [3, 6, 7]

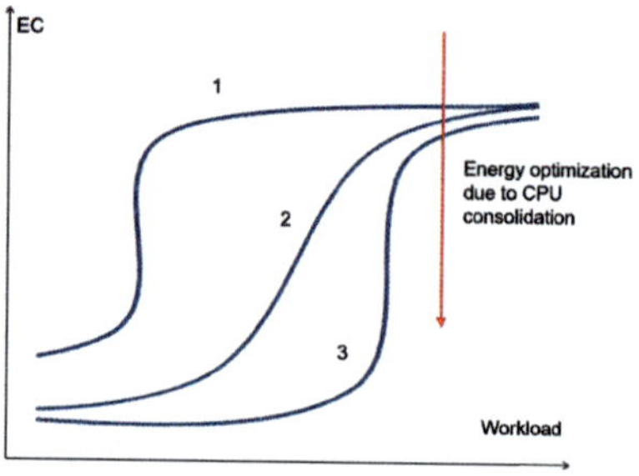

(d) Energy Optimisation due to CPU Consolidation by a growing Workload

Fig. 1. Migration and clustering effects

2 Best Practices for Energy-Efficient Constructions

In order to better understand the underlying principles and problem, the following example shall be considered.

Example 1

Within the HAEC project at TU Dresden [3, 6] the best practices for energy efficient constructions were investigated. The cluster architecture examined by the HAEC team consists of a large set of CPUs with multiple radio interconnections (each link of 100 Gbps) as well as of multiple optical interconnections (each link of 1 Tbps) (cf. Fig. 1b). Due to the large count of participating nodes (n) the quantity of interconnections (C) grows quadratic. The given architecture solves the problem of high-rate inter-chip communication. However, the computing resources consume a large amount of energy even when idling (cf. Fig. 1c). The more complex the servers, the higher their idle power consumption. With the aim of achieving energy efficiency, switching off idle or under-utilised resources are of a big interest. HAEC's proposed algorithm of the workload consolidation by distributed computing is as follows:

1. Estimation the incoming workload.
2. Live time VM migration (migration of containers).
3. Workload consolidation enables the downtime for the under-utilised physical servers.
4. This migration involves image encoding and transfer from a source server to a destination server.

Surely, the energy gain depends on migration latency (MT) and energy consumption (MC). These parameters are dependable on the VM size and the available network bandwidth (BW). The energy costs of a VM migration is a function of MT and MC. Interestingly, in some publications members of the HAEC team claim MC increases super-proportionally as the available BW decreases linearly. Design of scalable multi-processor servers with multiple CPUs and GPUs per board and multiple boards per server requires efficient inter-processor communication for interaction and load balancing between the executed jobs. In such cases, flexible virtualisation of execution engines is required. Migration policies taking energy, load, service quality and other parameters (such as memory or networking) as well as jobs structure into account [3,6,7]. As a future work, the scheduling mechanisms are necessary for detailed control and decision-making concerning VM placement and migration to achieve load-proportional energy consumption and to reduce absolute energy consumption and CO_2 footprint in total.

3 Performance Parameters and Energy Optimisation Due to Load Consolidation

Within this Section the mostly important QoS parameters of the clusters and data centres are discussed, among them the parameters are as follows [8–11]: Cluster availability classes 1 through 7, Cluster performance P_{Cluster}, Cluster PW (Performance-2-Power), Cluster average downtime T_{down}, Cluster PUE (Power Usage Effectiveness), and Cluster ERE (Energy Reuse Efficiency). An explanation of such important computing cluster and data centre parameters [8] like average downtime T_{down} are summarised in Table 1.

The classes 1 through 7 of service availability are given based on the sidereal year under consideration of leap years. The given representation is based on [8–11] and depicts the services from the 90%-reliable clusters and data centres up to the error resistant systems. The Cluster Speedup shows acceleration and performance increasing with n available CPUs in relation to normal performance of 1 CPU:

$$A_n = \frac{P_{\text{Cluster}}}{P_1} \tag{1}$$

Table 1. Availability boundaries and classes [8]

#	Yearly avail. in %	Yearly T_{down}	Unit	Available classes
1	90	36.52	Days	1 – reliable up to 90%
2	95	18.26	Days	
3	98	7.30	Days	
4	99	3.65	Days	2 – stable
5	99.5	1.83	Days	
6	99.8	17.53	h	
7	99.9	8.77	h	3 – available
8	99.95	4.38	h	
9	99.99	52.59	Min	4 – highly available
10	99.999	5.26	Min	5 – insensitive to errors
11	99.9999	31.56	Sec	6 – fault tolerant
12	99.99999	3.16	Sec	7 – error resistant

Note: "yearly" refers to an avg. sidereal year with 365.256 days

where P_1 is the performance on single CPU (ca. 100 GFLOPS for a convenient desktop PC). The parameter PW (Performance-2-Watt) represents the cluster efficiency by reaching of the specified performance:

$$PW = \frac{P_{\text{Cluster}}}{PIT} \tag{2}$$

where PIT is the nominal electrical power of the cluster.

3.1 Generalised Law for Cluster Speedup

Sun-Ni's Law [12] as the Generalised Law for Cluster Speedup was developed in the early 1990s. It refers the memory bounds and increasing workload within a cluster and then leads to a generalisation of both, Amdahl's and Barsis-Gustafson's models [10,11], considering that:

$$1 = e + p \Leftrightarrow p = 1 - e \tag{3}$$

where p is the parallelism grade.

If n CPUs are available within a cluster for a workload W, then the parallel portion of the workload W_{new} will be increased in $G(n)$ times as a reflection of the increasing of the memory consumption:

$$W_{\text{new}} = e \cdot W + (1 - e) \cdot G(n) \cdot W \tag{4}$$

where W_{new} is the parallel portion of the workload W. Then the speedup which considers the memory bounds can be calculated as follows:

$$A_n = \frac{e \cdot W + (1 - e) \cdot G(n) \cdot W}{\frac{e \cdot W}{1 + (1 - e)} \cdot G(n) \cdot \frac{W}{n}} \tag{5}$$

and thus:

$$A_n = \frac{e + p \cdot G(n)}{e + p \cdot \frac{G(n)}{n}} \tag{6}$$

where p is the parallelism grade and $p = 1 - e$.

Depending on the growth factor $G(n)$, there are three possible cases:

1. $G(n) = 1$, like Amdahl's:

$$A_n = \frac{e + (1-e)}{e + (1-e) \cdot \frac{1}{n}} = \frac{1}{e + \frac{(1-e)}{n}} = \frac{1}{(1-p) + \frac{p}{n}} \tag{7}$$

2. $G(n) = n$, like Barsis-Gustafson's with Eq. (3) inserted:

$$A_n = \frac{e + (1-e) \cdot n}{e + (1-e) \cdot \frac{n}{n}} = e + (1-e) \cdot n = 1 + (n-1) \cdot p \tag{8}$$

3. $G(n)\, m > n$, for Sun-Ni's Law:

$$A_n = \frac{e + (1-e) \cdot m}{e + (1-e) \cdot \frac{m}{n}} = \frac{1 - p + p \cdot m}{1 - p + p \cdot \frac{m}{n}} \tag{9}$$

Sun-Ni's Law's formula is used for the case that the workload grows faster than memory. The speedup is slightly higher than Barsis-Gustafson's, i.e. the ratio $\frac{m}{n} = 1.1 \ldots 1.5$.

Data centre virtualisation technologies have attracted a lot of attention to enable various cloud computing services and to facilitate VM migration. VM migration can help service providers to achieve the goals of saving energy, enhancing resource efficiency and quality of service (QoS). In order to ensure the QoS, the migration time and the downtime of VM should be considered while implementing the VM migration.

3.2 Workload Consolidation and VM Migration

Computing resources consume a large amount of energy even when they are idle. The workload consolidation and VM migration leads to the decreasing of idle times, overall costs and energy overhead (waste) but optimisation of the Migration Time (MT), Migration Power (MP), Migration Energy Consumption (EC) must be also taken into account [3,6,7,10,11]. The following VM migration models are examined and represented in the literature (cf. [9] and [10]). Linear regression is used as a rule [3,6,7]:

$$\begin{aligned} EC &= MP \cdot MT \\ &= a \cdot V_{\text{mig}} + c \\ &= a + b \cdot V_{\text{load}} + c \cdot BW \end{aligned} \tag{10}$$

where $\{a, b, c\}$ are constants, V_{mig} is the volume for a migrating VM, V_{load} is the workload volume, and BW is the bandwidth of the inter-CPU-links (e.g., Fibre Channel with 16 Gbps).

$$MT = a + b \cdot INST + c \cdot V_{\text{L3}} \tag{11}$$

where $\{a, b, c\}$ are constants, $INST$ is the volume of the processed instructions during a VM migration, and $VL3$ is the volume of the processed lines of L3 cache during a VM migration. Therefore, flexible migration is possible when considering the power, load, network, and storage usage as well as jobs structure. Only under such conditions QoS increasing by a growing workload within the cluster can be reached. The energy optimisation due to CPU consolidation is depicted in Fig. 1d. The depicted curves represent: 1) without VM migration, 2) and 3) with migration (less energy by more workload).

4 Energy Efficiency Parameters and Energy Recycling

4.1 Energy Efficient Computing and Applications

In addition to the performance parameters such as TFLOPS and PFLOPS, the experts speak more and more about energy efficiency in Cluster Computing, namely about the values [4–11], such as PUE (Power Usage Effectiveness) and ERE (Energy Reuse Efficiency), which are defined as follows:

$$PUE = \frac{P_{\text{IT}} + P_{\text{A}}}{P_{\text{IT}}} \tag{12}$$

$$ERE = \frac{P_{\text{IT}} + P_{\text{A}} - P_{\text{R}}}{P_{\text{IT}}} \tag{13}$$

where P_{IT} is the necessary power for IT and computing in the cluster, P_{A} is the waste heat power, and P_{R} is the recycled power. The first one describes effectiveness of the power usage within a data centre cluster under considering of the waste heat P_{A}. The second represents the efficiency of the energy recycling P_{R} under considering of the P_{A} value. The branch's best [3–6] computing centres and clouds delivers $PUE = 1.02 \ldots 1.12$. By significant improvement [5] via cooling and energy recycling ERE can be estimated in the vicinity of $ERE = 1.02 \ldots 1.06$.

4.2 Best Practices for Energy Efficiency: Hot Water Cooling

The solutions for Green IT [1–7], energy efficiency for computing clusters and data centres have been in the press since more than a decade, for example: Google's Data Centres in The Dalles, OR and Douglas County, GA in 2010 (cf. Fig. 2a and 2b), CoolMUC-2 in Munich in 2016, Mäntsälä in Finland in 2017, or hot water cooling at TU Dresden's Centre for High Performance Computing in 2017.

Currently, CLOUD&HEAT Technologies GmbH [4,5] belongs to the few Cloud providers following the Green IT best practices in Germany, and maybe worldwide, too:

- innovative and highly efficient hot-water cooling system,
- fully water-cooled hardware with waste heat reuse (to heat of buildings directly with servers running in the basement instead of a district heating connection), and
- integrated hydraulic system to reuse server waste heat by ca 55°C on the processor surface and heat disposal.

(a) Cooling Towers at Google's Data Centre in The Dalles, OR (Picture © by Google) [4, 6]

(b) Central cooling plant in Google's Data Centre in Douglas County, GA (Picture © by Google/Connie Zhou) [4, 5]

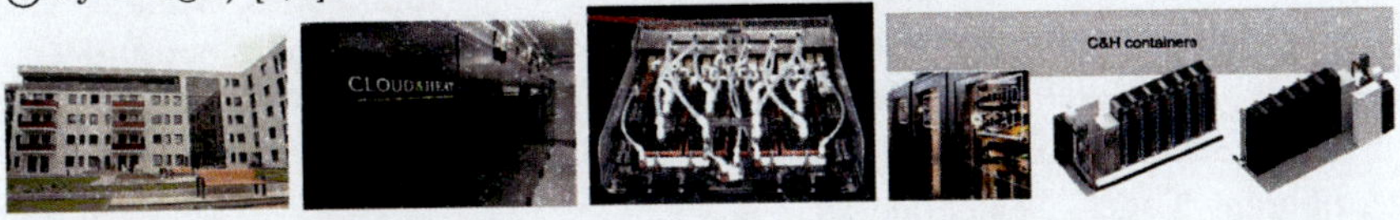

(c) Building with a Data Centre in the Basement

(d) Data Centre providing Building Heat

(e) Water-cooled 19"-Rack

(f) Water-cooled Data Centre Containers

Fig. 2. Data centre heat utilisation (Figs. c–f © by [5])

Whole data centres can be collapsed to an energy-efficient box, or specialised water-cooled 19"-rack, or container for a lot kW of recycled waste heat which can be utilised to cover a building's heating demands instead of district heating (cf. Fig. 2c and 2d). The mentioned hardware solution consists of a freely scalable number of server blades and storage units: configurable cores per rack, configurable storage with redundant storage servers etc. The systems are redundantly dimensioned for all essential components for failure-tolerant operation.

5 Blockchain and Mining of Crypto-Currencies

Blockchain is a cryptographical concept supporting decentralised applications with demand of consistency, integrity, and accountability assurances. Such decentralised applications include, but are not limited to, payment systems, financial online transactions, or crypto-currencies. By design, the Blockchain is a peer-to-peer (P2P) concept. The economic success of this crypto-technology will be evident in the next 10 up to 20 years. The deployment of Blockchain technology speaks mainly for a decentralized financial system. The advantages of such a solution seem obvious [4,5]:

- sustainability, general transparency and commitment,
- accelerated economic workflows and digitisation processes (so-called "IT in the digital age"), and
- it is also well suited to supporting current crypto-currencies (Bitcoin, Ethereum, Ripple, Litecoin, ZCash, Monero, Stellar, etc.).

5.1 Blockchain Architecture

Decentraliszed, cryptographically secured and unified blocks, their chains and transactions are grouped under a general, global public ledger (account), the structure of which is as follows: The Blockchain, as a networked Public Ledger, consists of participating nodes that represent an efficient P2P communication model. Typical features of the Blockchain are [13–15]:

- redundancy and synchronisation,
- cryptographic hash procedures for integrity assurance and attack safety,
- decentralised management and control of the Blockchain,
- network subscribers are also referred to as Nodes (Full-Nodes, Miners, Validators [13–15]) and run redundantly with mutual synchronisation, and
- often also considered a (anti-)feature: large block volumes can cause "Big Data" problem.

The different Nodes are defined as:

- *(Generic) Node*: All Blockchain participants (computers, smartphones, tablets, clusters, ...) are qualified as Node, if they have installed the corresponding software.
- *Full-Node*: A Node with a full (local) copy of the Blockchain; it checks for so-called "consensus rules".
- *Miner* and *Validator*: The individual participants or mining pool (with high resource requirements regarding hardware and energy consumption); they finalise blocks, where in a Miner "generates" a new block by calculating valid hashes, and a Validator proves the validity of the new block. Externally, they each act like a large participant; however, in fact many small blocks are generated for fractions of the system units (e.g., for small payments within a crypto-currency).

However, the following problems may occur during Blockchain operations:

- enormous energy consumption due to "mining" of valid blocks (processing of the hash blocks via its algorithmic complexity),
- exponential memory growth (including capacity migration between USB media, smartphones, PC, storage media such as SAN/NAS, as well as cloud storages), and
- privacy issues may arise even though cryptographic data security is guaranteed.

A solution may be not to process the complete Blockchain with all the transactions, but only use of excerpts of the Blockchain without a prehistory.

5.2 Mining of Crypto-Currency and Resource Consumption

At least eight factors are necessary and must be considered when calculating the profitability of the mining [13–15]:

1. Investment (basis device costs): hardware investment for a mining rig or cluster.
2. Electricity Costs: in real life currency per kWh (e.g. €/kWh).
3. Energy Consumption: electrical power of Mining Rig in KW.
4. Hash Rate: hash values computed each second.
5. Mining Difficulty: factor has always a new actual value.
6. Pool Fees: how much in % belongs to the joined pool.
7. Block Reward: monetary or ideological compensation for each new computed block (or for other units).
8. Unit Price: exchange course for the BTC (or other units like XMR, ETH, ZEC, or LTC).

There are different multi-purpose and specialised devices for mining available, so-called Mining Rigs. Standard PCs or smartphones can also be used, but considering the energy consumption problems, they provide rather more disadvantages than justifiable use. The following types of devices can be deployed to generate the hash values in the mining process:

- CPU mining: powerful processor is required,
- GPU mining: powerful graphics card is required, or
- ASIC mining: using Application-Specific Integrated Circuits (cf. Table 2).

Table 2. Features of an AntMines S15 mining device according to Amazon

Feature	High-performance mode	Low-energy mod
Hash rate	28 TH/s	18 TH/s
Electric power	1,596 W	900 W
Power efficiency	57 J/TH	50 J/TH
NW connection	Ehternet	
Weight, dimensions	7 kg @ $(240 \times 178 \times 296)$ mm^3	

Example 2

Using AntMiner as mining rig, which can calculate 28 TH/s with a power consumption of 1,596 W: The average daily gain from this performance is 0.00115956 Bitcoin. The contribution to the joined mining pool is 1.5%. The current value of a Bitcoin is 6, 640 €[3]. In Germany, the average electricity price is 0.3137 €/kWh[4]. When looking for the expectable investment to conduct a week's mining one shall consider:

- the expenditures E for the used electricity,
- gain/proceeds from the Bitcoins GB,
- efficiency Eff (in J/TH), and
- gain or losses (in €) GL per week of such mining.

The purchase costs are negligible. As a result, one can calculate:

$$\begin{aligned}
E &= 1.596\,\text{kW} \times 168\,\text{h/week} \times 0.3137\,\text{€/kWh} \\
&= 268.128\,\text{kWh/week} \times 0.3137\,\text{€/kWh} \\
&= 84.1117\,\text{€/week} \\
GB &= 0.00115956\,\text{Bitcoin/d} \times 7\,\text{d/week} \times 6,640\,\text{€/Bitcoin} \\
&= 53.8963\,\text{€/week} \\
GL &= 53.8963\,\text{€/week} - 84.1117\,\text{€/week} \\
&= -30,2154\,\text{€/week} \qquad \text{(losses due to negative value)} \\
Eff &= 1,596\,\text{W}/28\,\text{TH/s} = 57\,\text{J/TH}
\end{aligned}$$

A simple cross-multiplication of Eff shows that AntMiner rig would need to at least calculate 43.7 TH/s in order to cover the cost and turn a marginal profit.

The AntMiner rig has an extra energy saving mode. In that mode, the rig calculates only 18 TH/s with a reduced power consumption of 900 W. The process results in gain of 0.00074543 Bitcoin per day. All other parameters remain unchanged. Then the new calculation derives:

$$\begin{aligned}
E &= 0.9\,\text{kW} \times 168\,\text{h/week} \times 0.3137\,\text{€/kWh} \\
&= 47.4314\,\text{€/week} \\
GB &= 0.00074543\,\text{Bitcoin/d} \times 7\,\text{d/week} \times 6,640\,\text{€/Bitcoin} \\
&= 34.6476\,\text{€/week} \\
GL &= 34.6476\,\text{€/week} - 47.4314\,\text{€/week} \\
&= -12.7838\,\text{€/week} \\
Eff &= 900\,\text{W}/18\,\text{TH/s} = 50\,\text{J/TH}
\end{aligned}$$

Once again, the rig produces a net-loss.

The interesting question fora private individual is as follows: How high does the price Pr_{kWh} for 1 kWh have to be, so that mining is profitable for a given situation?

[3] As of 18 April 2020.

[4] National average calculated by the German Federal Grid Agency. Given price represents average as of 17 April 2020.

– High-Performance mode:

$$GB = 53.8963\,€/\text{week} = 268.128\,\text{kWh/week} \times Pr_{\text{kWh}}$$
$$\Leftrightarrow Pr\text{kWh} = 0.2010\,€/\text{kWh}$$

– Low-Power mode:

$$GB = 34.6476\,€/\text{week} = 151.2\,\text{kWh/week} \times Pr_{\text{kWh}}$$
$$\Leftrightarrow Pr_{\text{kWh}} = 0.2291\,€/\text{kWh}$$

This practical example shows that in a lot of cases mining of the crypto-currencies like BTC, ETH, ZEC, XMR, etc. leads to "no reward"-scenarios due to the large energy consumption in combination with the unsustainable electricity price. Additionally, the initial purchase costs for the rigs cannot be considered negligible. In contrast, positive results can be reached if the powerful CPU+GPU or ASIC chip clusters are used with access to low-cost renewable energy[5], or can gain energy via waste recycling (both with minimal Pr_{kWh}). A cluster with an optimal ERE value can use such recycled energy (cf. Sect. 4). The gained PR can then be targeted for crypto-currency mining.

Example 3

Considering a no-name XMR mining operation using a similar container concept like it was described earlier [5,13–15]. The mostly important container features are —as usual— the electrical power and water proximity. E.g., $P_{\text{IT}} + P_{\text{A}} = 2.4\,\text{MW}$ for overall electrical input, $P_{\text{A}} = 0.25\,\text{MW}$ for waste heat, as well as $P_{\text{R}} = 0.15\,\text{MW}$ can be recycled additionally. Therefore:

$$PUE = \frac{2.4\,\text{MW}}{2.15\,\text{MW}} = 1.116$$
$$\text{and} \quad ERE = \frac{2.4\,\text{MW} - 0.15\,\text{MW}}{2.15\,\text{MW}} = \frac{2.25\,\text{MW}}{2.15\,\text{MW}} = 1.047.$$

For 1 ASIC chip:

$$E = 47.4314\,€/\text{week}$$
$$GB = 0.00521801\,\text{Bitcoin/d} \times 7\,\text{d/week} \times 6,640\,€/\text{Bitcoin}$$
$$= 242.5331\,€/\text{week}$$
$$GL = 195.1017\,€/\text{week} \qquad \text{(profit!)}$$

for a heat waste recycling cluster with

$$P_{\text{R}} = 0.15\,\text{MW}, \quad P_{\text{Chip}} = 900\,\text{W}, \quad N_{\text{Chips}} = \frac{P_{\text{R}}}{P_{\text{Chip}}} = 166,666.\overline{6}.$$

[5] Such as hydro-electric power in Norway, or geothermal power in Iceland. Both locations are favoured by many crypto-currency mining operations.

Hence, with 166, 667 ASIC chips and re-utilised waste heat:

$$\begin{aligned} GB &= 242.5331\,€/\text{week} \times N_{\text{Chips}} \\ &= 40,422,264.1777\,€/\text{week} \\ GL &\approx 40.4 \cdot 10^6\,€/\text{week} \qquad \text{(pure profit!)} \end{aligned}$$

The calculation results in a pure profit as the expenditure E can be set as $E = 0$ since the investment into the power costs is calculated against the primary task of the cluster rather than the "waste-product" of Bitcoin mining.

A 40'-standard container can include ca 100,000 ASIC chips, which are divided into multiple rows of 19"-racks including liquid cooled servers, pump groups for high and low temperature levels, heat-exchangers and controlling units, as well as the hydraulic pipe systems [4,5] (cf. Fig. 2e and 2f). Such CLOUD&HEAT data centres and computing clusters are deployed no further than 30 km away from the cloud users [4,5] because of a targeted latency of $\Delta t < 1$ ms. These energy-efficient containers with data centres and computing clusters can be implemented in two variants: 1) $PIT = 500$ kW in a 20' shipping container, 2) $PIT = 1$ MW in a 40' shipping container [4,5].

6 Conclusions and Outlook

1) Advanced efficient constructions for computing centres are required.
2) Workload consolidation and intelligent VM migration are necessary.
3) Optimisation of the cooling processes (guaranteeing PUE) and energy recycling (guaranteeing ERE and reduction of the CO_2 footprint) are increasingly important.
4) Optimised ERE within data centres and computing clusters provides advanced and, simultaneously, resource consuming distributed applications, such as pharmacology simulations, Big Data analytics, Blockchain, or crypto-currencies mining.

Acknowledgement. The authors thank the colleagues and students at BA Dresden (namely, L. Zipfel and N. Liebich), CLOUD&HEAT Technologies GmbH (namely, J. Struckmeier and M. Feldmann), IBH IT-Services GmbH (namely, T. Horn), and current and former colleagues at the Chair of Computer Networks at TU Dresden (namely, A. Schill, D. Gütter, and K Rybina).

References

1. Zipfel, L.: Green IT problems: welcome word. In: Workshop Green IT at BA Dresden, October 2018
2. Horn, T.: Energieeffiziente RZ in Dresden und Sachsen. In: Workshop Green IT at BA Dresden, October 2018
3. Schill, A.: Energy-efficient service execution. In: Workshop Green IT at BA Dresden, October 2018

4. Struckmeier, J.: Cloud&heat datacenter for AI, AR and big data: secure, scalable and sustainable. In: Workshop Green IT at BA Dresden, October 2018
5. CLOUD&HEAT Technologies GmbH (Online). https://www.cloudandheat.com/
6. HAEC Project at TU Dresden (SFB 912 for Highly Adaptive Energy-Efficient Computing; Online). https://tu-dresden.de/ing/forschung/sonderforschungsbereiche/sfb912
7. Rybina, K., Dargie, W., Umashankar, S., Schill, A.: Modelling the live migration time of virtual machines. In: OTM Conferences 2015. Springer, Switzerland (2015)
8. Jurzik, H., Ahmke, D.: Verfügbarkeitswerte und Verfügbarkeitsklassen, p. 7 (2007)
9. Luntovskyy, A., Gütter, D.: Moderne Rechnernetze – Theoriebuch: Protokolle, Standards und Apps in kombinierten drahtgebundenen, mobilen und drahtlosen Netzwerken. Springer Fachmedien Wiesbaden GmbH (2020). ISBN 978-3-658-256173
10. Luntovskyy, A., Spillner, J.: Architectural Transformations in Network Services and Distributed Systems: Service Vision. Case Studies. Springer, Heidelberg (2017). ISBN 9-783-6581-484-09
11. Luntovskyy, A.: Performance and energy efficiency in distributed computing. In: 20th International Multi-Conference on Advanced Computer Systems (ACS-2016). Hard and Soft Computing for Artificial Intelligence, Multimedia and Security. Springer, Heidelberg (2016)
12. Sun, X.H., Ni, L.: Scalable problems and memory-bounded speedup. J. Parallel Distrib. Comput. **19**, 27–37 (1993)
13. Luntovskyy, A., Gütter, D.: Cryptographic technology blockchain and its applications. In: Proceedings of UkrMico-2018, Odessa, Ukraine. Springer (2019)
14. Survey on Crypto-platforms. https://hackernoon.com/top-blockchain-platforms-to-watch-out-in-2019-aa80e336a426/
15. Blockchain as a Service: Teamwork ohne Daten-Grenzen (2020). https://www.t-systems.com/blockchain/

Implementation of Nussinov's RNA Folding Using the Kokkos Library

Mateusz Gruzewski(✉) and Marek Palkowski(✉)

Faculty of Computer Science and Information Systems, West Pomeranian University of Technology in Szczecin, Zolnierska 49, 71210 Szczecin, Poland
mateusz-gruzewski@zut.edu.pl, mpalkowski@wi.zut.edu.pl
http://www.wi.zut.edu.pl

Abstract. Dynamic programming of bioinformatics codes requires efficient programming on multi- and many-core platforms. In this paper, we propose an implementation of Nussinov's RNA folding using the innovative Kokkos framework. Kokkos Core allows us to quickly code parallel programs, including bioinformatics tasks, for major HPC platforms. The Kokkos abstracts enable multi-threaded patterns like Posix Threads, OpenMP, or CUDA without significant code modifications. Our implementation consists of two steps. First, we parallelize Nussinov's RNA folding code using well-known compilers like Traco and PLUTO. Second, we re-implement manually output to the Kokkos C++ modern and multi-platform code using templates and lambdas. We explain and practice the framework features for memory and threads management in terms of RNA folding programming. We carry out experiments to study the performance of Nussinov's codes within Kokkos. We analyze speed-up as a ratio of serial run-time to the time taken by the parallel codes for various RNA sequences lengths. Related and future work are outlined.

Keywords: Computational biology · The Nussinov algorithm · RNA folding · Kokkos · Parallel computing

1 Introduction

Optimization of bioinformatics programs is a current challenging task for developers. One of the basic dynamic programming and computational biology is Nussinov's approach for secondary structure prediction [7]. The Nussinov RNA folding is an interesting benchmark because is an example of "nonserial polyadic dynamic programming" (NPDP). The term "nonserial polyadic" stands for family of dynamic programming codes exposing non-uniform data dependences, which is more difficult to be optimized [4]. Other bioinformatics for RNA folding or DNA sequences alignment like Zuker's RNA folding, Smith-Waterman's approach, or John McCaskill's algorithm for probabilistic folding evolves more complex data patterns.

In this paper, we introduce the Kokkos implementation for Nussinov's approach. Our target is performing parallel and portable code dedicated to CPUs,

M. Choraś et al. (Eds.): CORES 2021, IP&C 2021, ACS 2021, LNNS 255, pp. 310–317, 2022.
https://doi.org/10.1007/978-3-030-81523-3_30

GPUs, and Intel Xeon Phi. Kokkos supports modern templated C++ programming with abstracts and lambdas. It allows us to create universal code for various platforms. The code can be easily modified to run with OpenMP, Posix Threads, and CUDA.

There are many approaches implementing parallel and cache-efficient Nussinov's RNA folding [1,3,5,9,10] on CPUs, GPUs, co-processors, and FPGA platforms.

Li et al. proposed an interesting manual solution for Nussinov's RNA folding algorithm. Using lower and unused part of Nussinov's, they changed column reading to more efficient row reading. Diagonal scanning exposes parallelism in the output code. The experimental study was presented on the CPU and GPU architectures. However, the codes and algorithms are different depending on the architecture.

Zhao et al. improved the *Transpose* method and performed the experimental study of the energy-efficient codes. However, authors do not present multi-platform implementations, hence we do not consider this approach in our analysis [11].

Automatic compilers like Pluto and Traco are state-of-art tools, however, they apply parallel tiling strategies on CPUs only. The PPCG tool is an automatic approach for GPUs but focused only on CUDA. Cetus and Par4All are multi-platform and source-to-source compilers but the tools generate separated codes for CUDA and OpenMP.

Intel TBB is an advanced library with many parallel patterns using templated C++ with abstracts and lambdas, tasking, and many more. Unfortunately, TBB is limited to multi-core CPUs.

Authors of PluTo, Bondhugula and et al. presented dynamic tiling for the Zuker's optimal RNA secondary structure prediction [6]. 3-d iterative tiling for dynamic scheduling is calculated using reduction chains. Operations along each chain can be reordered to eliminate cycles in an inter-tile dependence graph. The authors consider only the CPU platforms.

Wonnacott et al. introduced 3-d tiling of "mostly-tileable" loop nests of RNA secondary-structure prediction codes in paper [10]. But, the paper does not consider any parallel code. The approach is also limited to the NUMA CPU architectures.

Summing up, limitations of these approaches are that authors provide state-of-art performance only to one HPC platform or different codes for various platforms. The portability of code allows bioinformatics developers to use one code for various graphic cards, CPUs, or other HPC hardware. In this paper, we engage the Kokkos library to provide multi-threaded portable codes with speed-up.

The paper is organized as follows. Section 2 introduces Nussinov's RNA folding code and wave-fronting strategy. The next section presents the Kokkos approach. Nussinov's code implemented with the Kokkos features is presented in Sect. 4. Next, we describe the experimental study. The last section summarizes the paper.

2 Nussinov's RNA Folding Algorithm

Nussinov made one of the first attempts at folding RNA in a computationally efficient way as the base pair maximization approach in 1978 [7]. An RNA sequence is a chain of nucleotides from the alphabet G (guanine), A (adenine), U (uracil), C (cytosine). Given an RNA sequence, the Nussinov algorithm solves the problem of RNA non-crossing secondary structure prediction by means of computing the maximum number of base pairs for subsequences, starting with subsequences of length 1 and building upwards, storing the result of each subsequence in a dynamic programming array.

Let N be a $n \times n$ Nussinov matrix and $\sigma(i,j)$ be a function which returns 1 if RNA[i], RNA[j] are a pair in the set (AU, UG, GC)[1] and $i < j-1$, or 0 otherwise. Then the following recursion $N(i,j)$ is defined over the region $1 \leq i \leq j \leq n$ as

$$N(i,j) = max(N(i+1,j-1) + \sigma(i,j), \max_{1 \leq j \leq n} (N(i,k) + N(k+1,j))) \quad (1)$$

and zero elsewhere [10].

The equation leads directly to the C/C++ code with triple-nested loops presented in Listing 1 [6].

Listing 1. Nussinov loop nest.

```
for (i = N-1; i >= 0; i--) {
 for (j = i+1; j < N; j++) {
  for (k = 0; k < j-i; k++) {
   S[i][j] = MAX(S[i][k+i] + S[k+i+1][j], S[i][j]); //s0
  }
  S[i][j] = MAX(S[i][j], S[i+1][j-1] + bond(RNA,i,j));//s1
 }
}
```

Nussinov's code can be parallelized by well-known wave-fronting strategies where statements are delayed to remove dependences between them in the same time partition. The optimizing compiler like Pluto or TRACO can be applied to skew statements and generate the code presented in Listing 2. The second loop nest is parallel and marked with the OpenMP pragma [8] meanwhile the outermost loop nest scans time partitions in serial.

Listing 2. Nussinov loop nest.

```
for (i = 1; i <= N - 1; i++){
  #pragma omp parallel for private(j, k) shared(i)
  for (j = i; j <= N - 1; j++)
  {
    for (k = 0; k <= i - 1; k++)
      S[-i+j][j] = max(S[-i+j][k-i+j] + S[k-i+j+1][j], S[-i+j][j]);
    S[-i+j][j] = max(S[-i+j][j], S[-i+j+1][j-1] + bond(RNA, -i+j, j));
  }
}
```

[1] order of letters is unimportant in a pair.

Listing 3. A trivial example of the Kokkos C++ code.

```
int main(int argc, char **argv)
{
  Kokkos::initialize(argc, argv);
  {
    Kokkos::View<double *, Kokkos::HostSpace > hostData{"hostData", N};
    Kokkos::View<double *, Kokkos::CudaSpace > deviceData{"deviceData", N};
    for (int i = 0; i < N; ++i)
    {
       hostData(i) = i;
    }
    Kokkos::deep_copy(deviceData , hostData);
    Kokkos::parallel_for(
      Kokkos::RangePolicy <Kokkos::Cuda, int >{0, N},
      KOKKOS_LAMBDA(int i) {
      deviceData(i) = deviceData(i) * deviceData(i);
    });
    Kokkos::fence();
    Kokkos::deep_copy(hostData , deviceData);
  }
  Kokkos::finalize();
}
```

3 The Kokkos Library

Kokkos [2] evolves quickly and provides a programming model in C++ for coding performance portable applications on different kinds of hardware, such as GPUs, Intel Xeon Phis, or many-core CPUs. Kokkos supports modern templated C++ programming and promotes good programming practices. It implements abstractions for both parallel executions of code and data management.

Developers can choose one of the execution spaces. Kokkos can use POSIX Threads, OpenMP and NVIDIA's CUDA programming model for graphics processing units (GPUs) and serial (no thread parallelism). The library supports fundamental parallel algorithms (execution patterns): parallel_for, parallel_reduce, parallel_scan, and tasks. To implement Nussinov's approach we applied parallel_for.

Next, we have two options for the execution policy. *Range policies* are used to execute an operation once for each element in a range. Team policies are used to implement hierarchical parallelism. A thread team is a collection of one or more parallel "threads" of execution.

Kokkos is designed to target complex node architectures with N-level memory hierarchies. Memory spaces are the places *where* data resides. Memory layouts adapt memory structures in a given algorithm and can perform memory optimization. Memory traits specify how a data structure is accessed in an algorithm.

Listing 3 present a simple Kokkos program. It uses Kokkos::CUDA and Kokkos::parallel_for. Arrays are presented by Kokkos::View which describes type, sizes, and location (host, GPU). Arrays can be copied between devices using Kokkos::deep_copy. Kokkos::fence synchronizes threads after the parallel loop.

Listing 4. The Nussinov's code using the Kokkos RangePolicy.

```
Kokkos::View<int**> gpu::nussinov::rangePolicy(std::string_view rnaChain,)
{
  using Policy = Kokkos::RangePolicy <std::size_t >;
  const int N = rnaChain.length();
  const Kokkos::View<int**> output { "output", N, N };
  const Kokkos::View<char*> deviceRnaChain { "deviceRna", N};
  std::memcpy(deviceRnaChain.data(), rnaChain.data(), rnaChain.length()
         * sizeof(char));

  for (int t2 = 1; t2 <= N - 1; ++t2) {
  Kokkos::parallel_for(
     "nussinovFor",
     Policy { 0, static_cast <std::size_t >(N - t2) },
     KOKKOS_LAMBDA(std::size_t index) {
     const int t4 = t2 + index;
     output(-t2 + t4, t4) = 0;
     int maxValue = output(-t2 + t4, t4);
     for (int j = 0; j < t2; ++j) {
       maxValue = std::max(output((-t2 + t4), j + (-t2 + t4)) +
                  output(j + (-t2 + t4) + 1, t4), maxValue);
     }
     output((-t2 + t4), t4) = std::max(maxValue , output((-t2 + t4) +1, t4-1)
                                + bond((-t2 + t4), t4));
  });
  Kokkos::fence();
  }
  return output;
}
```

4 Implementation of Nussinov's RNA Folding Within Kokkos

We chose Kokkos to implement studied bioinformatics task because we target to portable performance. Listing 4 presents code with the Range Policy. The GPU threads execute an operation once for each element in a range independently. Nussinov's array and RNA chain are transferred between host and CUDA. They are represented by means of the Kokkos abstracts, views. Next, the outermost loop nest is serial. The second loop nest is parallelized with the Kokkos lambda. The lambda body is executed within one thread.

Listing 5 presents alternative solution with TeamPolicy. Groups of thread perform calculations of the statement groups independently. The result is calculated by threads with the *Max* reduction within one team.

Listing 5. The Nussinov's code using the Kokkos TeamPolicy.

```
template <unsigned int... Traits >
Kokkos::View<int**> gpu::nussinov::teamBased(std::size_t teamSize , std::
string_view rnaChain)
{
  using Policy = Kokkos::TeamPolicy <>;
  using Team = Policy::member_type;
  const int N = rnaChain.length();
  Kokkos::View<int**> output { "output", N, N };
  const Kokkos::View<char*> deviceRnaChain { "deviceRna", N};
  std::memcpy(deviceRnaChain.data(), rnaChain.data(), rnaChain.length() *
                                                          sizeof(char));
  for (int t2 = 1; t2 <= N - 1; ++t2) {
    const int groupsCount = (N - t2 + (teamSize - 1)) / teamSize;
    Kokkos::parallel_for(
    "nussinovFor",
    Policy { N - t2, Kokkos::AUTO },
    KOKKOS_LAMBDA(const Team& team) {
    Kokkos::parallel_for(
      Kokkos::TeamThreadRange(team, teamSize),
      KOKKOS_LAMBDA(int i) {
      const int t4 = t2 + (team.league_rank() * teamSize) + i;
      if (((-t2 + t4) < output.extent(0)) && (t4 < output.extent(0))) {
        output(-t2 + t4, t4) = 0;
        int maxValue = output(-t2 + t4, t4);
        Kokkos::parallel_reduce(
          Kokkos::ThreadVectorRange(team, t2),
          KOKKOS_LAMBDA(int j, int& value) {
          value = std::max(output((-t2 + t4), j + (-t2+ t4)) +
                                 output(j + (-t2 + t4) + 1, t4),value);
      },
    Kokkos::Max<int> { maxValue });
    output((-t2 + t4), t4) = std::max(maxValue ,
                 output((-t2 + t4) + 1, t4 - 1) + bond((-t2 + t4), t4));
    }
    });
  });
  Kokkos::fence();
  }
  return output;
}
```

5 Experimental Study

To carry out experiments, we used a machine with a processor Intel i5-6000K (3.2 GHz, 4 cores, 4 threads, 6 MB Cache) and with a GPU card GeForce GTX 1060 (1280 CUDA cores, 6 GB memory, CUDA 10.0). The optimized codes were compiled by means of the nvcc with the Kokkos wrapper and Clang.

Experiments were carried out for eight RNA sequences lengths of the problem defined with parameter N. For test data, we used randomly generated RNA sequences.

Table 1 shows the time execution of the Kokkos GPU code with both execution policies and compared to the Kokkos::serial code and the OpenMP wavefronting implementation of Nussinov's algorithm.

The experimental study shows that the RangePolicy policy allows us to achieve slightly better performance than using TeamPolicy. The parallel implementation does not perform operations that could be more effective using the

Table 1. Time execution of the Nussinov RNA folding codes implemented applying Kokkos.

RNA length, N	Time execution [s]			
	i5 Serial	i5 OpenMP	GPU RangePolicy	GPU TeamPolicy
1000	0.2714	0.3422	0.0735	0.1303
2000	2.6398	3.8188	0.4321	0.6161
3000	11.9649	12.5858	1.2925	1.6029
4000	36.3067	33.7493	2.8039	3.2939
5000	83.6751	54.3392	5.5891	6.1095
6000	159.8300	97.2908	9.9438	10.7284
7000	277.7970	165.616	16.8424	18.1222
8000	437.4100	270.247	27.6812	30.4544

TeamPolicy strategy for this HPC platform. Furthermore, the allocated number of threads to blocks may not be optimal. The main advantage of using TeamPolicy is the ability to merge global memory access and shared memory usage, providing quick access to data used by all threads in the groups. It is hard to predict which the global memory elements will be used repeatedly by the group for the NPDP programs like Nussinov's approach because of non-uniform dependences. Hence, we will focus on more efficient RangePolicy employing the transpose or tiling techniques in further research for this class of problems. Summing up, we noted the significant performance of the portable Kokkos code for the studied GPUs comparing to the CPUs speed-up.

6 Conclusion

In this paper, was presented the Nussinov RNA secondary structure prediction codes using the Kokkos tool. The code is portable for different kinds of HPC architectures. We used GeForce GTX 1060 to check performance on GPUs. The original code has been parallelized within the TRACO compiler. Next, the code has been manually transformed with the Kokkos lambdas, templates, and abstracts. The output is a universal and modern C++ STL code.

Portability has the efficiency price but allows developers to run their multi-core computational biology tasks employing various platforms without any special effort for code adjusting. It is significant for the algorithm prototyping without efficient manual implementations. In the future, we plan to study other RNA folding approaches like the Zuker algorithm and MEA. The challenge is also the output code improvement with data management strategies like *Transpose* or tiling dedicated to the bioinformatics NPDP programs. Finally, we are going to generate the Kokkos codes automatically within the TRACO compiler.

Acknowledgments. Special thanks to Tomasz Michalak for practical support with the Kokkos framework.

References

1. Chang, D.J., Kimmer, C., Ouyang, M.: Accelerating the Nussinov RNA folding algorithm with CUDA/GPU. In: The 10th IEEE International Symposium on Signal Processing and Information Technology, pp. 120–125 (2010). https://doi.org/10.1109/ISSPIT.2010.5711746
2. Edwards, H.C., Trott, C.R., Sunderland, D.: Kokkos: enabling manycore performance portability through polymorphic memory access patterns. J. Parallel Distrib. Comput. **74**(12), 3202–3216 (2014). https://doi.org/10.1016/j.jpdc.2014.07.003. http://www.sciencedirect.com/science/article/pii/S0743731514001257. Domain-Specific Languages and High-Level Frameworks for High-Performance Computing
3. Jacob, A., Buhler, J., Chamberlain, R.D.: Accelerating Nussinov RNA secondary structure prediction with systolic arrays on FPGAs. In: Proceedings of the 2008 International Conference on Application-Specific Systems, Architectures and Processors, ASAP '08, pp. 191–196. IEEE Computer Society, Washington, DC (2008). https://doi.org/10.1109/ASAP.2008.4580177
4. Liu, L., Wang, M., Jiang, J., Li, R., Yang, G.: Efficient nonserial polyadic dynamic programming on the cell processor. In: 25th IEEE International Symposium on Parallel and Distributed Processing, IPDPS 2011, Anchorage, Alaska, USA, 16–20 May 2011 - Workshop Proceedings, pp. 460–471 (2011)
5. Liu, L., Wang, M., Jiang, J., Li, R., Yang, G.: Efficient nonserial polyadic dynamic programming on the cell processor. In: IPDPS Workshops, pp. 460–471. IEEE, Anchorage (2011)
6. Mullapudi, R.T., Bondhugula, U.: Tiling for dynamic scheduling. In: Rajopadhye, S., Verdoolaege, S. (eds.) Proceedings of the 4th International Workshop on Polyhedral Compilation Techniques, Vienna, Austria (2014)
7. Nussinov, R., Pieczenik, G., Griggs, J.R., Kleitman, D.J.: Algorithms for loop matchings. SIAM J. Appl. Math. **35**(1), 68–82 (1978)
8. OpenMP Architecture Review Board: OpenMP application program interface version 4.0 (2013). http://www.openmp.org/mp-documents/OpenMP4.0.0.pdf
9. Palkowski, M., Bielecki, W.: Parallel tiled Nussinov RNA folding loop nest generated using both dependence graph transitive closure and loop skewing. BMC Bioinform. **18**(1), 290 (2017). https://doi.org/10.1186/s12859-017-1707-8
10. Wonnacott, D., Jin, T., Lake, A.: Automatic tiling of "mostly-tileable" loop nests. In: IMPACT 2015: 5th International Workshop on Polyhedral Compilation Techniques, At Amsterdam, The Netherlands (2015)
11. Zhao, C., Sahni, S.: Cache and energy efficient algorithms for Nussinov's RNA folding. BMC Bioinform. **18**(15), 518 (2017)

Ontology-Driven Approach to Research and Educational Organization Information Representation

Larysa Globa[1(✉)], Rina Novogrudska[2], Maryna Popova[2], Bohdan Zadoienko[1], and Yu Junfeng[3]

[1] Institute of Telecommunication Systems of National Technical University of Ukraine "Igor Sikorsky Kyiv Polytechnic Institute", Kyiv, Ukraine
lgloba@its.kpi.ua

[2] National Center "Junior Academy of Sciences of Ukraine", 38-44 Dehtyarivska str., Kyiv, Ukraine
rinan@ukr.net, pm@man.gov.ua

[3] Information research institute of Shandong academy of science, Shandong, China
yujf@sdas.org

Abstract. The paper presents an approach to the representation of various types of information that is accumulated at the level of different scientific and educational organizations with the purpose of further assessment of the quality of their functioning. This approach is based on ontological modelling. Ontology allows structuring and organizing scientific and educational organizations information aiming its representation and analyses. General ontology of scientific and educational organizations information is divided into several ontologies in order to represent all aspects and important indicators of organizations performance. Ontological system elements are described. Practical implementation of proposed ontology and its filling with information was performed using cognitive IT-platform "POLYHEDRON".

Keywords: Research and educational organization · Quality · Assessment · Activity · Parameters · Information · Ontology

1 Introduction

Scientific and educational activities are one of the most influential factors in the development of each country's economy. An objective analysis of the effectiveness of organizations functioning in terms of world and domestic science, education, economics, culture, improving the quality of human life is extremely important. It is important to obtain and analyse data in an open and transparent manner using objective methods of obtaining organizations performance indicators.

Government agencies are complex organizational systems. They have a developed hierarchical structure consisting of many departments and subordinate organizations. Information technology is the basis of information and analytical

M. Choraś et al. (Eds.): CORES 2021, IP&C 2021, ACS 2021, LNNS 255, pp. 318–329, 2022.
https://doi.org/10.1007/978-3-030-81523-3_31

activities for the evaluation of such objects, and information processing is one of the most complex parts of the evaluation process. This is due to the large number of complex information flows circulating in such objects; high dynamics of change in their structure, which is due to legislative changes and the presence at each level of subjective factors; the need to take into account when processing both quantitative and qualitative information.

Information processing in such organizations is performed by experts who, based on their own experience and with the help of simplified mathematical methods, assess the general state of performance. It is necessary to identify the key points of such activities:

- the evaluated information is often incomplete, part of the data is presented in quantitative form, part – in qualitative;
- variability of requirements for information processing and system model;
- the work of assessing the situation is performed by a person at all levels of the hierarchy of the object, so even in the same type of units with the same input data can be obtained different estimates;
- due to the fact that information processing is carried out by a person, such a system loses efficiency;
- based on the final reports submitted to the relevant evaluation unit, it is difficult to track and identify those parameters that have the greatest impact on performance indicators.

As a result, the task of developing an information system that will automatically assess the quality of the functioning of scientific and educational organizations according to state-accepted standards is urgent. It is proposed to use ontology-driven approach to such system development. Ontologies are used as the basis of such system and serves for scientific and educational organizations information representation.

The paper describes the approach to ontology development for the representation of scientific and educational organizations information. Nowadays there are no com-plex approaches to the automatic assessment of the scientific and educational organizations functioning. Proposed approach bases on qualitative representation of information that is already accumulated at the level of such organizations with its further usage for their assessment. Such an approach involves the design of the ontological model structure, the description of ontological model elements, and ontology direct development, as well as filling it with data using the appropriate environment for the ontologies development.

The structure of the paper is following: Sect. 2 gives analyses of related works and backgrounds for the research, description of various approaches to the quality of organizations functioning assessment, basic notions of ontological modelling. Section 3 describes the structure of ontology of scientific and educational organizations in-formation representation. In section 4, elements of developed ontology are shown. Section 5 depicts main features of cognitive IT-platform "POLYHEDRON" that is proposed to be used for ontology development. Section 6 presents conclusions and plans for future work.

2 State of Art and Related Works

2.1 Approaches to Quality of Organizations Functioning Assessment

Recently, the improvement of methods and techniques for assessing the effectiveness of scientific and educational organizations is becoming increasingly important in Ukraine, Belarus, Kazakhstan, Russia and foreign countries [1]. All these techniques include the analysis of the following main parameters: qualitative publications; attracting additional funding or grants; availability of young specialists, candidates of sciences, graduate students; activity in the international market of scientific and educational cooperation. It should be noted that effective developments in the field of scientific evaluation are set out in the "Leiden Manifesto for Scientometrics" [2]. The process of assessing the quality of organizations in Ukraine is regulated by three national regulations [3].

In the process of organizing scientific and educational activity, with its subsequent assessment, several main processes are involved: organization of organization functioning; determination of indicators by which it is possible to identify the level of performance of a certain type of activity in organizations; availability of criteria for evaluation of such indicators, by means of which it is possible to perform a qualitative assessment of the organization's achievements and to establish whether its activity meets the requirements; organization of evaluation process of the organization directly with the determination of the sequence of actions, the circulation of relevant documents and the organizations involved [4,5].

Improving the quality of assessment of the functioning of educational and research organizations is possible due to the systematization and structuring of heterogeneous poorly structured information that is accumulated at the level of such organizations. The approach to developing a system for assessing the quality of educational and scientific organizations includes: analysis of methods for assessing the quality of such organizations both in Ukraine and abroad, analysis of existing regulations governing the process of such assessment at the state level, development of a system of indicators based on parameters (indicators) that characterize the process of scientific and educational activity and the criteria by which to assess such parameters; development of knowledge representation ontology for systematization and structuring of information already accumulated at the level of scientific and educational organizations and organizations, development of an intelligent method of automated assessment of the quality of organizations, direct development of system modules.

This approach to the development of a system for assessing the quality of educational and scientific organizations will optimize the flow of work involved in the evaluation process and automate the transfer of information that is directly involved in such a process.

2.2 Ontological Modelling

Today there is a tendency to the development of artificial intelligence and their penetration into any sphere of human life. Specialized technologies based on artificial intelligence are used to develop information systems of various kinds and purposes [6,7]. Artificial intelligence technologies also offer a wide range for the presentation of knowledge and data [8,9]. Among such models of knowledge representation we can distinguish: production model, semantic model, logical model, frame model and ontological model. The choice of the type of knowledge representation model depends on the requirements for the designed system. Thus, it is advisable to use the production model in the case of developing a system aimed at implementing expert opinions with a rigid structure of "cause and effect"; frame models are used to implement a scenario approach in order to create systems focused on forecasting and modelling future events; logical models are used to represent information in the form of formal mathematical structures, which should be used when necessary to optimize or automate certain processes in the design of information systems; ontological models can be used both for the purpose of presenting information in the process of system design and for further operation of such information. Presented study proposes to use the ontology as one of the models of knowledge representation to organize information of scientific and educational organizations for its structuring and systematization, as well as for its further processing and use.

In the era of Big Data, the management and integration of big on-board data provide tremendous opportunities for generating new knowledge. The ontologies, that represent domain knowledge in the form of concepts, their classes, and the interconnections between them, are a versatile tool for overcoming barriers to integrate data and knowledge from diverse sources of information, thus facilitating the search for new knowledge [10,11].

There are following advantages of ontological models usage:

- structuring and systematization of domain information,
- organization of semantic search,
- automated integration with the semantic web,
- a large number of existing software and web applications to develop ontological models.

It is ontology that can be used as a universal tool for integrating data and knowledge from diverse sources of information [12].

Using ontologies allows you to structure, organize and classify information [13]. An important characteristic of ontology is that its structure a priori fits into the paradigm of the semantic web that makes it possible to automatically adjust and use the developed ontology in the Internet environment. Such a feature is in demand in terms of the development of modern information systems, the main requirement of which is to provide online access to users in real time.

Thus, it is proposed to use ontology to represent information that correlates to scientific and educational organizations functioning aiming at it further analyses and assessment.

3 Ontology of Scientific and Educational Organizations Information Representation

The functioning of scientific and educational organizations is associated with certain specific features that are not characteristic to other types of organizations. For example, important indicators of the scientific and educational processes are:

- number of publications;
- citations;
- various scientific indexes and ratings;
- participation in international projects and programs;
- number of performed research projects;
- trained specialists of different qualification levels in different specialties and specializations;
- etc.

That is why the information produced during the operation of such organizations also should reflect all aspects of such functioning. This feature must be taken into account when developing the ontology.

As well we can define several basic processes that are involved in organization scientific and educational activity:

- organization of organization activity;
- definition of indicators by which it is possible to identify the level of performance of a certain type of activity in an organization;
- the existence of criteria for evaluating such indicators (using them it is possible to evaluate the achievements of organization and to determine whether its activity meets the requirements);
- organization of the organization's evaluation process itself.

On this basis, it is proposed to divide general ontology into several ontologies in order to represent all aspects of information accumulated by organizations [14]. As a result of this selection, the overall ontology will be a system of ontologies, each describing the appropriate sub process within both the activities of the organization itself and the process of its evaluation. Such system as a whole will allow both to structure and organize the information accumulated by organizations, and to organize context-independent structures for its further processing, analyses and assessment. Thus general ontology of SEIR is a set of several components (see Fig. 1):

- Ontology of scientific organizations – reflects all possible processes that occur during different processes within the activities of organizations.
- Ontology of organizations activities organization – describes general concepts that relate to the organization of scientific and educational activity as a whole.

- Ontology of organization activities parameters – allows to describe in detail the parameters and indicators on all aspects of the organization activities.
- Ontology of organizations activities evaluation – introduces concepts that make it possible to evaluate the effectiveness of scientific and educational activity of a particular organization in the framework of its state review or establishment of its scientific and educational level.
- Ontology of organizations activities evaluation criteria – sets criteria for evaluating the performance of the organization.
- Ontology of evaluation process – describes the assessment process itself by the relevant authorities.

Ontology developed in such a way will not only enable the representation of information accumulated by scientific and educational organizations in the process of their functioning, but also organize structures for its further analysis and processing.

In the process of ontology development, there is a need to describe its elements. The ontology includes the following elements [15]:

Ontological model = <classes, attributes, relations, types of attribute values, constraints on attribute values, instances of classes>

Such set of elements allows to define all needed data while ontology development. In proposed research al such element for each of depicted above ontologies were defined. Let us as the example describe in more detailed way the *Ontology of organizations activities organization (OOAO).*

4 Ontology of Organizations Activities Organization

Ontology of organizations activities organization includes a set of classes that describe general concepts related to the organization processes of the institution (see Fig. 2). These classes are related by relations of different types. Different properties of each concept are described on the basis of attributes of concepts and the restrictions imposed on area of their values.

Figure 2 describes elements of OOAO. Classes and sub classes are shown as ellipses and relations between classes as arrows. The are two types of relations: "associative relations" that allow performing meaningful searches through the ontology information space and "part-to-whole" relations that allow establishing relations between classes at the hierarchy level. Relations of "part-to-whole" type are depicted with dotted arrows.

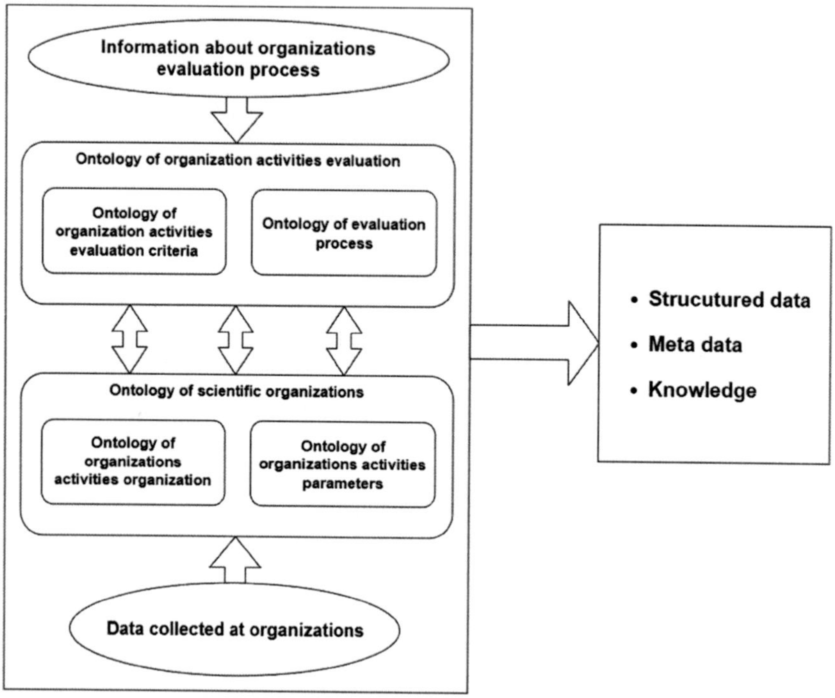

Fig. 1. Ontology of scientific and educational organizations information representation

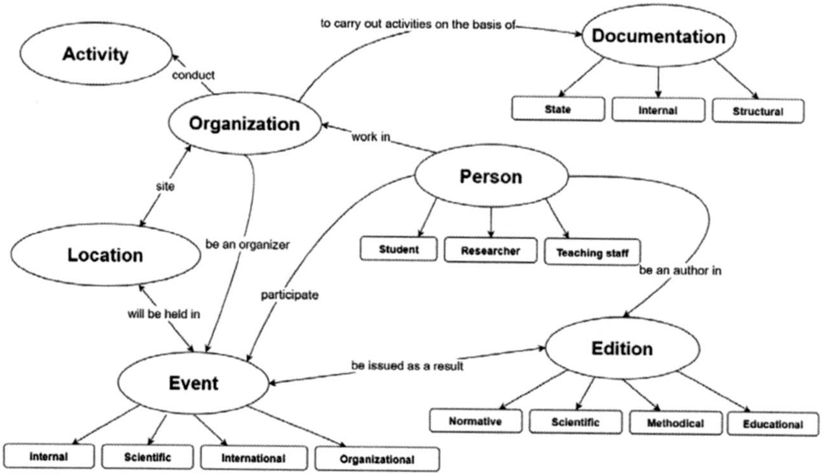

Fig. 2. Fragment of Ontology of organizations activities organization.

Table 1 gives detailed description of OOAO classes, their attributes, subclasses and relations with other classes of ontology.

Table 1. Description of OOAO elements.

Class	Description	Class attributes	Child classes	Relation
Person	Concepts related to the subjects of the organization: researchers, teachers, staff and members of organizations	Personal data, academic degree, academic title, position, field of activity in the organization, awards, publications	Scientific staff, teaching staff, support staff, administra-tion, students	to be an author to be a student, work in.
Organzation	The class describes different types of organizations, such as research institutes, academies, universities, ministries, communities and associations, and more	Name (full and abbreviated), location, type, details of the organization		Conduct, To be located, Carry out activities on the basis of
Event	The class describes various activities that take place in organizations in the process of scientific and educational activities: conferences, seminars, workshops, etc.	Name, venue, start date, end date, subject, name and details of the organizer responsible for the event	Organizational, Internal, International Scientific	
Activity	The class describes all activities that take place in a scientific organization, such as research, projects, international research programs, the educational process, and others)	Name, type, venue, participants, result	State, Internal, Structural	
Location	The class allows to describe the geographical and administrativeterritorial localization of all elements of the ontology	Name, geographical type, administrativeterritorial affiliation, country of location, full address		Event, organization
Edition	The class describes all types of publications (presented in printed or electronic formats) that are used and produced in the process of the organization	Title, type, type, short description, date and language of the publication, number pages, characteristics of the publication	Regulatory, Methodical, Educational, Scientific	Be issued results
Documentation	The class describes the different types of documentation that regulate the functioning of the organization	Title, type, authors, number of pages approved by approved by		

5 Ontological Solutions Development Platform

The practical implementation of proposed ontology and its filling with information was performed using cognitive IT-platform "POLYHEDRON" (CIT "POLYHEDRON").

CIT "POLYHEDRON" is designed to support the processes of linguistic and se-mantic analysis of large volumes of spatially distributed unstructured information (Big Data), their structuring, establishing contextual links between processed documents, forecasting and support of rational choice processes with the subsequent formation of information-analytical WEB [16,17].

CIT "POLYHEDRON" means provide the implementation of the following information processes (see Fig. 3):

- Linguistic and semantic analysis of network information resources, which have a significant number of interdisciplinary relationships and are created based on the use of various information technologies and standards;
- Transdisciplinary integration with other network information systems and WEB-oriented applications;
- Taxonomization of narratives of arbitrary documents and reflection of their structure, including intertextual relations;
- Creation of ontological interactive documents;
- Detection of latent information in the analysed information resources;
- Deep Learning (Machine Learning);
- Support for Semantic Web formats and protocols;
- Big Data processing.

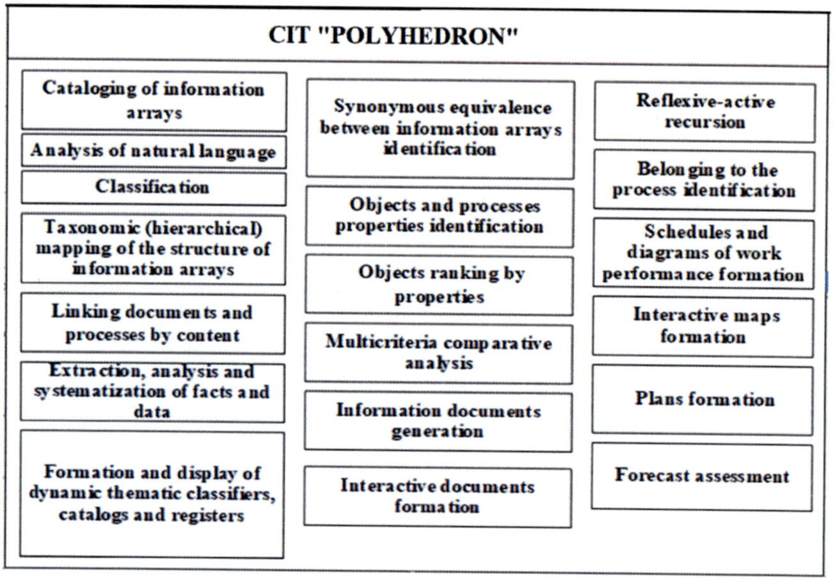

Fig. 3. Components of CIT "POLYHEDRON".

CIT "POLYHEDRON" is a modular system, and different modules may have different degrees of integration with each other [18]. CIT "POLYHEDRON" provides many software tools for storing and processing knowledge through the development of ontologies. CIT "POLYHEDRON" provide the mechanism for ontology checking and verification, thus using it is possible to verify if developed ontology is correct.

Figure 4 shows a fragment of an ontology of scientific and educational organizations information representation developed using CIT "POLYHEDRON".

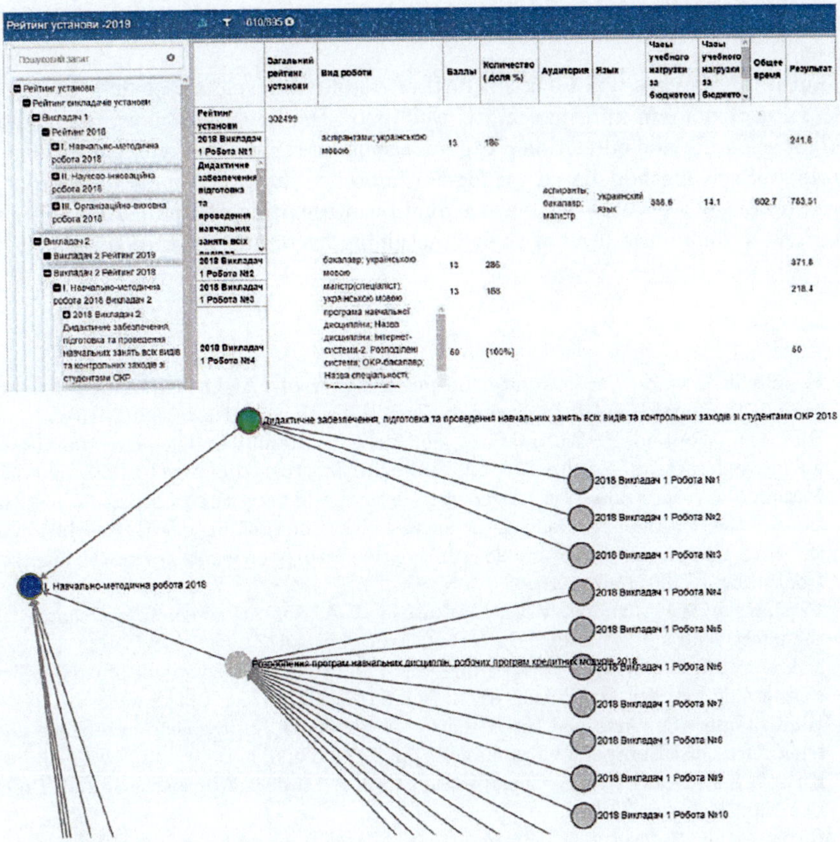

Fig. 4. Fragment of ontology for scientific and educational organizations information representation developed using the CIT "POLYHEDRON".

6 Conclusions

The paper presents an approach to the representation of information accumulated in various scientific and educational organizations on the basis of ontological modelling. The ontological representation allows evaluating the quality of scientific and educational organizations functioning on the basis of national principles for such assessment. The information how to organize the evaluation process as well as generally accepted evaluation criteria are also stored in the ontology. The associative relations between objects that are determined in ontology allow linking the organization activities indicators with the evaluation criteria for assessment process automation.

The limitations of using the proposed approach lie in filling the ontology with input data of a certain subject domain (namely, data on functioning of scientific and educational organizations). In the general case proposed approach can be used for any subject domain with minor modification of the proposed ontological model.

Future researches will focus on further ontology development and its filling with subject domain information, as well as on the novel method implementation for scientific and educational organizations functioning assessment based on ontology. Such method based on logical inference and reasoning through proposed ontology will allow evaluating quality of scientific and educational organizations to define the level of their functioning according to state standards.

References

1. Feigelman, M.: How to evaluate the performance of FANO institutions. Private view of the participant in the process. Trinity Option **13**(182), 1–2 (2015)
2. Hicks, D., Wouters, P., Waltman, L., de Rijcke, S., Rafols, I.: The leiden manifesto for research metrics. Nature **520**, 429–431 (2015). https://doi.org/10.1038/520429a
3. Methodology for evaluating the effectiveness of the activities of scientific institutions of the National Academy of Sciences of Ukraine (in Ukrainian), Resolution of the NAS Presidium of Ukraine 241, July 2018. http://www.nas.gov.ua/legaltexts/DocPublic/P-180711-241-1.pdf
4. Varshavsky, A.E., Ivanov, V.V., Markusova, V.A.: On an adequate assessment of the effectiveness of scientific. Bull. Rus. Acad. Sci. **81**(7), 587–593 (2011)
5. Mikhailov, O.V.: Criteria and parameters of an objective assessment of the quality of scientific activity. Bull. Rus. Acad. Sci. **81**(7), 622–625 (2011)
6. White Paper: On Artificial Intelligence - A European approach to excellence and trust. Brussels: European Commission, pp. 1–27 (2020)
7. Russell, S.J.; Norvig, P.: Artificial Intelligence: A Modern Approach. 3rd ed. Prentice Hall, Hoboken (2010)
8. Harmelen, F., Lifschitz, V., Porter, B.: The Handbook of Knowledge Representation, p. 1034. Elsevier Science, San Diego (2007)
9. Hermann, H.: Knowledge Representation and the Semantics of Natural Language, p. 647. Springer, Berlin, Heidelberg, New York (2006)
10. Sowa, J.: The role of logic and ontology in language and reasoning. In: Poli, R., Seibt, J. (eds.) Theory and Applications of Ontology: Philosophical Perspectives, pp. 231–263. Springer, Dordrecht (2010). https://doi.org/10.1007/978-90-481-8845-1_11
11. Tanwar, P., Prasad, T., Dutt, K.: A tour towards the various knowledge representation techniques for cognitive hybrid sentence modeling and analyzer. Int. J. Inform. Commun. Technol. **7** (3), 124–134 (2018). https://doi.org/10.11591/ijict.v7i3.pp124-134
12. Changrui, Y., Yan, L.: Comparative research on methodologies for domain ontology development. In: Huang, D.S., Gan, Y., Gupta, P., Gromiha, M.M. (eds.) Advanced Intelligent Computing Theories and Applications. With Aspects of Artificial Intelligence, ICIC 2011. Lecture Notes in Computer Science, vol. 6839, pp. 349–356. Springer, Berlin, Heidelberg (2012). https://doi.org/10.1007/978-3-642-25944-9_45
13. Uschold, M., et al.: Making the case for ontology. Appl. Ontol. **7**, 373–373 (2012). https://doi.org/10.3233/AO-2012-0110

14. Novogrudska, R.L., Globa, L.S., Koval, O.V., Senchenko, V.R.: Ontology model of intelligent modeling system for marine facilities identification. In: International Conference on Radio Electronics & Info Communications (UkrMiCo), pp. 1–5 (2017). https://doi.org/10.1109/UkrMiCo.2017.8095426
15. Globa, L., Novogrudsk, R., Oriekhov, O.: Method of heterogeneous information resources structuring and systematizing for Internet portals development. In: Eurocon 2013, pp. 319–326 (2013). https://doi.org/10.1109/EUROCON.2013.6625003
16. Stryzhak, O., Prykhodniuk, V., Podlipaiev, V.: Model of transdisciplinary representation of GEOspatial information. In: Ilchenko, M., Uryvsky, L., Globa, L. (eds.) Advances in Information and Communication Technologies. UKRMICO 2018. Lecture Notes in Electrical Engineering, vol. 560, pp. 34–75. Springer, Cham (2019). https://doi.org/10.1007/978-3-030-16770-7_3
17. Dovgyi, S., Stryzhak, O.: Transdisciplinary fundamentals of information-analytical activity. In: Ilchenko, M., Uryvsky, L., Globa, L. (eds.) Advances in Information and Communication Technology and Systems. MCT 2019. Lecture Notes in Networks and Systems, vol. 152, pp. 99–126. Springer, Cham (2021). https://doi.org/10.1007/978-3-030-58359-0_7
18. Popova, M., Novogrudska, R.: Transdisciplinary system for student youth educational and research activities support. In: 2020 IEEE 15th International Conference on Computer Sciences and Information Technologies (CSIT), Zbarazh, Ukraine, 2020, pp. 320–323 (2020). https://doi.org/10.1109/CSIT49958.2020.9321894

Analogy of the Degree of Interest in a News Article by Gaze Information

Takuya Ogawa(✉), Keiichi Endo, Hisayasu Kuroda, and Shinya Kobayashi

Graduate School of Science and Engineering, Ehime University, Matsuyama, Japan
ogawa@koblab.cs.ehime-u.ac.jp

Abstract. There is a huge amount of news information on the Internet. This gives users the advantage of having access to a large amount of information. However, users are not always interested in all the news that is distributed. Therefore, when users read the news on the Internet, they need to select the information they are interested in. We have developed NEAR, a news application for smartphones, with the aim of solving the problem of information overload. However, in order for the app to learn the user's interest trends, it needed to have the user sort through a lot of news, which was a burden for the user. Therefore, we propose a method that focuses on the user's gaze information as a method for learning interests while reducing the burden of article selection by the user. By focusing on gaze, it is thought that it will be possible to analogize the degree of interest on a word-by-word basis rather than on an article-by-article basis, and it is expected that the number of article selection required for learning will be reduced.

Keywords: Information distribution system · User-aware · Gaze

1 Introduction

The Internet is filled with a vast amount of news information, but not all of it is of interest to users. Therefore, when users read the news, they need to find the information they are interested in among the vast amount of news information. When the amount of information increases and it becomes difficult for users to sort the information, it is called information overload. We developed PINOT (Personalized INformation On Television screen), an information distribution system for individuals, to solve the problem of information overload [1]. PINOT displays news information in ticker format, and the user can pause, fast-forward, and rewind on the displayed information. PINOT inferred the user's interests from their actions and provided news that the user was interested in. In response to the widespread use of smartphones, we developed NEAR, a news application for smartphones [2]. NEAR now supports the provision of information by voice, making it possible to provide information in many more situations. However, in order to learn the user's interests, the user needed to sort through many articles, which was a burden for the user.

M. Choraś et al. (Eds.): CORES 2021, IP&C 2021, ACS 2021, LNNS 255, pp. 330–338, 2022.
https://doi.org/10.1007/978-3-030-81523-3_32

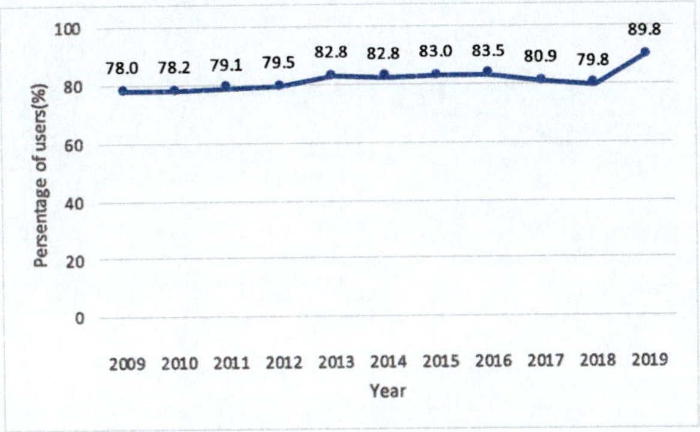

Fig. 1. Internet usage rate

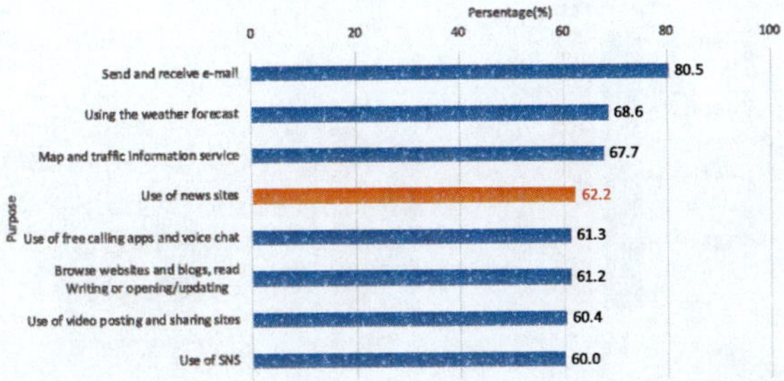

Fig. 2. Purpose of Internet use

2 Background

2.1 Spread of the Internet

In recent years, many people have been using the Internet. The percentage of people who used the Internet at least once during the past year is shown in the Fig. 1 [3]. This figure shows that 89.8% of people use the Internet in 2019.

One of the Internet services that people use is to obtain Internet news. Figure 2 shows a selection of Internet users' purposes and uses of the Internet, the percentage of which exceeds 60% [4]. The percentage of people who use the Internet to obtain news information is 62.2%. From this figure, we can see that many people obtain news information via the Internet.

There is a huge amount of news information on the Internet. This gives users the advantage of having access to a large amount of information. However, users are not always interested in all the news that is distributed. Therefore, when users read the news on the Internet, they need to select the information they are interested in. The more information there is on the Internet, the more difficult it is to select only the information that interests you. This problem is called information overload.

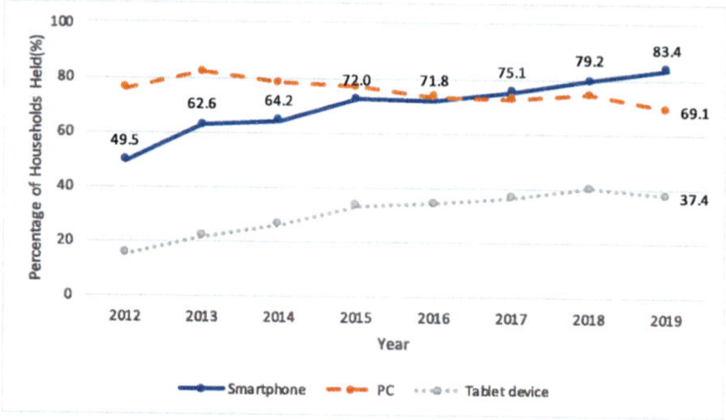

Fig. 3. Ownership rate of information terminals (household)

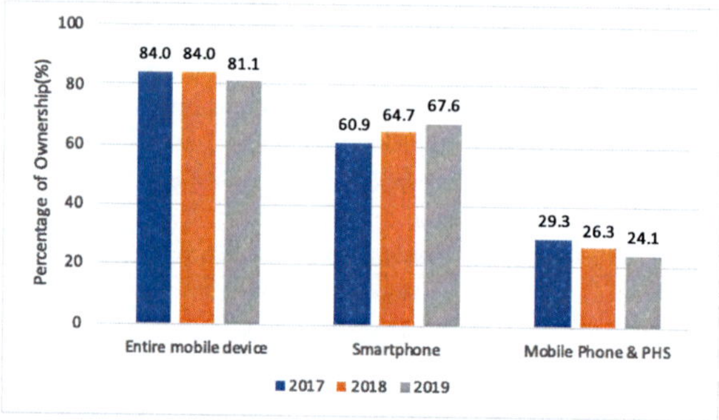

Fig. 4. Mobile device ownership (individual)

2.2 Smartphone Penetration

In recent years, smartphones have become popular among the general public, and many people are using them as information terminals. Figure 3 shows the changes in household ownership of smartphones, PCs, and tablet devices. As Fig. 3 shows, as of 2019, 83.4% of households in Japan owned a smartphone, which is more than the percentage of households that owned a computer [3].

Figure 4 shows the individual ownership rate of mobile devices. As of 2019, 67.6% of individuals own a smartphone, indicating that many individuals own a smartphone.

Many people also use Internet services via smartphones. Figure 5 shows a graph showing which terminal is used to access the Internet. This figure shows that 63.3% of people use the internet using their smartphones as of 2019. It can be said that smartphones are used in a very high percentage of cases compared to other devices.

Since many people use the Internet on their smartphones, it is expected that many people will use their smartphones to get the news. In fact, according

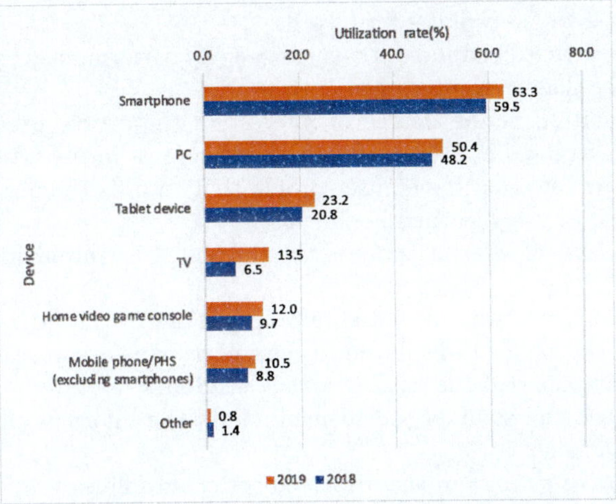

Fig. 5. Percentage of terminals using the Internet

to a Pew Research center survey [5], 57% of Americans who get Internet news frequently in 2019 will use a mobile device, compared to 30% who use a computer.

The amount of news available in news apps is enormous, and not all of it is useful to users. Since each user has different interests and targets of interest, the users themselves need to select the news when they want to get the news information. Sorting through the huge amount of news to find something of interest is burdensome and time-consuming for the user. There is a need to solve the information overload in smartphone news applications to reduce the burden on users to obtain useful information.

2.3 NEAR

We have developed NEAR, a news application for smartphones, with the aim of solving the problem of information overload. The application learns which articles the user has viewed, and automatically provides news based on the user's interests. In addition, the system supports voice-based information provision, making it possible for users to obtain news information even when they are unable to gaze at the smartphone screen. And the ability to control the playback of audio information by operating the smartwatch makes it possible for users to obtain news information without having to operate their smartphones at all.

The procedure for analogy and learning of user interests in NEAR is shown below.

Step 1 Obtain article information from the information distribution server
Acquire article information of news articles from the information distribution server set by the user in the application.

Step 2 Break article headlines into words
By dividing article headlines into words, we can perform analogy and learning based on the degree of interest in each word.
Step 3 Calculation of the degree of interest in the article headline text For each word extracted in Step 2, the degree of interest in the article headline is calculated by referring to the user profile that records the degree of interest in each word of users learned in the past.
Step 4 If the level of interest is above the threshold, it will be displayed on the screen
If the degree of interest calculated in Step 3 is higher than the threshold, the headline of the article is displayed to the smartphone as an article of interest.
Step 5 Operate smartphone against article headlines
The user taps the smartphone to manipulate the information displayed on the smartphone.
Step 6 Analogize interest in the displayed article headlines
The presence or absence of user interest in the displayed article headline is inferred analogously from the user's operation of the smartphone.
Step 7 Update user profiles
Update the content of the user profile based on the results of the analogy obtained in Step 6.

Step 3, Step 6, and Step 7 are explained in detail.

Step 3 Calculation of the degree of interest in the article headline text
1. Get the level of interest for each words
For each word $\omega_n (n = 1, 2, ..., N)$ extracted in Step 2, the degree of interest $i(\omega_n)(0 \leq i(\omega_n) \leq 1)$ of each word is obtained by referring to the user profile. N is the total number of extracted words. If the extracted word is a new word that is not included in the user profile, $i(\omega_n) = 1$.
2. Calculate the degree of interest in textual information
The following expression (1) is used to calculate the degree of interest $I(W)$ in the article headline W. This $I(W)$ is the average value of the degree of interest $i(\omega_n)$ of each extracted word.

$$I(W) = \frac{\sum_{n=1}^{N} i(\omega_n)}{N} \tag{1}$$

Step 6 Analogize interest in the displayed article headlines
When the degree of interest in the article headline text exceeds a threshold, the article is displayed on the smartphone screen as an article of interest. The interest is analogized by the tap operation performed by the user on the article headline displayed in the article headline list. An article headline that is tapped by a user is judged to be an "article of interest". Among the article headlines seen by the user, the article headlines that are not tapped are judged as "article of not interested".
Step 7 Update user profiles
Updating the user profile means to calculate the new degree of interest $i(\omega_n)$

for each word ω_n extracted from the article headline W based on the analogy of the interest in the article headline W, and to rewrite the profile with the new one. The degree of interest is calculated for each word $\omega_n(n = 1, 2, ..N)$, using the following expression (2).

$$i(\omega_n) := \alpha \cdot i(\omega_n) + (1 - \alpha) \cdot J \tag{2}$$

In this expression, J takes the value 1 if it is judged as "article of interested" or 0 if it is judged as "article of not interested" in the analogy of interest in the article headline. The amount of change in the degree of interest in a word depends on the threshold α. The smaller the value of α, the larger the proportion of new degrees of interest when updating the degree of interest of a word. On the other hand, if α is large, the percentage of old degree of interest is large.

By iteratively updating the user profile, the value of the degree of interest of words that frequently appear in the article headline of interest becomes higher, and the value of the degree of interest of words that frequently appear in the article headline of disinterest becomes lower. As a result, a user profile is created that matches the user's interests.

In NEAR, learning is performed based on whether the article headline is selected by the user or not. However, since the judgment of interest is made on an article-by-article basis, it is not reliable as the degree of interest in a word. Therefore, a huge amount of article selection is required to create user profiles, which is a burden for users.

3 Proposed Method

As a method to obtain the degree of interest in more detail, we consider a method for learning interest that takes the user's gaze into account. If there is a relationship between a user's eye movement when reading text information and the degree of interest in the text information, it is possible to analogize the user's interest using eye movement information.

In this study, we investigate the characteristics of eye movement when users look through news information, with the aim of analogizing the degree of interest in sentences and the degree of interest in words appearing in sentences, targeting Internet news articles.

3.1 Assumed Gaze Movement

The following is a list of possible gaze movements for article headlines.

- Read the headline to the end and display the detailed text or not
  When a user reads a headline to the end, user understands the content of the sentence and makes a choice whether to read detailed information or not. In this case, learning is performed by uniformly changing the degree of interest for all words in the sentence according to the presence or absence of interest. In this part, there is no difference from the traditional learning method.

- Read the headline to the middle and display the detailed text or not
  If user reads the headline until the middle of the sentence and then chose whether to read the details or not, it is likely that user decided whether to read the details or not based on the last word user read. In this case, by changing the degree of interest of the last word read more than usual, it is thought that learning can take into account the strength of interest of each word.
- Don't read the headline
  In this case, the degree of interest is not changed. However, with the conventional method, it is difficult to determine whether the user has seen the article or not, and those that did not display detailed information were learned as uninteresting articles. Therefore, this method can deter false learning.

Next, the following is a list of possible gaze movements for detailed informations.

- Read back a specific sentence
  When the user reads back the sentence, it can be expected that user is interested in the sentence. By changing the degree of interest in the read back sentences more than usual, learning is performed in consideration of how the user reads the article.
- Skip the sentence from the middle
  When skipping from the middle, the user can judge that user is not interested in the text based on the information up to that point. Therefore, "not interested" learning is performed for the sentences read by the user. At this time, erroneous learning can be suppressed by not learning for sentences that have not been read.

3.2 System Overview

The configuration of the system used in this paper is shown in Fig. 6. This system displays Internet news on the screen and acquires the movement of the user's gaze when reading the news from the camera image. There are two types of screens presented to the user: a screen that displays a list of news headlines and a screen that displays detailed information on any news. On the list screen, the movement of the gaze when selecting multiple heading sentences is acquired. On the detailed information screen, the movement of the line of sight for a long sentence that explains a specific content is acquired. With this system, the movement of the gaze when the user reads the news is investigated. After that, we will investigate which article you were interested in by hearing survey. By identifying the presence or absence of interest in each article and associating it with the gaze information, we investigate the difference in the way the line of sight moves depending on the presence or absence of interest. In addition, if a characteristic movement of the gaze is observed for a specific word in the sentence, the presence or absence of interest in that word is also investigated.

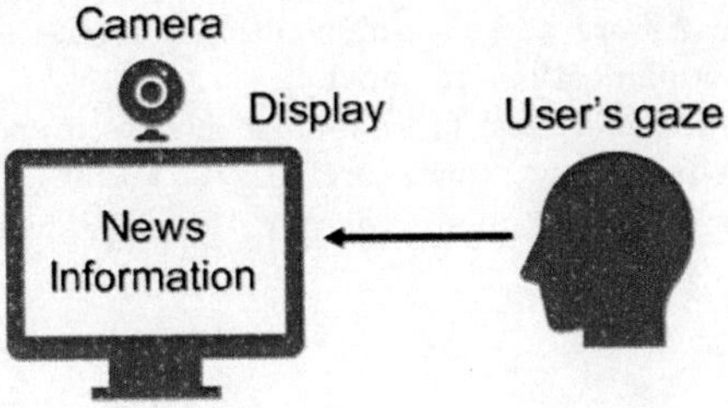

Fig. 6. Configuration of the system

3.3 Future Plan

Currently, we are developing a system for gaze estimation. After the completion of the gaze estimation system, we will conduct an experiment by combining it with a news display system. Based on the results of a multi-person experiment, we investigate the characteristic gaze movements and their relationship to the presence or absence of interest. After that, we will develop a system that learns whether the user is interested or not based on gaze information.

4 Conclusion

In this paper, we described the information overload in Internet news. We also mentioned that NEAR was developed to solve the problem of information overload. NEAR solved the problem of information overload by learning the user's interests and providing only the information the user wanted to read. However, since the learning of interest is done on an article-by-article basis, it cannot be said that the degree of interest in a word is accurately learned. Therefore, we proposed a system to analogize the degree of interest from gaze information. By taking gaze into account, it becomes possible to analogize the degree of interest on a word-by-word basis, which allows for more accurate interest learning. In the future, after developing a system that can detect the user's gaze, we will record the user's gaze movements while reading the news and investigate the relationship between the user's interest in the article and the user's gaze.

References

1. Ono, S., Inamoto, T., Higami, Y., Kobayashi, S.: Development of the smartphone application to choose and display news for individuals based on an operation history. In: 2016 Proceedings of Multimedia, Distributed, Cooperative and Mobile Symposium (2016). (in Japanese)
2. Ogawa, T., Fujihashi, T., Endo, K., Kobayashi, S.: A proposal of personalized information distribution system using smart watch and wireless headphones. In: Asian Conference on Intelligent Information and Database Systems (ACIIDS2020) (2020)
3. Ministry of Internal Affairs and Communications, Japan: 2020 White Paper on Information and Communications in Japan

4. Ministry of Internal Affairs and Communications, Japan: 2019 White Paper on Information and Communications in Japan
5. Pew Research Center: Americans favor mobile devices over desktops and laptops for getting news. https://www.pewresearch.org/fact-tank/2019/11/19/americans-favor-mobile-devices-over-desktops-and-laptops-for-getting-news/. Accessed 26 Feb 2021

Check for updates

Examination of Water Temperature Interpolation Method for Prediction

Yu Agusa(✉), Keiichi Endo, Hisayasu Kuroda, and Shinya Kobayashi

Graduate School of Science and Engineering, Ehime University, Matsuyama, Japan
g863001b@mails.cc.ehime-u.ac.jp

Abstract. In aquaculture, fishery workers and researchers strong require to visualize water temperature. This is because knowing the water temperature makes it possible to take measures against fish diseases and predict the occurrence of red tide. In collaboration with fishery researchers, we constructed a multi-depth sensor network consisting of 17 water temperature observation devices over the Uwa Sea in Ehime Prefecture. In addition, we developed a Web system that can instantly visualize the water temperature measured on this network and provide it to fishery workers in the Uwa Sea. However, the water temperature information provided by this Web system is only current or past information, and fishery workers are requiring the provision of near-future water temperature forecasts for about 3 days. Therefore, in this research, as a new function of this Web system, we will implement a function to predict the water temperature in the near future from the past water temperature information and provide the forecast information. This paper examines the steps to be solved for the prediction of seawater temperature, which is the current problem in this system. Moreover, among the steps, this paper reports on the solution method and its evaluation experiment for improving the accuracy of water temperature interpolation.

Keywords: IoT · Big data · Water temperature · Forecast · Interpolation

1 Introduction

The world population is about 7.7 billion in 2019, and is expected to reach 9.7 billion in 2050 and 10.9 billion in 2100 [1]. The critical problem caused by the increase in the world population is food shortage. Aquaculture is considered an important food supply source because it produces stably.

Ehime Prefecture is one of the leading fisheries prefectures in Japan. (annual production in 2018: about 87.7 billion yen (3rd place in Japan) [2]). The production is particularly supported by the aquaculture in Uwa Sea (annual production in 2018: 65.3 billion yen [3]).

In Ehime University, we developed the system that annoucned the amount of plankton and descoloration to fishery workers to help aquaculture in Uwa

M. Choraś et al. (Eds.): CORES 2021, IP&C 2021, ACS 2021, LNNS 255, pp. 339–350, 2022.
https://doi.org/10.1007/978-3-030-81523-3_33

Sea [4]. Moreover, the system that annoucned the red tide (the phenomenon caused by overgrowth of plankton and causes great damage to fishery production) occurrence information sent by the Fisheries Research Center in Ehime to fishery workers was developed [5].

This paper introduces a system that announces water temperature information to fishery workers (Seawater Temperature Announcement System [6]). In addition, this paper examines the steps to be solved for the prediction of seawater temperature, which is the current problem in this system. Moreover, among the steps, this paper reports on the solution method and its evaluation experiment for improving the accuracy of water temperature information.

2 Background

2.1 Importance of Water Temperature

In aquaculture, especially in the Uwa Sea, water temperature in the sea area around the site is very important for the following two reasons. First, water temperature zones that cause fish diseases (for example, Red sea bream iridovirus infection [7]) exist. According to interviews with fishery workers, they must reduce the amount of feed to fish when fish disease occurs. If they can get the current information about the water temperature in their aquaculture ground, they can observe the health condition of their fish and reduce the damage caused by fish diseases. Second, fishery workers can predict the tidal flow in the future based on the water temperature in their aquaculture ground and around. The degree of influence of tidal flows on aquaculture fisheries varies depending on the time, thickness and range of inflow, and it is necessary to adjust the amount of feed according to the tidal flow conditions.

Moreover, fishery researchers also focus to water temperature because they predict the occurrence of red tides using water temperature in an ocean physics perspective.

2.2 Proposed System

Issues

This section describes the situation before developing the proposed system from the viewpoint of fishery workers and researchers in Ehime Prefecture.

Previously, fishery workers had no means of knowing the water temperature at various depths of their aquaculture grounds. Therefore, they were estimating the water temperature from the air temperature. However, in the Uwa Sea, the actual water temperature was significantly different from them estimates because the water temperature changed drastically by a high-temperature tide called "Kyucho" and a low-temperature tide called "Sokoirijio" [8].

Fishery researchers surveys the water temperature every month in the Uwa Sea, at 24 fixed points, against this issue. In addition, five water temperature observation devices that enabling measurements on multiple depths (min: 1 m,

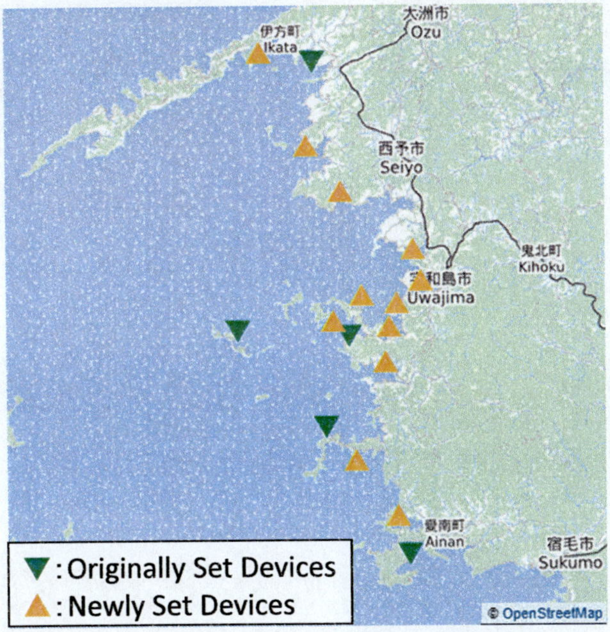

Fig. 1. Location of water temperature observation devices

max: 60 m) set (Fig. 1 shows the location) to obtain the value every hour. However, Kyucho may occur at intervals shorter than half a month, which cannot be grasped by fixed-point surveys. In addition, the number of observation devices that is insufficient for observing marine phenomena in the Uwa Sea. Moreover, fisheries workers could not view the information obtained.

Overview
Proposed system is shown in Fig. 2.

First, we configured a sensor network that contains 17 (as of October 2020) observation devices (Fig. 1 shows the location of devices). By configuring this network, fishery workers and researchers can monitor tidal flow conditions in the upper and lower layers in the Uwa Sea. When constructing this network, we developed the observation devices that is inexpensive, and sufficiently accurate as information obtained, because the commericially available devices is very expensive (more than 1.5 million yen per a unit) [9].

Next, we developed a system called "Seawater Temperature Announcement System" that provided the information collected on using this network system to fishery workers and researchers. The means for fishery workers to browse the measured information was established by implementing this system.

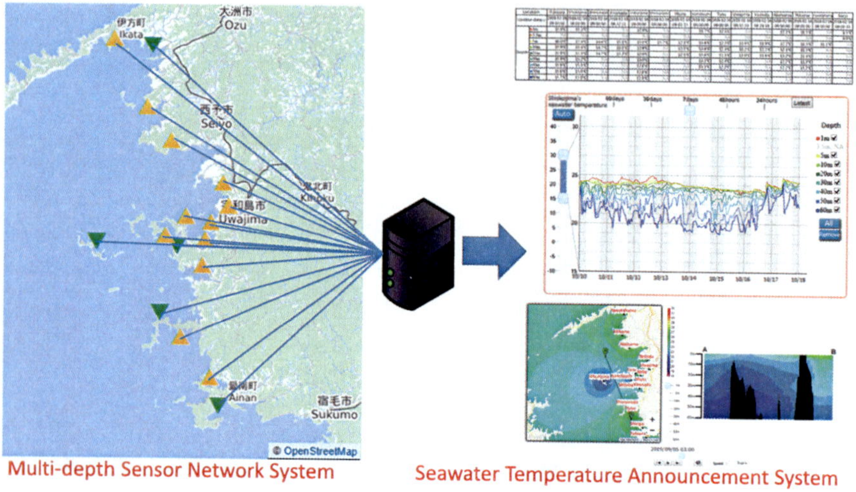

Fig. 2. Proposed system

Table 1. Commercial products vs our products

	Commercial products	Our products
Cost	High	Low
Measurement interval	30 min or 1 h	30 min
Number of measurement trials	Once	3 times
Resolution	0.001 °C	0.0625 (1/16) °C
Data send method	Email	Communicate with server

2.3 Water Temperature Observation Device

There are roughly two types of water temperature observation devices: commercially available ones and our originally ones. The Table 1 summarizes the differences between the two.

In the former, the water temperature is measured once every 30 min or 1 h. In addition, an e-mail containing measurement data is sent to the specified e-mail address from a send-only e-mail address that differs for each measurement point.

In the latter, the water temperature is measured three times in a row every 30 min (It has been verified that the measured values are not always the same) to suppress missing measurements and improve the accuracy of water temperature information. Moreover, unlike the former, the mail does not arrive, but the communication module connects to the server and saves the measurement data on the server. Compared to the former, the resolution is inferior because it is cheaper, but from interviews with fishery workers, it is known that it is sufficient to know the water temperature in units of 0.1 °C, so this resolution is fully satisfied the requirement.

2.4 Seawater Temperature Announcement System

Overview

This system provides fishery workers and researchers with the latest data and measurement history about seawater temperature in the Uwa Sea.

This system consists of two subsystems: Storage System and Visualization System. The storage system immediately extract the data and store it on the server after measuring the water temperature. The visualization system processes the data saved by the storage system into data in the form of tables, graphs, etc. and displays it on a web page.

Storage System

The storage system can be used to decode the e-mails sent from the commercial devices and to extract the data upon receiving them. When the extracted data are the data measured for the first time by a device, this system generates a text file named the same as the sender's e-mail address and writes the data in this file on the server. Otherwise, this system adds the data to a text file named the same as the sender's e-mail address on the server. Using the above method, the data can be stored on the server.

Visualization System

Fishery workers and researchers can browse the data accumulated on the server using the storage system by using the visualization system. To satisfy all aforementioned requirements, we realize the announcement system as a Web application, as it can represent the seawater temperature information regardless of platforms. When this system is launched (accessed), the data stored on the server can be read. Thereafter, this system allows creating such as dynamic tables and graphs based on the registered data and displaying them on the Web page.

Users can switch between the displayed information, including the current status and the past status, by pushing button.

Functions

We list the requirements determined based on interviewing fishery workers and fishery researchers as follows: The display system should visualize the seawater temperature considering the spatial spread of marine area.

Concerning their requirement, we implemented five functions defined as follows:

Function (a) Displaying the location of the observation point on the map.
Function (b) Displaying the measurement data in a tabular form.
Function (c) Displaying the current state of the seawater temperature as a graph.
Function (d) Displaying the seawater temperature variation over time as a graph.
Function (e) Displaying the seawater temperature variation over time in three dimensions using a distribution chart.

Figure 3 shows an example of display in Function (e). This display format is the result of pursuing a format that is easy for fishery workers to see through interviews with them.

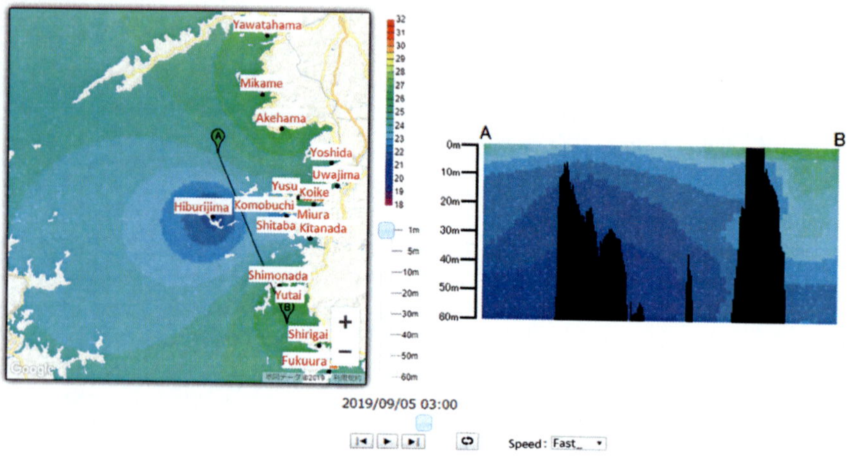

Fig. 3. Example of display in the visualization system

By implementing these functions, the situation has changed from the situation where fishery workers cannot browse the collected water temperature information to the situation where they can browse the latest water temperature information at any time using the Web system. Moreover, for fisheries researchers, visualization of water temperature information has reduced the time and effort required for analysis.

2.5 Purpose

The information provided through our system is the current state and time changes of water temperature in the Uwa Sea area. However, through interviews with fishery workers, we feel the need for forecast information for the near future, which is about one to two weeks, in addition to these information.

Water temperature prediction has already been realized by the Web system called "Akashio Net", and the service by the venture company of researchers at Kyoto University [13].

The former provides only water temperature forecasts for a single layer (depth 5 m). The latter is capable of predicting the water temperature of 36 layers from the surface to the seabed, using data from the Japan Meteorological Agency and the US National Oceanic and Atmospheric Administration.

On the other hand, in this study, we predict multi-layer water temperature by machine learning using the measured values by the observation equipment installed on the sea and the meteorological data that affect the water temperature. As mentioned in Sect. 2.2.1, the water temperature of the Uwa Sea changes drastically due to the occurrence of tidal flows near the sea surface and near the seabed, so fishery workers are demanding highly accurate water temperature information at multiple depths. Therefore, this research is expected not only by fishery workers but also by fisheries researchers.

3 Seawater Temperature Prediction

3.1 Issues

The issues we are concerned about this research are as follows:

1. Improving accuracy of water temperature information
   - Implementation and evaluation of outlier detection and removement method
   - Improving accuracy of water temperature interpolation around a measurement point
2. Examination, implementation and evaluation of water temperature prediction method

This paper reports on the examination of improvement methods for interpolation.

3.2 Improvement of Interpolation Accuracy

Purpose

Our system interpolates the water temperature around each measurement point when running the function shown in Fig. 3. In the current method, we adopted the inverse distance weighted method (IDW) for the plane, and the linear interpolation method for the depth direction. Figure 4 shows a simple interpolation image.

IDW is a method to estimate the data at a certain point by weighting the reciprocal of the square of the distance to the measurement point to the measurement data [14]. In this function, the estimated value at point s is calculated using the following formula:

$$\mu(s) = \frac{\sum_{i=1}^{n} w_i(s)\mu_i}{\sum_{j=1}^{n} w_j(s)}, w_i(s) = \frac{1}{d(s, s_i)^2}$$

Symbol n denotes the number of measurement points; symbol μ_i corresponds to the real value of the i-th measurement point; function $d(s, s_i)$ describes the distance between points and measurement point s_i, and function $w_i(s)$ means the weight of measurement point s_i at point s in the equation.

Linear interpolation is a method of estimating the estimated value at one point between two measured points as approximately linear expression [15]. In

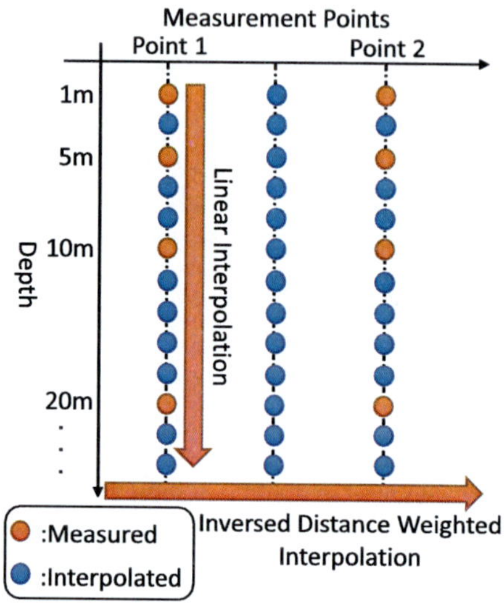

Fig. 4. Current interpolation method

this function, the estimated value at water depth d ($d_1 < d < d_2$) between the measured values $\lambda(d_1)$ and $\lambda(d_2)$ is calculated by the following formula:

$$\lambda(d) = \lambda(d_1) + \frac{\lambda(d_2) - \lambda(d_1)}{d_2 - d_1}(d - d_1)$$

However, the interpolation accuracy (maximum error) in this method exceeds the initial target of 0.5 °C. Moreover, fishery workers require more strictical accuracy so that errors in units of 0.1 °C are unacceptable.

In this research, we examine, implement, and evaluate improvement method for the interpolation method.

Proposed Method

In this research, we are planning to adopt kriging method as alternative method to IDW, which is an interpolation method on a plane. In Kriging, we create a variogram and a covariance function to find a quantitative statistical dependency (spatial autocorrelation) that depends on the autocorrelation model before calculating the estimate. Thereafter, at the time of estimation, the obtained spatial autocorrelation is weighted together with the distance from the estimation point [16].

Kriging is an advanced method of IDW, and the paper that reports that the accuracy is improved compared to IDW by adopting kriging exists [17]. However, it cannot be said that the successful cases in that report can be applied to our research, so we need to verify.

Analysis of Spatial Structure

In the analysis of the spatial structure, which is the first phase of kriging, we create variogram clouds showing the relationship between the spatial distance of each sample pair and the degree of dissimilarity (the squared diffrence).

Figure 5 is a variogram cloud formed from water temperature data measured on April, 2020.

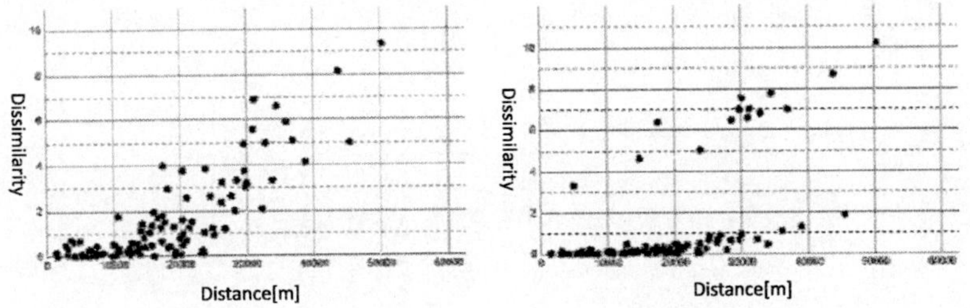

(a) Data measured at 8 pm on April 16, 2020

(b) Data measured at 0 am on April 13, 2020

Fig. 5. Variogram clouds

Most of the variogram clouds formed from past water temperature data, including Fig. 5a, are characterized by the dissimilarity diverging for sample pairs over 20000 to 30000 m regardless of the season. Therefore, it is considered appropriate to use a sample for interpolation within 20000 m from the estimated position.

In addition, as shown in Fig. 5b, even sample pairs with a spatial distance of 20000 m or less may have a high degree of dissimilarity. We consider that this is because sudden temperature changes occurred in some sea areas due to kyucho.

In fact, looking at the water temperature on April 12, 2020, a rapid temperature rise when kyucho occurred was observed (Fig. 6). When a kyucho occurs, the interpolation method using the sample values cannot be used, so it is necessary to consider an estimation method different from kriging. In this evaluation, interpolation by kriging is performed from the variogram during the time when no kyucho is occurring. In addition, considering these variogram, we selected the variogram model to be approximated is the sphere model, when we implemented kriging method.

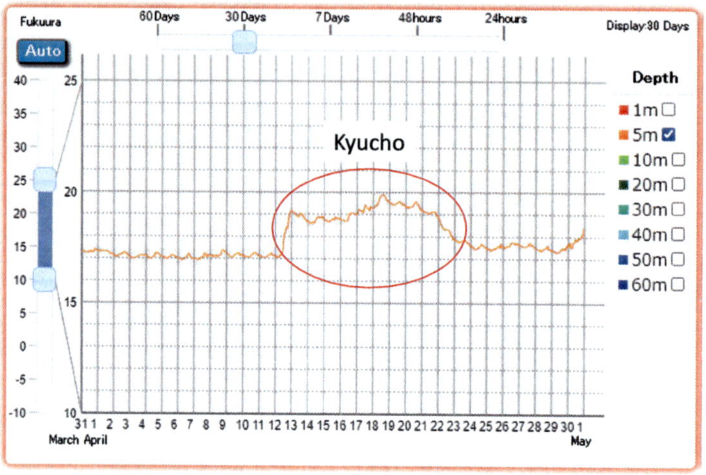

Fig. 6. Observation of kyucho

4 Evaluation

4.1 Purpose

This paper evaluates how much the interpolation method we implemented has an improvement effect compared to the previous method.

4.2 Method

In this evaluation experiment, the measured value of the exclusion point and the interpolation value when interpolation is performed by excluding one of all points are compared. We used sample data is the data at 10 o'clock on December 24, 2020 at a water depth of 5 m. Figure 7 shows the location of sample points.

4.3 Results

Table 2 shows the evaluation results. Of the sample areas, the improvement effect is particularly apparent in areas where observation points is dense, such as Yusu and Yoshida.

5 Future Works

In this paper, we showed that the estimation results based on the sample data at a specific time. In the future, we will verify whether improvement effects can be seen from the estimation results from long-term data such as one day and one month. Further, we will present more reliable evaluation data by using not only the measurement data of the observation device but also the measurement data of the survey ship as the actual measurement data to be compared with the interpolation data.

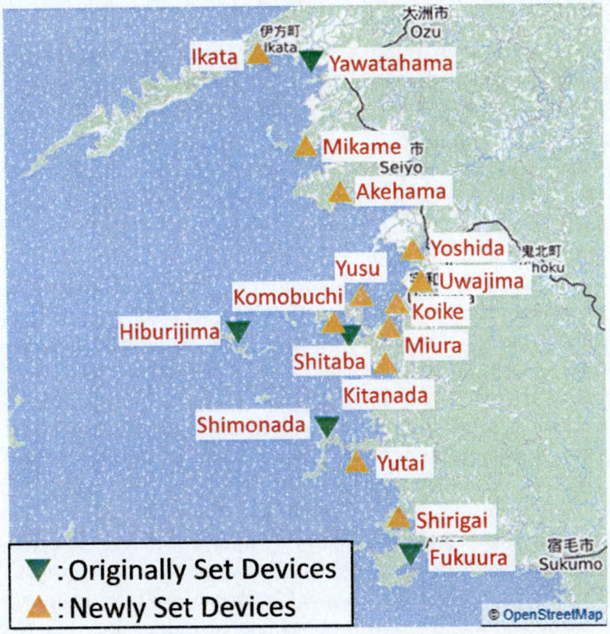

Fig. 7. Location of sample points

Table 2. Result of evaluation

	Previous method (IDW)	Proposed method (Kriging)
Fukuura	1.42	1.43
Shirigai	0.64	0.91
Yutai	0.52	0.51
Shimonada	0.39	0.48
Kitanada	0.18	0.20
Hiburijima	0.26	0.15
Shitaba	0.12	0.13
Komobuchi	0.02	0.01
Yusu	0.14	0.03
Uwajima	0.39	0.24
Yoshida	0.21	0.01
Akehama	0.23	0.37
Mikame	0.25	0.13
Yawatahama	0.46	0.39
Ikata	0.22	0.22

References

1. United Nations: World Population Prospects: The 2019 Revision, p. 1 (2019)
2. Ehime Prefectural Government Office: Ehime Prefecture Fisheries Status in 2018 (Ehime ken gyogyo no chii (Heisei 30)) (2018). https://www.pref.ehime.jp/h37100/toukei/documents/1-1.pdf. Accessed 15 May 2020
3. Ehime Prefectural Government Office: Changes in fishery production by sea area in 1964 to 2018 (Kaiiki betsu gyogyo sansyutsu gaku no suii (Sokatsu : Showa 39 - Heisei 30)). https://www.pref.ehime.jp/h37100/toukei/documents/3-1-2.pdf. Accessed 17 May 2020
4. Ando, K., Okamoto, T., Endo, K., Kuroda, H., Higami, Y., Kobayashi, S.: Development of marine information collection support system for red tide and fish disease estimation. In: Proceedings of the IPSJ 78th National Convention, no. 1, pp. 937–938, March 2016
5. Suehiro, E., Takaichi, R., Fujihashi, T., Endo, K., Kuroda, H., Kobayashi, S.: Development of marine information distribution system for quick countermeasures against red tide damage. In: Proceedings of the IPSJ 80th National Convention, no. 1, pp. 549–550, March 2018
6. Agusa, Y., Fujihashi, T., Endo, K., Kuroda, H., Kobayashi, S.: Development of seawater temperature announcement system for improving productivity of fishery industry. In: Intelligent Information and Database Systems, pp. 714–725. Springer (2019). (Proceedings of the 11th Asian Conference on Intelligent Information and Database Systems (ACIIDS 2019), Yogyakarta, Indonesia, 8–11 April 2019)
7. Nakajima, K., Kurita, J.: Red sea bream virus disease. Virus **55**(1), 115–126 (2005)
8. Takeoka, H.: Fisheries support by advanced altitude information (Kodo kaikyo joho ni yoru suisangyo shien). Ehime J. **31**(5), 84–87 (2017)
9. Araki, K., Fujihashi, T., Endo, K., Kuroda, H., Kobayashi, S.: Development of seawater temperature observation device for ocean forecast information service. In: Proceedings of the IPSJ 80th National Convention, no. 1, pp. 553–554, March 2018
10. IPSJ-SIG-IS: Information System Effectiveness Evaluation, Guidelines for Quantitative Evaluation, version 1.1, November 2012
11. IPSJ-SIG-IS: Information System Effectiveness Evaluation, Guidelines for Qualitative Evaluation, version 1.0, September 2013
12. Ehime Prefectural Government Office: Trends in the number of management entitles by type of fishery in 1964 to 2018 (Omo toshite itonanda gyogyo syuruibetsu keieitaisu no suii (Showa 39 - Heisei 30), p. 2 (2019). https://www.pref.ehime.jp/h37100/toukei/documents/2-1-2.pdf. Accessed 17 May 2020
13. Ocean Eyes: OceanEyes Service Overviews. https://oceaneyes.co.jp/en/overview-2. Accessed 27 Oct 2020
14. Meijering, E.: A chronology of interpolation: from ancient astronomy to modern signal and image processing. Proc. IEEE **90**(3), 319–342 (2002)
15. Bailey, T.C., Gatrell, A.C.: Interactive Spatial Data Analysis. Prentice Hall, Hoboken (1996)
16. Wackernagel, H.: Multivariate Geostatistics: An Introduction with Applications. Springer, Heidelberg (1995)
17. Jedermann, R., Palafox-Albarrán, J., Barreiro, P., Ruiz-García, L., Robla, J.I., Lang, W.: Interpolation of spatial temperature profiles by sensor networks. In: SENSORS, pp. 778–781. IEEE (2011)

Author Index

M. Choraś et al. (Eds.): CORES 2021, IP&C 2021, ACS 2021, LNNS 255, pp. 351–352, 2022.
https://doi.org/10.1007/978-3-030-81523-3